WHAT EVERY VETERAN SHOULD KNOW

87th EDITION

BOOK PRICE - $30.00 – INCLUDES SHIPPING

WITHDRAWN

This is the 87th Edition of the book "WHAT EVERY VETERAN SHOULD KNOW," a service officer's guide since 1937.

Monthly supplements are available from the publisher, which provide updates to the information contained in this book. A one-year subscription to the supplement service is $38.00.

The material herein covers veterans' benefits, rights, privileges, and services over which the Department of Veterans Affairs has jurisdiction. All references to "VA" pertain to the Department of Veterans Affairs. Revised and new laws passed by the 118th Congress as of December 31, 2022, are incorporated in this text.

This guide was prepared solely for convenient reference purposes, and does not have the effect of law. Although diligent effort has been made to ensure its accuracy, in the event of any conflict between this book and any regulation, the latter is, of course, controlling.

Veterans and the dependents of deceased veterans are advised to contact their local veterans organization service officers, or the nearest Veterans Administration Office, for help in completing any valid claim.

VETERANS INFORMATION SERVICE
P.O. Box 111
East Moline, Illinois 61244-0111
Telephone: (309) 757-7760
Fax: (309) 278-5304
Email: help@vetsinfoservice.com
www.vetsinfoservice.com

ASK FOR YOUR **FREE** SAMPLE COPY OF THE
MONTHLY SUPPLEMENT,
DESIGNED TO KEEP THIS BOOK UP-TO-DATE.
ONE-YEAR SUPPLEMENT SERVICE: $38.00

WHAT EVERY VETERAN SHOULD KNOW

VETERANS INFORMATION SERVICE
P.O. Box 111
East Moline, Illinois 61244-0111

Phone: (309) 757-7760
Fax: (309) 278-5304
E-mail: help@vetsinfoservice.com
www.vetsinfoservice.com

A NOTE TO THE READER:

Revised and new laws passed by the 117th Congress as of December 31, 2022, are incorporated in the text of this book. Chapter 1 of the book contains a brief summary of some of the important legislative changes affecting veterans' benefits during 2022.

You will find a Table of Contents, listing the main subject headings. For your convenience there is also a complete index in the back of the book, as well as an Edge Index on the back cover.

We hope you enjoy this edition of **WHAT EVERY VETERAN SHOULD KNOW**. Any comments or suggestions for future improvements are welcome.

© 2023
What Every Veteran Should Know
Library of Congress Cataloging in Serials
ISSN 1532-8112
ISBN 978-0-9600887-3-7

Publisher:

Veterans Information Service
P.O. Box 111
East Moline, IL 61244-0111

TABLE OF CONTENTS

VA PHONE NUMBERS

Main VA Phone Numbers

Where to Call	Toll-Free Number
MyVA411 (main information line for VA)	800-698-2411
Telecommunications Relay Service Using TTY	744

VA Health Care

Where to Call	Toll-Free Number
PACT Act Questions	800-698-2411
Health Care Benefits	877-222-8387
My HealtheVet Help Desk	877-327-0022
Civilian Health and Medical Program of the Department of Veterans Affairs (CHAMPA)	800-733-8387
CHAMPVA Meds by Mail	888-385-0235 or 866-229-7389
Foreign Medical Program (FMP)	877-345-8179
Quit Vet (Get Help from a Counselor to Quit Smoking)	855-784-8838
MISSION Act	844-698-2311, Option 1
Spina Bifida Health Care Benefits Program	888-820-1756
Caregiver Support Line	855-260-3274

VA Benefits

Where to Call	Toll-Free Number
VA Benefits Hotline	800-827-1000
GI Bill Hotline	888-442-4551
Students Outside the U.S.	+1-918-781-5678
National Pension Call Center	877-294-6380
Support for SGLI or VGLI	800-419-1473
All Other VA Life Insurance Programs	800-669-8477
Special Issue Hotline (Blue Water Navy Act, Gulf War, Agent Orange, and Other Information)	800-749-8387

Burials and Memorials

Where to Call	Toll-Free Number
National Cemetery Scheduling Office	800-535-1117
Headstones and Markers	800-697-6947

Other VA Support

Where to Call	Toll-Free Number
Women Veterans Hotline	855-829-6636
eBenefits Technical Support	800-983-0937
Debt Management Center (Collection of Nonmedical Debt)	800-827-0648
Vets Center Call Center	877-327-8387
White House VA Hotline	855-948-2311
Veterans Crisis Line	800-273-8255

QUICK GUIDE—
APPLYING FOR BENEFITS

The following are guides to applying for the most commonly used benefits and services for veterans.

Applying for Benefits Under The PACT Act

The full name of the PACT Act is The Sergeant First Class (SFC) Heath Robinson Honoring Our Promise to Address Comprehensive Toxics (PACT) Act. The PACT Act is a health care and benefits expansion. It expands eligibility for VA health care for veterans with toxic exposures, and veterans of the Vietnam, Gulf War and Post-9/11 eras.

If you are a Gulf War or Post-9/11 veteran you can get free VA health care for any condition related to your service for up to 10 years from the date of your most recent discharge or separation. You can enroll at any time during this period and get care as well.

You can also file a disability claim for a new presumptive condition. You can do this online or by mail, or with the help of a trained professional. If the VA denied your disability claim in the past and it now considers your condition presumptive, you can submit a Supplemental Claim, and the case will be reviewed again.

If you have a pending claim for a condition that's now presumptive, you don't have to do anything. If the VA added your condition after you filed your claim, it will still consider it presumptive and will send a decision notice when completing their review. Follow the below instructions for applying for health care and disability benefits.

VA was set to begin processing PACT Act-related benefits in January 2023. If you apply for benefits within the next year and the VA grants your application, they'll likely backdate your benefits to the date of the bill signing, meaning they would pay you the amount you would have received from August 10, 2022, to the date they grant your application.

Apply for VA Health Care

Collect the following:
- Discharge papers (DD214 member-4 or equivalent)
- Your most recent tax return
- Social security numbers for yourself and your dependents
- Account numbers for insurance programs you're enrolled in
- If applicable, your VA Disability Rating Decision Application
- Apply by completing the health care application form (VA Form 10-10EZ) in one of a few ways.
  - Visit online (VA.gov)
  - Call 877-222-8387 and press 1
  - Visit in person at a VA medical center
  - Print out and mail the completed form to the Health Eligibility Center (2957 Clairmont Rd. Suite 200 Atlanta, GA 30329)
  - Call 877-222-8387 and press 0 if you haven't heard back from the VA within one week of submitting your application
- If approved, take steps to access your benefits.
  - Review your priority assignment group and personalized benefits handbook, mailed to you by the VA after enrollment.
  - Contact your local VA Medical Center to set up an appointment.

- o Make an appointment to obtain your Veterans Health Identification Card (VHIC).
- You are eligible for VA health care if:
  - o VA rated you for a service-connected disability.
  - o You received a purple heart, or you are a former POW.
  - o You are a combat veteran who separated within the last five years.
  - o You are eligible for Medicaid.
  - o Your family income is below the income threshold.
  - o You served in Vietnam, Southwest Asia during the Gulf War or Camp Lejeune during certain periods.

To Apply Online (The Equivalent of the VA Form 10-10 EZ):

If you're not signed into your account on the VA's health care website, you'll need to sign into your account. If you sign in, it can make the process faster because parts of the application can be pre-filled with account details. If you do it this way, you can also save your progress and finish filling out the form later. You have 60 days from the date you start or update your application to submit the form. After 60 days, your form isn't saved, and you'll have to start over.

If you need help filling out an application, you can work with an accredited representative such as a Veterans Service Officer.

You may also qualify for vision and dental benefits as part of your VA health care benefits.

When you apply for VA health care, you're assigned to 1 of 8 priority groups. Then, the symptoms that need immediate help can get signed up quickly. Veterans with a service-connected disability are assigned the highest priority. Veterans who earn a higher income and don't have a service-connected disability are assigned a lower priority level.

If you're applying in person, you fill out the 10-10EZ and then bring it to your nearest VA medical center or clinic.

If you're using a power of attorney, you need to submit a copy of the Power of Attorney with the application

If you sign with an X, you must have two witnesses who also sign and print their name on the form.

Getting Started with Mental Health Services

If you need help right now, you can call, text, or chat online with responders at the Veterans Crisis Line. Many responders are also veterans. The line is always open. To connect call 1-800-273-8255 and Press 1.

You can also text a message to 838255. You can go online at veteranscrisisline.net and click chat. If you need emergency mental health care, you can go directly to your local VA medical center, regardless of discharge status or enrollment in other VA health care.

If you don't need immediate health care but would like to learn more about VA mental health care, you may be eligible for services.

All former servicemembers can access emergency VA mental health care.

Apply for a Disability Rating

A rating is between 0% and 100%. A rating may give you access to VA benefits like health care and compensation.

Collect the documents you'll need to file a claim, which includes:

- Discharge papers (DD214 member-4 or equivalent)
- Military medical records
- Hospital reports and VA medical records related to your disability
- Private medical records and hospital reports related to your disability
- Any other medical evidence related to your injury or illness
- Apply by submitting all documents in one of two ways—submitting a claim online or completing a claim form in person at a VA Regional Office
- Track your status at va.gov/claim-or-appeal-status
- If requested, provide more information or documentation
- Attend a VA medical exam if requested
- If you're approved, review your award letter for your rating

You may be able to file a claim if you are a veteran, a servicemember preparing to separate from the military, or you're a survivor or family member. Veterans Service Organizations can file a claim on your behalf as well.

Apply for VA Pension as a Veteran

The Veterans Pension Program provides monthly payments to wartime veterans who meet certain age or disability requirements and who have income and net worth within certain limits. If you are eligible, you'll need the following information:

- Social Security number or VA file number
- Military history
- Financial information and financial information of dependents
- Work history
- Medical information
- Bank account direct deposit information

You can apply online or by mail. You need to fill out an Application for Pension (VA Form 21P-527EZ).

If you're applying by mail, you'll send it to:

Department of Veterans Affairs
Pension Intake Center
PO Box 5365
Janesville, WI 53547-5365

You can also bring your application to a nearby VA regional office or work with an accredited representative to complete it.

You can start by submitting an intent to file while you gather documents to send in with your pension application. If you submit an intent to file before you file your claim, you may get retroactive payments.

You can submit an intent to file by phone by calling 800-827-1000. You can also download and submit an Intent to File a Claim for Compensation and/or Pension, Survivors Pension, and/or DIC (VA Form 21-0966).

Using Community Care

You may be eligible to see a community provider if you meet one of six eligibility criteria:

- Service is unavailable: You need a service not available at VA.
- Facility unavailable: You live in a state or territory without a full-service VA medical facility.

- Grandfathered choice: You were eligible for community care under the Veterans Choice Program distance criteria.
- Medical interest: Your VA provider determines it's in your best medical interest to be referred to a community provider.
- Quality standards: You need care from a VA medical service that VA determines is not providing care that complies with VA's quality standards.
- Access standards: You meet the access standards for average drive time or wait time for a VA appointment.

You may be eligible to see a community provider if your average drive time from home to a VA medical facility that provides the service you need takes more than:
- 30 minutes travel time to receive primary care, mental health, and non-institutional extended care services including adult home day care
- 60 minutes travel time to receive specialty care services

You may be eligible if the wait time for an appointment at a VA medical facility that provides the service you need exceeds:
- 20 days for primary care, mental health, and non-institutional extended care services
- 28 days for specialty care from the date of request with certain exceptions

Accessing Urgent Care
VA offers eligible veterans an urgent care benefit for treating minor injuries and illnesses at retail and urgent care providers who are part of VA's contracted network.

The urgent care benefit is offered to veterans and urgent care and same-day services VA provides through its medical facilities.

To be eligible for the urgent care benefit, you must be enrolled in the VA healthcare system, and you must have received care through VA from either a VA or a community provider within the last 24 months. To find an urgent care facility, use the VA Facility locator at www.va.gov/find-locations. You can also call your local VA medical facility to find an in-network urgent care provider. When you arrive, tell the provider you would like to use your VA urgent care benefit and confirm they're part of the VAs network.

Apply for Education Benefits
Education benefits include the Post-9/11 GI Bill, the Montgomery GI Bill, career counseling, and on-the-job training. To get started, decide on the type of education benefit you want to use, and learn about schools that offer VA-approved programs. You can utilize the online GI Bill Comparison tool to compare school statistics and your out-of-pocket costs.

Once you decide on a program, you will need your discharge papers and your orders if activated from the Guard or Reserves. You can submit your application online or complete an application in person at a VA regional office. You will work with your school's VA certifying official.
If you have questions, you can call the Education Call Center at 888-442-4551.
Some of the types of education benefits include:
- Post-9/11 GI Bill: You must have qualified active service on or after 9/11/01. It includes payment for 40-100% of tuition and stipends for housing and books.
- Montgomery GI Bill—Active Duty: This requires qualifying active service and a high school diploma or GED. The benefit includes a monthly stipend for tuition and fees.
- Montgomery GI Bill—Selected Reserve: This requires a six-year commitment to Selected Reserve, and it's usable only while serving. There's a monthly stipend for tuition and fees.

- Survivors and Dependents' Educational Assistance: This is for a spouse or dependent of a qualifying veteran, and it includes a monthly stipend for tuition and fees.

Apply for Burial in a VA National Cemetery and Memorial Products

To start, decide whether a burial will be in a VA national cemetery, state, or tribal veterans' cemetery. You might also decide on a private cemetery. From there, decide on the desired memorial products.

For burial in a VA national cemetery, a private cemetery, or for a Presidential Memorial Certificate, you will need to submit the following with your application, if available:
- Discharge documents (DD214 member-4) or equivalent
- Orders if activated by Executive Federal Order from the National Guard or Reserves
- Pre-need determination approval letter if previously completed

Apply for burial in a VA national cemetery by:
- Faxing discharge documents to 866-900-6417 or
- Scanning discharge documents to NCA.Scheduling@va.gov and
- Calling the National Cemetery Scheduling Office at 800-535-1117

Apply for a government-furnished headstone, marker, or medallion for placement in a private cemetery by submitting a:
- VA Form 40-1330, Claim for Standard Government Headstone or Marker or
- VA Form 40-1330M, Claim for Government Medallion for Placement in a Private Cemetery

Apply for a Presidential Memorial Certificate by submitting a VA Form 40-0247, Presidential Memorial Certificate Request Form.

Apply for a burial flag by submitting a VA Form 27-2008, Application for United States Flag for Burial Purposes.

A funeral director can assist a family by gathering and providing the VA's documents, calling the National Cemetery Scheduling Office to schedule the burial, acquiring a burial flag, requesting Military Funeral Honors from the Department of Defense, and serving as a liaison between the family and VA.

Request Your Military Service Records (Including DD214)

The best way to begin to request your military service records is to use the milConnect website (https://milconnect.dmdc.osd.mil/milconnect/). You will need to sign into milConnect to get your service records. To do this, you will need a Premium DS Logon account. You can't use your My HealtheVet or ID.me credentials. You will have to go to milConnect to sign up, register, and upgrade your account to premium.

Once you have done that, you can take the following steps to request your military service records:
- From the homepage where you are signed in, click or tap on Correspondence/Documentation
- Select Defense Personnel Records Information (DPRIS) from the drop-down menu
- Choose the Personnel File tab
- Select Request My Personnel File
- Fill out the form
- In the Document Index section, check the boxes next to the documents you want to request

- Click or tap the Create and Send Request button

You can request the following by taking the above steps:
- DD214
- DD215
- Report of Separation
- Other release papers
- Orders and endorsements
- Performance reports
- Awards and decorations
- Qualifications, licenses, and certificates
- Security clearance

Once you submit your request, you'll get an email letting you know the request is being processed. You'll receive a second email when your request is complete, and your files are ready to be viewed and downloaded. You can also check your request's status by signing into milConnect (https://milconnect.dmdc.osd.mil/milconnect/) and going to the Personnel File tab in the Defense Personnel Records Information section.

If you prefer, you can also request your military records in other ways. These options include:
- Mail or fax a Request Pertaining to Military Records (Standard Form SF 180) to the National Personnel Records Center (NPRC)
- Write a letter to the NPRC at 1 Archives Drive, St. Louis, Missouri 63138
- Visit the NPRC in person
- Contact your state or county veteran's agency

If you want to request someone else's military records because you're planning a burial for a veteran in a VA national cemetery, you can contact the National Scheduling Office at (800) 535-1117.

You can also request records by mailing or faxing a request to NPRC at the address above or taking any of the other steps named above. If you are not next of kin for the veteran and the veteran was discharged more than 62 years ago, you can order a copy of their military records from the National Archives. If the veteran was discharged less than 62 years ago, you might be able to request limited information from their Military Personnel File.

Applying for Disability Compensation
How to prepare to file a claim for disability compensation:
- Find out if you're eligible
- Gather any evidence and supporting documents you'll submit
- Make sure your claim is filled out completely
- Find out if you need any supplemental forms with your claim

Evidence can include:
- VA medical records and hospital records relating to claimed illnesses or injuries
- Private medical records and hospital reports that relate to your claimed illnesses or injuries
- Supporting statements, you want to provide from friends, family members, clergy members, law enforcement, or people you served with

You should also know that you have up to a year from the date the VA receives your claim to turn in any evidence. If you start your application and need time to gather more supporting documents, you can save your application and come back later to finish it. The VA will recognize the date you started your application as your date of claim if you complete it within 365 days.

The best way to apply for disability compensation is by obtaining an eBenefits account and applying online. When applying for compensation benefits, you must have access to the following information:

- Discharge or separation papers (DD214 or equivalent)
- Medical evidence (doctor and hospital reports)
- Dependency records (marriage and children's birth certificates)
- Alternatively, you may print and mail-in VA Form 21-526EZ, Application for Disability Compensation and Related Compensation Benefits, or call VA at 1-800-827-1000 to have the form mailed to you.

Service members may be able to receive disability compensation benefits sooner if you apply before your discharge from service through the Benefits Delivery at Discharge (BDD) or Quick Start pre-discharge programs.

How to File an Appeal

Starting in 2019, elements of the Veterans Appeals Improvement and Modernization Act of 2017 (Public Law 115-55) went into effect. This modernized the current claims and appeals process and includes three review options for disagreements with decisions. It requires improved notification of VA decisions, provides earlier claim resolution, and ensures you receive the earliest effective date possible.

Supplemental Claim

If you are adding new and relevant evidence, you can choose to file a Supplemental Claim. A reviewer looks at the evidence and determines whether it changes the decision. You can file a supplemental claim anytime if you have new evidence. The recommendation is that you file within one year from the date of your decision letter. To file a supplemental claim, you will use VA Form 20-0995: Supplemental Claim.

Higher-Level Review

If you request a higher-level review, you're asking for someone to review the same evidence. A senior reviewer will look at your case and determine if the decision can be changed. You can't submit any evidence, but either you or your representative can speak with the reviewer on the phone and tell them why you think the decision should be changed.

To follow this option, you would use VA Form 20-0996: Higher-Level Review. There is a deadline of one year from the date on your decision letter to request a higher-level review.

Please note, the below steps are only applicable for VA decisions received before February 19, 2019.

- After the Veterans Benefits Administration decides on your claim, you can appeal some or all of the decision if you don't agree with it. You can have a representative help you with your appeal. This person could be a lawyer, claims agent, or Veterans Service Officer.
- The first step in the process requires filing a Notice of Disagreement (NOD). When you file a NOD, you begin the appeals process. This needs to be filed within one year of the date on the letter notifying you of the claim decision. The NOD is VA Form 21-0958.
- You can fill out the NOD and mail it to the address on the VA claim decision notice you receive. You can also bring it to your local regional office.
- An accredited representative can help you file a NOD.
- After the VA receives your NOD, it's reviewed by a Decision Review Officer. If they don't determine there's enough evidence to grant your appeal, they send you findings in what's called a Statement of the Case. At that point, you can decide

- whether to continue your appeal to the Board of Veterans' Appeals.
- When you receive the Statement of the Case in the mail, you also get VA Form 9. VA Form 9 needs to be returned to the Board of Veterans' Appeals within 60 days if you want to continue your appeal.
- You can submit new evidence at any time in the process.
- The case is sent to the Board of Veterans Appeals when the Decision Review Officer finishes the review.
- If you have a serious illness, are in financial distress, or have sufficient cause, you can apply for your appeal Advanced on Docket. If you're older than 75, your appeal automatically receives this status.
- You have the option to request a hearing with a Veterans Law Judge, but the judge doesn't make a decision at the hearing.
- The board reviews your appeal and provides a decision on each issue. There are three ways it can be decided—allowed, remanded, meaning more evidence is needed or denied.
- If you disagree with the Board's decision, you can appeal to the Court of Appeals for Veterans Claims.

How to Apply for a VA Home Loan Certificate of Eligibility (COE)

The first step to get a VA direct or VA-backed home loan is to apply for a Certificate of Eligibility. This indicates and confirms to the lender that you qualify for the VA home loan benefit. The information you'll need to apply for your COE is listed below:

- Veteran: You will need a copy of your discharge or separation papers (DD214)
- Servicemember: Servicemembers need a statement of service signed by a commander, adjutant, or personnel officer. It should include your Social Security number, date of birth, the date you entered duty, duration of any lost time, and the name of the command providing the information
- Current or Former Activated National Guard or Reserve Member: You'll need your discharge or separation papers (DD214)
- Discharged member of the National Guard Who Was Never Activated: You'll need your Report of Separation and Record of Service (NGB Form 22) for each period of service OR your Retirement Points Statement (NGB Form 23) and proof of the character of service
- Surviving Spouse of a Veteran Who Died on Active Duty or Who Had a Service-Connected Disability: You'll need the veteran's discharge documents (DD214) if available, AND you'll need to fill out VA form 26-1817 (Request for Determination of Loan Guaranty Eligibility—Unremarried Surviving Spouses), OR if you don't receive DIC benefits, you'll need to fill out VA Form 21P-534EZ and have a copy of your marriage license and the veteran's death certificate.

You can then use the eBenefits service online to apply for your COE. You can also apply through the Web LGY system—ask a lender about this option. If you want to apply by mail, you can complete a Request for Certificate of Eligibility (VA Form 26-1880) and mail it to the address on the form. This option will take the longest.

How to File for Increased Disability Compensation

To apply for increased disability compensation, a veteran can use VA Form 21-526-EZ (Application for Disability Compensation and Related Compensation Benefits).

Veterans can also sign into the VA's website using their DS Logon, My HealtheVet, or their ID.me account. They can then begin the disability application. When the button is clicked to begin the process online, it is declaring intent to file. This reserves a potential effective date for when you could begin getting benefits. You have a year from the day you submit your intent to file to complete an application.

The following are the steps to file a claim for increased disability compensation:

- When you file for a disability increase, you'll need to provide evidence to support your claim. This evidence can include VA medical and hospital records, private medical and hospital records, or support statements.
- The VA advises that you check to see if you're already receiving the maximum disability rating for your claimed condition before filing for an increase.
- After applying, you receive a confirmation message, which should be printed for your records.
- The VA then processes the applications in the order they're received.
- Once it's received, you receive a notice in the mail with the decision.

Applying for VA Survivor Benefits

To begin, learn more about the different survivor benefits and services that may be available. Collect the necessary paperwork including:

- Veteran's discharge paperwork—DD214 or other separation documents
- Veteran's death certificate
- Proof of your relationship to the veteran (marriage or birth certificate)

Apply by submitting all required documents, including the applicable VA Form and the veteran's death certificate to the required servicing location. You can mail it to the VA office that serves your state.

You can also work with an accredited Veterans Service Organization or directly to your VA Regional Office for assistance.

For Life Insurance Claims, call 800-669-8477.

Surviving family members can learn whether they are eligible for survivor benefits by contacting the Office of Survivors' Assistance at 202-461-1077 or contacting the VBA directly at 800-827-1000.

Benefits may include:

- Internment in a National Cemetery
- Memorialization and Legacy Programs
- Financial Burial Benefits—VA Form 21P-530
- Dependency and Indemnity Compensation (VA Form 21P-534EZ)
- Parents' Dependency and Indemnity Compensation (VA Form 21P-535)
- Survivors' Pension (VA Form 21P-534EZ)
- Dependent's Educational Assistance Program (VA Form 22-5490_
- Fry Education Scholarship
- VA Home Loan Benefits
- Civilian Health and Medical Program of the Department of Veterans Affairs (CHAMPVA)

Types of Veteran ID Cards

There are different types of veteran ID cards you can use to show you're a veteran. The following are those cards and instructions on getting them.

Department of Defense ID Card

A Department of Defense ID Card is used to show your military status and to get access to services at military bases. You can also use the card to get veteran discounts. If you have a DoD ID Card, you don't need another type of photo ID to prove you're a veteran.

You'll need to fill out an Application for Identification Card/DEERS Enrollment (DD Form 1172-2). Then, turn your completed application into a Real-Time Automated Personnel Identification System (RAPIDS) office for processing.

11

Veteran Health Identification Card
When you're enrolled in VA health care, you can get a Veteran Health Identification Card (VHIC) that you use to check into your appointments at VA medical centers. You can also use a VHIC to get veterans' discounts at restaurants, stores and other businesses. If you have a VHIC, you don't need to request another type of photo ID card to prove you're a veteran. To get this card, you need to be enrolled in VA health care.

Veteran ID Card
A Veteran ID Card (VIC) is a form of photo ID you can use to get military discounts, so you don't have to bring your military discharge papers with you wherever you go.

To be eligible, you must have served on active duty in the Reserves or the National Guard, including the Coast Guard. You must have also received an honorable discharge or a general discharge. If you received other-than-honorable, bad conduct or dishonorable character of discharge, you aren't eligible for a Veteran ID Card.
You can apply for the card online using your DS Logon, My HealtheVet, or Id.me account.

Veteran's Designation on a State-Issued Driver's License or ID
Currently, all 50 states and Puerto Rico offer a veteran designation printed on state-issued driver's licenses or IDs. Most states will ask you to provide a copy of your discharge papers, such as your DD214, to get this designation. Some states may also require other documents. You should talk to your local DMV about what you need to apply for a veteran's designation in your state.

Filing a Pre-Discharge Claim
If you have an illness or injury, you believe was caused or made worse by active-duty service, you can file a claim for disability benefits 180 to 90 days before leaving the military. This may help speed up the claims decision process so that you can get benefits sooner.

You must be a service member on full-time active duty, have a known separation date, and that date must be in the next 180 to 90 days.

If you have 180 to 90 days left, you'll apply through the Benefits Delivery at Discharge (BDD) program. To do so, you'll need to submit your service treatment records along with your application. You'll need to be available for 45 days from the date you apply for a VA medical exam and finish all phases of the VA/DoD medical separation examination process before your military release.

2022 UPDATES

Following is a summary of some of the important legislative changes and VA policy changes affecting veterans' benefits during 2022.

New Measure Signed into Law to Make Sure GI Bill Recipients Receive Their Full Housing Stipend

On the day special pandemic protections for student veterans were set to expire, President Joe Biden signed into a law a measure ensuring recipients of the GI Bill will receive their full housing stipends through next semester, even if their classes go online.

The REMOTE Act, finalized by Congress, will affect around 57,000 students who currently use the GI Bill for their college classes. In some instances, individuals could have lost thousands of dollars in housing support over the next six months.

The issue was with how post-9/11 GI Bill benefits are paid out to students who attend college classes remotely and not in person.

Students using veterans' education benefits receive tuition money plus a monthly housing stipend. Individuals enrolled in in-person; traditional classes receive the full financial benefit.

Students in online-only classes get half the housing stipend.

The difference between a full and half a stipend for housing can be hundreds of dollars up to thousands of dollars, depending on the school and its location.

The new law pushes the authorization back until next summer.

VA Increases Access to Home, Community-Based Services for Eligible Veterans

The Department of Veterans Affairs Office of Geriatrics and Extended Care announced an expansion of the Home-Based Primary Care, Medical Foster Home and Veteran-Directed Care programs in an effort to make them available at all VA medical centers by the end of fiscal year 2026.

VA says it will add 58 medical foster homes and 70 veteran-directed care programs to veterans affairs medical centers (VAMCs) around the country. The Department also says it will add 75 home-based primary care teams to areas with the highest unmet need.

According to the VA's Policy Analysis and Forecasting Office, the number of veterans of all ages who are eligible for nursing home care is estimated to expand from the two million in 2019 to more than four million by 2039. As the population grows, VA says it wants to make sure it's able to provide the highest levels of care to veterans in the least-restrictive possible settings.

The programs provide an in-home or smaller care setting than traditional institutionalized long-term care. The smaller setting reduces the transmission of infectious diseases, and many veterans choose these programs over institutionalized care.

VA Sets New Threshold for Reporting Benefit and Medical Debt

The VA published a final rule in the Federal Register on Feb. 2 amending procedures for reporting debt to consumer reporting agencies. The revisions alter the VA's minimum requirements for reporting debt.

The rule also creates an opportunity for relief and helps veterans who are experiencing financial hardship.

The VA says this is particularly important as it has resumed debt collection activities as of October 1, 2021.

Veterans Will Be Getting Improved Breast Cancer Screenings Under New Legislation

Veterans, including individuals who may be at an increased risk for breast cancer because of service near burn pits, could have better access to breast cancer screenings under a set of bills that passed Congress recently.

Recently the House voted unanimously to approve the "Dr. Kate Hendricks Thomas Supporting Expanded Review for Veterans in Combat Environments," or SERVICE Act. The law would require the VA to conduct mammograms for all women who served near burn pits or other toxic exposures, regardless of age, family history or symptoms.

Also passing unanimously was the "Making Advances in Mammography and Medical options for Veterans Act."

Under this legislation, the VA would be required to create a strategic plan to improve breast imaging services within a year. It would create a three-year pilot program of telemammography for veterans who live in areas where the VA doesn't offer mammograms in-house and expand veteran access to clinical trials through a partnership with the National Cancer Institute.

VA Establishes Presumptive Service Connection for Rare Respiratory Cancers

As part of the Unity Agenda from President Biden, aimed at supporting the nation's veterans, the Department of Veterans Affairs recently announced the addition of nine rare respiratory cancers to its list of presumed service-connected disabilities due to military environmental exposures to fine particulate matter.

The list of respiratory cancers added to VA regulations through an Interim Final Rule and published in the Federal Register on April 26, 2022, include:

- Squamous cell carcinoma of the larynx
- Squamous cell carcinoma of the trachea
- Adenocarcinoma of the trachea
- Salivary gland-type tumors of the trachea
- Adenosquamous carcinoma of the lung
- Large cell carcinoma of the lung
- Salivary gland-type tumors of the lung
- Sarcomatoid carcinoma of the lung
- Typical and atypical carcinoid of the lung

VA determined through a focused review of scientific and medical evidence there is biological plausibility between airborne hazards and carcinogenesis of the respiratory tract — and the unique circumstances of these rare cancers warrant a presumption of service connection.

The rarity and severity of these illnesses and the reality these conditions present, in a situation where it may not be possible to develop additional evidence, prompted VA to take this action.

VA Disability Rating for Sinusitis

Sinusitis is also called a sinus infection. This condition occurs when there's inflammation or swelling of the tissues lining the sinuses. The condition can be chronic when it lasts for three months or longer. The condition may be linked to other conditions such as allergic rhinitis, asthma, and ear infections.

On August 2, 2021, the VA started processing claims for asthma, rhinitis, and sinusitis. The claims are now being processed on a presumptive basis for those veterans exposed during their military service, and when the condition appeared within 10 years of their service.

It's been found that particulate matter pollution is associated with chronic sinusitis, affecting veterans who were in the Southwest Asia theater of operations, Syria, Afghanistan, Djibouti and Uzbekistan.

To be successful in making a service connection, a veteran needs to show evidence of a current diagnosis, an in-service occurrence that caused or aggravated the disability, and medical information that connects the current diagnosed disability to what happened in-service.

There are also secondary service connections that apply to secondary disabilities. These are disabilities that come from a service-connected disability that worsened or caused a new or pre-existing injury or illness.

The VA ratings for sinusitis include a 0% rating if it's detected by X-ray alone.
There's a 10% rating awarded if the sinusitis is manifested in one or two incapacitating episodes a year, requiring prolonged antibiotic treatment of four to six weeks.

Thirty percent is awarded if there are three or more incapacitating episodes annually requiring antibiotics of four to six weeks. Fifty percent is awarded to veterans requiring surgery for sinusitis with chronic osteomyelitis.

New DD-214 Created for Guard and Reserve Troops

Troops in the National Guard and across reserve components of all military branches are getting a new standardized discharge form by 2025, according to Defense Department instruction released in May.

When the new policy goes into effect, members will receive the new DD Form 214-1 when they retire or separate from one of the reserve components of the military. Currently each service has its own reserve component separation form like the NGB22 or the NAVPERS 1070/615.

Groups representing reserve component troops like the Reserve Organization of America and the National Guard Association of the U.S. have been advocating for the change.

The new form comes after the fiscal 2020 defense policy bill directed DoD to study if it would be possible to standardize the forms.

All the different separation forms often created problems for reserve component troops when they were trying to access veterans' benefits.

Part-time troops are usually only given regular DD-214s when they complete an active-duty tour of 90 days or more. Before the Global War on Terrorism increase the pace of reserve component deployments, some troops went their entire career and even retired without getting one at any point after their initial training.

The new form will show when a retiring service member is eligible to start collecting non-regular retired pay, which members can't receive until they're 60 without qualifying service. Previously, retired service members had to track the date and do the math themselves, so they'd know how to submit their request to start their retirement pay and benefits.

PACT Act Becomes Law

The PACT Act is a newly signed law that expands VA health care and benefits for veterans exposed to burn pits and other toxic substances. The law provides veterans and their survivors with care and benefits.

The PACT Act is considered the largest health care and benefit expansion in VA history. The full name of the legislation is The Sergeant First Class (SFC) Heath Robinson Honoring Our Promise to Address Comprehensive Toxics (PACT) Act.
The Pact Act brings some key changes.

The law will expand and extend eligibility for VA healthcare to veterans with toxic exposures and veterans of the Vietnam, Gulf War, and post-9/11 eras.

The PACT Act adds more than 20 presumptive conditions for burn pits and other toxic exposures and adds more presumptive-exposure locations for Agent Orange and radiation.

The legislation requires the VA to provide a toxic exposure screening to every veteran enrolled in VA health care, and helps improve research, treatment related to toxic exposures and staff education.

Veterans and survivors can now apply for PACT Act-related benefits.

VA Launches Improved Access to Care Site

Based on feedback from veterans, the Department of Veterans Affairs launched a new Access to care site. According to the VA, it represents a simpler, user-friendly experience that will make it easier for veterans to make informed decisions as they plan their health care appointments.

As of July 19, veterans and the public can see veteran experience information for care delivered through VA facilities. You can also use the site to get a clear view of average wait times, with more detailed information on available health care services and specialties. The site is available at www.accesstocare.va.gov.

New VA Life Insurance Program Coming Jan. 2023 for Veterans with Service-Connection

In January 2023, VA plans to launch a new life insurance program, the Veterans Affairs Life Insurance or VALife. It will provide guaranteed acceptance whole life insurance coverage to veterans aged 80 and under, with any level of service-connected disability. Some veterans aged 81 and older may also be eligible.

Guaranteed acceptance is a whole life policy not requiring a medical exam and it doesn't ask health questions. It doesn't have a limited two-year window to sign up. Whole life insurance provides coverage for the entire life of a policyholder if their premiums are always paid. Premium rates are locked in for the life of the policy, and they don't increase as a policyholder ages.

Under Public Law 116-315, the new program meets the needs of service-connected veterans who may not have previously qualified for life insurance with VA.

Program of Comprehensive Assistance for Family Caregivers Expands to Veterans of All Eras

At the start of October, the VA announced the Program of Comprehensive Assistance for Family Caregivers was expanding, to include eligible veterans and family caregivers of all eras.

Previously, PCAFC was available only to eligible veterans who served on or after September 11, 2001. The expansion opens the program up to eligible veterans of all eras, including veterans who served after May 7, 1975, and before September 11, 2001.

The program provides caregivers of eligible veterans with financial assistance, resources, education, peer support, beneficiary travel and health insurance.

CHAPTER 2

BENEFITS FOR ELDERLY VETERANS

According to data from the U.S. Department of Veterans Affairs, around 34 percent of the country's veterans are age 70 and above. These veterans served in conflicts worldwide, including World War II, the Korean War, the Vietnam War, and even in the Persian Gulf War. As veterans age, the Department of Veterans Affairs (VA) will provide benefits and services that address a variety of issues including the changing health risks they face and financial challenges through VA benefits and health services.

The PACT Act (2022 Update)
The PACT Act is a new law expanding health care and benefits for veterans exposed to Agent Orange and other toxic substances. Based on the PACT Act, there are two new Agent Orange presumptive conditions. One is high blood pressure, also known as hypertension. The other is monoclonal gammopathy of undetermined significance or MGUS. You may also be eligible for disability compensation based on other Agent Orange presumptive conditions.

Please see Chapter 26 for more information.

Blue Water Navy Act of 2019
On January 1, 2020, the Blue Water Navy Act of 2019 went into effect. The Act was signed into law on June 25, 2019 and extends the presumption of herbicide exposure such as Agent Orange to Blue Water Navy veterans who served as far as 12 nautical miles from the shore of Vietnam and have since developed one of 14 conditions related to exposure. These 14 conditions believed to be diseases related to Agent Orange include:
- Chronic B-cell leukemia
- Hodgkin's disease
- Multiple myelomas
- Non-Hodgkin lymphoma
- Prostate cancer
- Respiratory cancer, including lung cancer
- Soft tissue sarcomas other than osteosarcoma, chondrosarcoma, Kaposi's sarcoma, or mesothelioma
- AL amyloidosis
- Chloracne
- Diabetes mellitus type 2
- Ischemic heart disease
- Parkinson's disease
- Peripheral neuropathy
- Porphyria cutanea tarda

- Peripheral

Veterans may be eligible for disability compensation and other benefits. Additionally, if you're a veteran who served in the Korean Demilitarized Zone (DMZ) between September 1, 1967, and August 31, 1971, you may qualify for compensation and benefits for yourself and your family members.

You can file an initial claim that hasn't been previously decided by VA by submitting form 21-526EZ, Application for Disability Compensation and Related Compensation Benefits.

VA will be using the new law to automatically review claims currently with the VA review process or under appeal. However, if you had an herbicide exposure claim with one or more presumptive conditions denied in the past, you are urged to file a new claim.

When you begin the claims process, be sure to provide or identify any new and relevant information regarding your claim, such as the dates the vessel you were serving on traveled through the offshore waters of the Republic of Vietnam or updated medical information.

Submit a VA Form 20-0995, Decision Review Request: Supplemental Claim.

If you are deemed eligible, you may receive a monthly disability compensation payment and free healthcare related to your disability. The amount of compensation you may receive is determined by your diagnosed condition and level of disability.

Another important component of the Blue Water Navy Act includes changes to the VA Home Loan program. VA now allows the no-down-payment option on guaranteed home loans, regardless of the mortgage amount for ALL Veterans. In addition, there is a reduction in the funding fee required for Reservists and National Guard borrowers and other changes.

Veterans over the age of 85 who have life-threatening illnesses will have priority in claims processing.

Thailand Military Bases and Agent Orange Exposure
Vietnam-era Veterans whose service involved duty on or near the perimeters of military bases in Thailand anytime between February 28, 1961, and May 7, 1975, may have been exposed to herbicides and may qualify for VA benefits.

Please see Chapter 26 for more information.

Parkinson's Disease
Veterans with suspected movement disorders are typically referred to a Parkinson's Disease, Research, Education and Clinical Center (PADRECC) or a Consortium Center by a consult from their VA primary care provider or general neurologist.

Veterans diagnosed with Parkinson's disease who served in-country or on the inland waterways of Vietnam between January 9, 1962, and May 7, 1975, are presumed exposed to Agent Orange or other herbicides and are eligible for presumptive service-connection.

VA Benefits
There are a wide variety of benefits available to all military veterans, including benefits for elderly veterans. Some of these benefits include disability compensation, pension, education and training, health care, home loans, insurance, vocational rehabilitation, and burial.

Aid & Attendance

Aid and Attendance (A&A) is a monthly pension that's paid if veterans meet certain conditions listed below. This VA program is designed to provide an additional amount of money for elderly veterans eligible for or are receiving a VA Pension benefit. Aid & Attendance is not a separate benefit but is instead provides additional monthly income over and above the Basic Monthly Pension. To be eligible for this, you must also meet the requirements for Basic Pension.

Eligibility conditions include:
- The veteran is required to have help performing daily functions including bathing, eating, or dressing
- The veteran is bedridden
- The veteran is a patient in a nursing home
- The veteran has eyesight limited to a corrected 5/200 visual acuity or less in both eyes, or concentric contraction of the visual field to 5 degrees or less

Housebound

Housebound is an increased monthly pension amount paid if a veteran is substantially confined to their immediate premises because of a permanent disability. This is for veterans with a disability rating of 100%, preventing them from leaving their home.

How to Apply for Aid & Attendance and Housebound

A veteran may apply for Aid and Attendance or Housebound benefits by writing to the Pension Management Center (PMC) that serves their state. Veterans can also visit their local, regional benefits offices to file a request.

Copies of included evidence should include, if possible, a report from an attending physician validating the need for Aid and Attendance or Housebound type care. The report should include sufficient detail to determine whether a disease or injury is producing physical or mental impairment, loss of coordination, or conditions affecting the ability to dress and undress, feed oneself, attend to sanitary needs, and to keep oneself ordinarily clean and presentable.

Regardless of whether the claim is for Aid and Attendance or Housebound, the report should indicate how well the applicant gets around, where the applicant goes and what the applicant is able to do during a typical day. It's also important to determine whether the claimant is confined to the home or immediate premises.

VA Health Care for Elderly Veterans

Geriatrics is health care for elderly Veterans with complex needs. Extended care – also known as long term care – is a program for Veterans of all ages who need the daily support and assistance of another individual. Elderly Veterans can receive geriatric and long-term care programs at home, at VA medical centers, or in the community.

Geriatrics and Extended Care Services

Geriatrics and Extended Care Services (GEC) is committed to optimizing the health and well-being of Veterans with multiple chronic conditions, life-limiting illness, frailty, or disability associated with chronic disease, aging, or injury. Below are some of the main options for care in these categories.

Adult Day Care

Adult Day Health Care is a veteran's program that can include daytime care, social activities, peer support, companionship, and recreation. It's designed for veterans who need skilled services, case management and help with daily living activities. Health services including care from nurses, therapists, social workers, and others may be available.

Adult Day Care can provide respite care for a family caregiver and help veterans and their caregivers gain skills to manage the veteran's care at home. The program may be provided at VA medical centers, State Veterans Homes, or community organizations.

Home-Based Health Care

Home Based Primary Care is health care services provided to veterans in their home, with supervision from a VA physician who manages a health care team delivering services. Home Based Primary Care is for Veterans who have complex health care needs for whom routine clinic-based care isn't effective.

The program is for veterans who need skills, services, case management, and help with activities of daily living. The program is also for veterans who are isolated or who have a caregiver experiencing burden.

Hospice Care

Hospice Care is comfort care provided to you and your family if you have a terminal condition, with less than six months to live, and are no longer seeking treatment other than palliative care. Hospice Care can be provided at home, in an outpatient clinic, or an inpatient setting.

Hospice is a benefit that the VA offers to qualify veterans in the final phase of their lives, typically six months or less. This multi-disciplinary team approach helps Veterans live fully until they die. The VA also works very closely with community and home hospice agencies to provide care in the home.

Other geriatric care options include homemaker and home health aide care, palliative care, respite care, skilled home health care, and tele-healthcare.

Residential Settings

Another option for elderly veterans in terms of long-term care is residential settings. These can include community residential care, medical foster homes, adult family homes and assisted living.

Residential settings may also include nursing homes, including community living centers and community nursing homes, and State Veterans Homes.

State Veterans Homes are facilities providing nursing homes, domiciliary or adult day care. They're owned, operated, and managed by state governments. To participate in the State Veterans Home program, the VA must formally recognize and certify a facility as a State Veterans Home. The VA then surveys all facilities each year to make sure they meet VA standards.

Eligibility for State Veterans Homes is based on clinical need and setting availability, and each individual state establishes eligibility and admissions criteria for the homes. Some State Veterans Homes may admit non-veteran spouses and gold star parents, while others may admit only veterans.

Paying for Long-Term Care

Home and Community-Based Services are part of the VA Medical Benefits Package. All enrolled veterans are eligible for these services. However, to get the service, you must have a clinical need for it, and the service must be available in your location.

VA Standard Benefits Package includes:
- Geriatric Evaluation to assess your care needs and to create a care plan

- Adult Day Health Care
- Respite Care
- Skilled Home Health Care

Some Home and Community Based Services may be prioritized based on your level of VA service-connected disability. Nursing Home and Residential Settings have different eligibility requirements for each setting. The VA does not pay for room and board in residential settings such as Assisted Living or Adult Family Homes. However, you may receive some Home and Community Based Services while living in a residential setting.

The VA will provide Community Living Center (VA Nursing Home) or community nursing home care IF you meet certain eligibility criteria involving your service-connected status, level of disability, and income.

Service-Connected Disability
A "service-connected disability" is a disability that is related to your active military service. Your disability is assigned a rating (0% to 100%) based on how severely it impacts your daily life. The greater your disability, the higher you're rating.

VA Disability Compensation
VA Disability Compensation is a monthly tax-free payment to Veterans who have a service-connected disability. The higher your rating, the higher your monthly payments will be.

However, having a disability caused by your active military service does NOT automatically start disability compensation payments. You need to APPLY for those benefits. It is important to do this because your service-connected disability status affects how much you pay for VA health care services, what programs you are eligible for, and your priority in receiving certain services

How Do I Apply for VA Disability Compensation?
There are several ways to apply for your service-connected disability status. You can:
- Complete and mail VA Form 21-526EZ to your Veterans Benefits Administration (VBA) regional office. To find your VBA regional office, visit the facility locator site, or
- Complete the application online using an eBenefits account, or
- Work with a representative, such as a Veterans Service Officer. You can search for a representative here or visit your local VBA office for assistance.

Your disability needs to be reviewed by staff at VBA and assigned a rating before you can receive payments.

Other Available Monthly Compensation
If you receive VA compensation for a service-connected disability, you may be eligible to receive additional monthly monetary benefits if you are ALSO:
- Require significant help with your personal care needs by another person because of your disability, or
- Are bedridden because of your disabilities

Veterans Pension
Veterans Pension is a sometimes-tax-free monetary benefit paid to low-income wartime Veterans who are ALSO:
- 65 or older, or
- Totally and permanently disabled, or
- Living in a nursing home receiving skilled nursing care, or

- Receiving Social Security Disability Insurance, or
- Receiving Supplemental Security Income Low-income guidelines are set by Congress. Additional income is allowed if you have dependents.

How much does it pay?
Your payments will raise your income to the Maximum Annual Pension Rate and not exceed it. Pension benefits are paid monthly.

How do I apply?
Complete and mail VA Form 21P-527EZ to your Veterans Benefits Administration (VBA) regional office. To find your VBA regional office:

- Visit the facility locator site, or
- Complete the application online using an eBenefits account, or
- Work with an accredited representative or agent, such as a Veterans Service Officer. You can search for representatives on the VBA website (benefits.va.gov) or visit your local VBA office for assistance.

World War II Veterans

World War II (WWII) was the most widespread war in history, with more than 100 million people serving in military units. About 16 million Americans served during WWII, and many veterans are now receiving VA benefits including Pension and Health Care. WWII Veterans who were a part of the Occupation Forces assigned to Hiroshima and Nagasaki, Japan soon after the detonation of Atomic-Bombs over those respective cities, and those American prisoners of war (POWs) who were housed close to those cities are sometimes called "AtomicVeterans."

VA Benefits
World War II Veterans may be eligible for a wide variety of benefits available to all U.S. military Veterans. VA benefits include disability compensation, pension, education and training, health care, home loans, insurance, vocational rehabilitation and employment, and burial. The following sections provide information tailored to the experiences of World War II Veterans to help you better understand specific VA benefits for which you may qualify.

Benefits for World War II Veterans Exposed to Ionizing Radiation

World War II-era Veterans might qualify for health care and compensation benefits if exposed to ionizing radiation during military service. Health care services include an Ionizing Radiation Registry health exam and clinical treatment at VA's War Related Illness and Injury Study Centers. You may also be entitled to disability compensation benefits if you have certain cancers due to exposure to ionizing radiation during military service.

Benefits for World War II Veterans Who Participated in Radiation-Risk Activities

World War II-era veterans may qualify for health care and compensation benefits if you participated in certain radiation-risk activities, such as nuclear weapons testing, during military service. These Veterans may be informally referred to as "Atomic Veterans." Health care services include an Ionizing Radiation Registry health exam and clinical treatment at VA's War Related Illness and Injury Study Centers. You may also be entitled to disability compensation benefits if you have certain cancers because of your participation in a radiation-risk activity during military service

Benefits for Survivors of Veterans with Radiation Exposure

Surviving spouses, dependent children, and dependent parents of Veterans who died due to diseases related to radiation exposure during military service may be eligible for survivors' benefits.

Korean War Veterans

Approximately 5.7 million Veterans served in the Korean War. Korean War Veterans are more prone to suffer from disabilities related to cold injuries from exposure to severe cold climates. Cold weather accounted for 16% of Army non-battle injuries and over 5,000 U.S. casualties of cold injury required evacuation from Korea during the winter of 1950-1951. In many instances, servicemembers could not seek or were unable to obtain medical care for cold injuries because of battlefield conditions.

VA Benefits

Korean War veterans may be eligible for a range of benefits available to all U.S. military veterans, including disability compensation, pension, education and training, health care, home loans, insurance, vocational rehabilitation and employment and burial.

Benefits for Veterans Who Experience Cold Injuries

Veterans who experienced cold injuries may have medical conditions resulting from a cold-related disease or injury. Examples of cold-related medical conditions include:

- Skin cancer in frostbite scars.
- Arthritis.
- Fallen arches.
- Stiff toes.
- Cold sensitization.

These cold-related problems may worsen as veterans grow older and develop complicating conditions such as diabetes and peripheral vascular disease, which place them at higher risk for late amputations.

Benefits for Korean War Veterans Exposed to Ionizing Radiation

Korean War-era veterans may qualify for health care and compensation benefits if exposed to ionizing radiation during military service. Health care services include an Ionizing Radiation Registry health exam and clinical treatment at VA's War Related Illness and Injury Study Centers. You may also be entitled to disability compensation benefits if you have certain cancers due to exposure to ionizing radiation during military service. Korean War-era veterans may qualify for health care and compensation benefits if you participated in certain radiation-risk activities, such as nuclear weapons testing, during military service. These Veterans may be informally referred to as "Atomic Veterans."

Health care services include an Ionizing Radiation Registry health exam and clinical treatment at VA's War Related Illness and Injury Study Centers. You may also be entitled to disability compensation benefits if you have certain cancers due to your participation in a radiation-risk activity during military service.

Vietnam Veterans

United States military involvement in the Vietnam War officially began on August 5, 1964; however, the first U.S. casualty in Vietnam occurred on July 8, 1959.

Approximately 2.7 million American men and women served in Vietnam. During the war, over 58,000 U.S. military members lost their lives, and 153,000 were wounded. There were 766 prisoners of war of which 114 died in captivity.

VA Benefits
Vietnam Veterans may be eligible for a wide variety of benefits available to all U.S. military Veterans. VA benefits include disability compensation, pension, education and training, health care, home loans, insurance, vocational rehabilitation and employment, and burial.

Disability Compensation for Veterans Exposed to Agent Orange
VA presumes that some disabilities diagnosed in certain Veterans were caused by exposure to Agent Orange during military service.

Children of Veterans with Exposure to Agent Orange
Children of Veterans exposed to Agent Orange who have a birth defect including spina bifida, a congenital birth defect of the spine, and certain other birth defects may be entitled to VA benefits. These include monetary benefits, health care, and vocational rehabilitation services.

Please see Chapter 26 for more information on exposure to Agent Orange.

What to Remember
✓ Elderly veterans may be eligible for a variety of benefits including disability compensation, pension, education, training, home loans, insurance and burial
✓ Under the Blue Water Navy Act of 2019, you may be eligible for benefits based on Agent Orange exposure
✓ Veterans over the age of 85 with life-threatening illnesses will have priority in Agent Orange claims processing
✓ Veterans' pensions are monetary benefits paid to low-income veterans who meet other criteria
✓ To apply for disability compensation, you can complete the application online or use VA Form 21-526EZ

CHAPTER 3

TRANSITIONING & YOUNGER VETERANS

Younger veterans are often referred to as Gulf War Veterans. More than 650,000 Service members served in Operation Desert Shield and Desert Storm from August 2, 1990, to July 31, 1991. For VA benefits eligibility purposes, the Gulf War period is still in effect. This means that anyone who served on active duty from August 2, 1990, to present is considered a Gulf War Veteran. For example, the Veterans Pension benefit requires service during a wartime period. Therefore, any veteran who served on active military service for any period from August 2, 1990, to the present meets the wartime service requirement.

The PACT Act (2022 Update)

The PACT Act expands VA health care and benefits for veterans exposed to burn pits and other toxic substances. The change expands benefits for Gulf War era and post-911 veterans.

Please see Chapter 26 for more information on the PACT Act and toxic exposures.

Solid Start Program

The VA, in collaboration with the Departments of Defense and Homeland Security, introduced VA Solid Start, which will proactively contact all newly separated service members at least three times during their first year of transition from the military. The program is part of Executive Order 13822, which was issued to improve mental health care and access to suicide prevention resources available to transitioning uniformed service members in the year following discharge, separation, or retirement.

Concierge for Care

In 2018, the VA announced the launch of the Concierge for Care program. This health care enrollment initiative is intended to help connect former service members shortly after they separate from service.

It was designed as research showed more than a third of veterans aren't aware of VA healthcare benefits. A quarter of veterans don't know how to apply for these benefits.

The Concierge for Care program allows VA staff members to directly contact recently separated service members to answer questions, process their health enrollment applications over the phone, and help schedule their first VA medical appointment if needed. The goal is to contact veterans within a month of discharge.

Veterans who served in a theater of combat operations can enroll and receive cost-free health care for medical conditions related to their military service during the five years after discharge.

Transitioning Service Members

As part of the U.S. Department of Veterans Affairs (VA) ongoing efforts to help transitioning service members navigate and understand VA's various benefits, the agency recently updated VA's briefing portion of the Transition Assistance Program (TAP) – an interagency initiative authorized as a voluntary program in 1991 under the National Defense Authorization Act and made mandatory under the VOW to Hire Heroes Act in 2011 to help service members adjust to civilian life.

The new VA briefing is more collaborative and stimulating, helping service members make informed decisions about their health care, employment, housing, and other benefits.

Because no two transitions are the same, the updates deliver elements relevant to service members based on where they are in their career and life. The redesign will encourage whole-health support for service members and their family members, to include relevant information about Veterans Service Organizations (VSOs) and allow time to identify local VSO representatives.

The updates incorporated suggestions made by veterans, VSOs, and post-9/11 veterans' groups, including taking a more holistic view of a service member's new life and the psychosocial aspects of the transition to civilian life.

VA will now work to integrate Transition Assistance Program (TAP) objectives into the military lifecycle fully, and as an added benefit, will begin implementing a post-transition veteran survey.

VA Benefits

Gulf War veterans may be eligible for a wide variety of benefits available to all U.S. military veterans, including disability compensation, pension, education and training, health care, home loans, insurance, vocational rehabilitation and employment, and burial.

Gulf War Veterans' Illnesses

Certain illnesses and diseases are presumed by the VA to be related to military service in designated areas of Southwest Asia and may entitle the veteran to VA disability compensation benefits. For Gulf War veterans, these presumptive diseases include:

- Medically unexplained illnesses often called "Gulf War Syndrome."
- Certain infectious diseases
- Amyotrophic lateral sclerosis (ALS) diagnosed in all veterans who had 90 days or more continuous active military service

Veterans discharged under conditions other than dishonorable who served in the Southwest Asia theater of military operations, which includes the areas specified by regulation, but not Afghanistan, may be entitled to disability compensation for certain undiagnosed illnesses, certain diagnosable chronic disability patterns, and certain presumptive diseases even though these disorders did not become manifest during qualifying service.

Veterans who served in Afghanistan on or after September 19, 2001, may be entitled to disability compensation for certain presumptive diseases.

To learn more please see Chapter 26: Toxic Exposures.

Blended Retirement

A new retirement system was introduced in 2017, which is known as the "Blended Retirement System" or BRS. The blending is of two primary sources of retirement income.

One of these is the existing annuity provision for people retiring after 20 or more years of service, in addition to the Thrift Savings Plan (TSP), which is a government-operated 401(k) retirement account allowing members to invest their own money in stocks or government securities and also receive a contribution to that account from their employer.

BRS will use the annuity formula currently in place: the average of the servicemember's highest 36 months of basic pay times 2.5% of their years of service -- but the 2.5% is adjusted downward by half of a percentage point, from 2.5 to 2%.

To make up for this reduction, the government will contribute to a member's TSP. After the first 60 days in the service, all members will be enrolled in TSP and receive an automatic government contribution of 1% of basic pay into their account each month. Additionally, the servicemember will be automatically enrolled to contribute 3% of their basic pay to the TSP each month.

After two years of service, the government will match the member's contributions up to an additional 4%.

So, after two years, members can get up to a 5% government contribution on top of what they contribute each month. Therefore, if a member contributes 5% of their basic pay, the government will match it, making a total contribution to the TSP of 10% of their basic pay.

BRS also includes a mid-career continuation pay at about 12 years of service as a further incentive to continue serving toward the traditional 20 years to qualify for monthly military retired pay.

When you retire (at age 60 for guard/reserve members), you will be given the option to take your full retirement pay, or you can take a lump-sum payment of either 25% or 50% of your gross estimated retired pay and receive a reduced monthly annuity until the age of 67 when your retirement pay goes back up to the full amount.

Current members who have less than 12 total years of service when the new plan was effective in 2018 will be able to switch over to the new system. There will be no back pay, but matching contributions will begin on enrollment.

If you joined active duty:
- Before January 1, 2006 - you will remain in your current retirement system
- After December 31, 2005, but before January 1, 2018 - you can choose either your current retirement system or the BRS
- After December 31, 2017 - you will be enrolled in the BRS

The opt-in/election period for the Blended Retirement System began January 1, 2018, and ended on December 31, 2018.

Reserve Component Member Eligibility

Reserve Component members with more than 4,320 retirement points will remain under their current retirement system. Reserve Component members with less than 4,320 retirement points as of December 31, 2017, will have the choice of whether to opt into the new Blended Retirement System or remain in the legacy retirement system. New accessions after January 1, 2018, will automatically be enrolled in the new Blended Retirement System. The opt-in/election period for the Blended Retirement System begins January 1, 2018 and concluded on December 31, 2018.

The mid-career continuation pay for reserve component members is 0.5x monthly basic pay (of active duty).

Transition Assistance Program

The Transition Goals, Plans, Success program, known as Transition GPS, was created by the Department of Defense and the Department of Defense and other partnering agencies. The goal is to provide in-depth services to servicemembers as they transition to work, life, and home after the military.

Earned Services

Transition GPS delivers information about the benefits and services servicemembers have earned. Some of the offerings include:

- Pre-Separation Counseling: Before leaving the military, servicemembers can receive individual assessments and one-on-one counseling with military service representatives who specialize in the transition process.
- Enhanced VA benefits briefings: Trained instructors lead training sessions to cover what the VA offers to servicemembers, veterans, family members, and services.
- Education and employment programs are offered to strengthen job skills and jump-start careers.
- VA benefits and services are offered to improve overall quality of life
- There are other benefits designed to help transitioning service members maintain a stable home environment.

Personalized Plans

The Transition GPS program is designed to make adjusting to post-military life easier, and all transitioning servicemembers are required to participate in the program. This includes Guard and Reserve members demobilizing after 180 days or more of active service. Services include individual transitioning planning to create a customized roadmap that will outline career goals and steps necessary to achieve them.

Employment workshops are designed to show transitioning service members what civilian employers are looking for in applicants. There are tailored tracks that include three optional workshops to prepare servicemembers for separation or retirement.

Through two briefings—VA Benefits I and II—learn what you need to know about VA benefits and services in a highly interactive, activity-based class attended by your peers.

Information is provided through these briefings:

VA Benefits I. This is a four-hour briefing providing information on education, health care, compensation, life insurance, and home loans, as well as vocational rehabilitation and employment benefits information and counseling. The program assists you in developing a personal plan of action for using VA benefits.

It's open to 50 attendees per class; spouses and family members are encouraged to attend. VA Benefits II. This two-hour supplemental briefing with video presentations provides an overview of the eBenefits portal and further information on VA health care benefits and services and the disability compensation process.

Pre-Discharge

If a veteran has an illness or injury they believe was caused or made worse by active-duty service, they can file a claim for disability benefits 180 to 90 days before leaving the military. This can speed up the claim process, so the veteran is able to get benefits sooner.

To file a claim through the Benefits Delivery at Discharge (BDD) program, all of the following must be true:

- You're a service member on full-time active duty including a member of the National Guard, Reserves or Coast Guard, and
- You have a known separation date, and

- Your separation date is in the next 180 to 90 days, and
- You're available to go to VA exams for 45 days from the date you submitted your claim, and
- You can provide a copy of your service treatment records for your current period of service when you file your claim.

You can't file a BDD claim or add more medical conditions to your initial claim if you have less than 90 days left on active duty. You can still start the process of filing your claim before discharge.

You can't use the BDD program if your claim requires special handling, even if you're on full-time active duty with more than 90 days left of service.

You can't use the BDD program if any of the following are true:
- You need case management for a serious injury or illness, or
- You're terminally ill, or
- You're waiting to be discharged while being treated at a VA hospital or military treatment facility, or
- You're pregnant, or
- You're waiting for the VA to determine your Character of Discharge, or
- You can't go to a VA exam during the 45-day period after you submit your claim, or
- You didn't submit copies of your service treatment records for your current period of service, or
- You added a medical condition to your original claim when you had less than 90 days left on active duty, or
- You need to have a VA exam done in a foreign country, unless the exam can be requested by the overseas BDD office in either Landstuhl, German, or Camp Humphreys, Korea.

Filing a Pre-Discharge Claim While Overseas
A veteran can file a disability claim through the Benefits Delivery at Discharge (BDD) program while stationed overseas. The veteran will need to ensure they have enough time to finish the medical exam in the country where they're located, and that may include visiting specialty clinics.

If a veteran is in Germany (Europe, Africa, or the Middle East) they should contact the Landstuhl BDD Office. They can do so by scheduling an appointment through VERA (the Visitor Engagement Reporting Application). They can also send an email to GermanBDD.vbapit@va.gov.

To complete the pre-discharge claim process, the veteran will need a copy of their service treatment records for their most recent service period, and they must be available for 45 days from the date they are applying for a VA medical exam. The veteran must also have enough time in-country to finish the medical exam, which can include visits to several specialty clinics.

In Korea (countries in the Pacific Theater) a veteran can contact the Camp Humphreys clinic. They can schedule an appointment through VERA or call the VA at DSN 757-2914. A veteran can also send an email to KoreaBDD.vbapit@va.gov.

Quick Start Program
Quick Start lets a servicemember submit a claim for disability compensation 1 to 59 days before separation, retirement, or release from active duty or demobilization. Submitting a claim before discharge makes it possible to receive VA disability benefits as soon as

possible after separation, retirement, or demobilization.

Servicemembers with 1-59 days remaining on active duty or full-time Reserve or National Guard service or servicemembers who don't receive BDD criteria requiring availability for all examinations before discharge may apply through Quick Start.

The program is available nationwide and open to all servicemembers on full-time active duty including members of the National Guard and Reserves.

Integrated Disability Evaluation System

The "Integrated Disability Evaluation System (IDES) Examination" is used to determine a Servicemember's fitness for duty. The Departments of Defense (DoD) and Veterans Affairs (VA) worked together to make disability evaluation seamless, simple, fast, and fair. If the Servicemember is found medically unfit for duty, the IDES gives them a proposed VA disability rating before leaving the service. IDES participants may be entitled to Vocational Rehabilitation and Employment services.

Post-9/11 GI Bill

The Post 9/11 GI-Bill can help pay for education and housing to those who:
- Have been discharged honorably.
- Have at least 90 days of aggregate service on or after September 11, 2001.
- Discharged with a service-connected disability after 30 days.

What Does the Bill Cover?
- Tuition and fees paid directly to the school are based on the maximum in-state tuition rate at a public institution of higher learning. That is the maximum coverage unless Yellow Ribbon comes into play.
- Monthly housing allowance.
- Living expense stipend based on E-5 BAH with dependents based on the zip code of your school.
- Annual books and supplies stipend up to $1,000 paid proportionately based on enrollment.
- Can transfer to eligible dependents.

If you have served on or called to active duty after September 11, 2001:
- **100%** benefit if you have at least 36 months of duty
- **100%** benefit if you have 30 days service and were discharged with a service-connected disability
- **90%** 30 to 36 months of duty
- **80%** 24 to 30 months of duty
- **70%** 18 to 24 months of duty
- **60%** 12 to 18 months of duty
- **50%** 6 to 12 months of duty
- **40%** 90 days to 6 months of duty

Readjustment Counseling Services

The VA operates 300 community-based counseling Vet Centers. Many providers at Vet Centers are Veterans of combat themselves. Vet Centers provide readjustment counseling and outreach services to all Veterans who served in any combat zone. Military sexual trauma counseling and bereavement counseling are also provided. Services are available for family members for military-related issues, and bereavement counseling is offered for parents, spouses, and children of Armed Forces, National Guard, and Reserves personnel who died in the service of their country. Veterans have earned these benefits through their service, and all are provided at no cost to the Veteran or family.

Substance Abuse Programs

Some Veterans who return from combat have problems with the use of alcohol, tobacco, or drugs. This can include use of street drugs as well as using prescription medications in ways they weren't prescribed. Such substance use can harm health, cause mood and behavior problems, hurt social relationships, and cause financial problems. Available treatments address all types of problems related to substance use, from unhealthy use of alcohol to life-threatening addictions.

A patient coming to VA can expect to find the following types of care:
- First-time screening for alcohol or tobacco uses in all care locations;
- Short outpatient counseling including a focus on motivation.
- Intensive outpatient treatment.
- Residential (live-in) care.
- Medically managed detoxification (stopping substance use safely) and services to get stable.
- Continuing care and relapse prevention.
- Marriage and family counseling.
- Self-help groups.
- Drug substitution therapies and newer medicines to reduce cravings.

How to Get Help
- Speak with your existing VA healthcare provider.
- Contact the OEF/OIF Coordinator at your local VA Medical Center.
- Contact your local Vet Center.
- Call 1-800-827-1000, VA's general information hotline.

Home Loans

VA helps Servicemembers, Veterans, and eligible surviving spouses become homeowners. As part of the mission to serve you, VA provides a home loan guaranty benefit and other housing-related programs to help you buy, build, repair, retain, or adapt a home for your own personal occupancy.

VA Home Loans are provided by private lenders, such as banks and mortgage companies. VA guarantees a portion of the loan, enabling the lender to provide you with more favorable terms.

Purchase Loans help you purchase a home at a competitive interest rate often without requiring a down payment or private mortgage insurance. Cash-Out Refinance loans allow you to take cash out of your home equity to take care of concerns like paying off debt, funding school, or making home improvements.
Interest Rate Reduction Refinance Loan (IRRRL): also called the Streamline Refinance Loan, can help you obtain a lower interest rate by refinancing your existing VA loan.
Native American Direct Loan (NADL) Program: Helps eligible Native American Veterans finance the purchase, construction, or improvement of homes on Federal Trust Land, or reduce the interest rate on a VA loan.
Adapted Housing Grants: help Veterans with a permanent and total service-connected disability purchase or build an adapted home or to modify an existing home to account for their disability.
Other Resources: many states offer resources to Veterans, including property tax reductions to certain veterans.

VA Caregiver Support—Caring for Seriously Injured Post 9/11 Veterans

The Program of Comprehensive Assistance for Family Caregivers offers enhanced support for Caregivers of eligible veterans seriously injured in the line of duty on or after September 11, 2001.

Enhanced services for eligible participants may include a financial stipend, access to health care insurance, mental health services and counseling, caregiver training, and respite care. Veterans eligible for this program must:

- Have sustained or aggravated a serious injury — including traumatic brain injury, psychological trauma, or other mental disorder — in the line of duty, on or after September 11, 2001; and
- Need personal care services to perform one or more activities of daily living and/or need supervision or protection based on symptoms or residuals of neurological impairment or injury.

Participating veterans may appoint one primary caregiver and up to two secondary caregivers who serve as back-up. The support available to caregivers will depend on their designation—primary or secondary—and the needs of the veteran.

For more information, contact your local Caregiver Support Coordinator. Caregiver Support Coordinators are available at every VA Medical Center to assist veterans and their caregivers with the application process. Additional application assistance is available at 1-877-222 VETS (8387).

What to Remember
✓ Younger veterans are often referred to as Gulf War veterans
✓ Certain illnesses are presumed to be related to military service in Southwest Asia
✓ Gulf War veterans' illnesses include certain infectious diseases, medically unexplained illnesses, and ALS
✓ The Post-9/11 GI Bill can help pay for housing and education
✓ Veterans may be eligible for VA home loans provided by private lenders
✓ The PACT Act is a new law expanding VA healthcare and benefits for veterans exposed to burn pits, Agent Orange, and other toxic substances.

CHAPTER 4

DISABILITY COMPENSATION FOR SERVICE-CONNECTED DISABILITIES

VA disability compensation offers a monthly tax-free payment to veterans who got sick or injured while serving in the military, and to veterans whose service made an existing condition worse. You may qualify for VA disability benefits for physical conditions, and mental health conditions that developed before, during or after service.

2022 Update: The PACT Act is a new law expanding access to VA health care and benefits for veterans exposed to burn pits and other toxic substances. See Chapter 26 for more information.

You may be eligible for VA disability benefits if you meet both of the following requirements:
- You have a current illness or injury that affects your mind or body, and
- You served on active duty, active duty for training, or inactive duty training.

At least one of these must be true:
- You got sick or injured while serving in the military and can link this condition to your illness or injury, called an in-service disability claim, or
- You had an illness or injury before you joined the military and serving made it worse, called a preservice disability claim, or
- You have a disability related to your active-duty service that didn't appear until after you end your service, known as a post service disability claim.

Presumed disabilities include chronic (long-lasting illnesses appearing within a year after discharge," an illness caused by contact with toxic chemicals or contaminants, or an illness caused by time spent as prisoner of war.

You may be able to get VA disability benefits for conditions including:
- Chronic back pain resulting in a current diagnosed back disability
- Breathing problems stemming from a current lung disease or condition
- Severe hearing loss
- Scar tissue
- Loss of range of motion
- Ulcers

- Cancers caused by contact with toxic chemicals or other dangers
- Traumatic brain injury (TBI)
- Post-traumatic stress disorder (PTSD)
- Depression
- Anxiety

How to File a Claim

You should find out if you're eligible for VA disability compensation and gather any evidence you're going to submit when you file your claim. Your claim should be filled out completely and you should have all your supporting documents ready.

The VA Fully Developed Claims Program can help you get a faster decision on your disability benefits.

You can submit a fully developed disability claim if you're applying for compensation for:
- An illness or injury caused by or one that got worse because of your active-duty service, or
- A condition caused by or made worse by a disability the VA has already determined is service connected

For a claim to be considered fully developed, a veteran need to:
- Submit their completed Application for Disability Compensation and Related Compensation Benefits (VA Form 21-526EZ,) and
- Submit all evidence (supporting documents), and
- Certify there's no more evidence the VA might need to decide the claim, and
- Go to any VA medical exams required for a claim decision. The VA lets the veteran know if they need any exams.

Evidence you'll need to submit along with your disability claim includes:
- All private medical records related to the claimed condition like reports from your own doctor, and
- Any records of medical treatment you received for the claimed condition while serving in the military, and
- Any military personnel records they have related to the claimed condition, and
- Information about any related health records that you don't have but that the VA can request on your behalf from a federal facility

If a veteran doesn't think their service records will include a description of their disability, they can also submit letters from family, friends, clergy members, law enforcement or the people they served with who can tell the VA more about their claimed condition, and how and when it happened.

Using the Fully Developed Claims program won't affect your benefits or how your claim is handled. If the VA determines they need non-federal records to make a decision, they'll remove the claim from the Fully Developed Claims program and process it as a standard claim. Once you start a Fully Developed Claim, you have up to a year to complete it. If the VA approves your claim, you'll be paid back to the day you started it.

Standard Disability Claims
To submit a standard disability claim, you'll need to:
- Submit your completed Application for Disability Compensation and Related Compensation Benefits (VA Form 21-526EZ), and
- Let the VA know about related records not held by a federal agency, and give the VA any information needed to get them, and
- Go to any medical exams the VA schedules, if they decide they're needed to make

a decision on your claim.

You can start your online application, or you can file a claim by mail, in person, or with the help of a trained professional.

The process is slower for getting a decision on a standard disability claim because the VA gathers the evidence for your claim. If you want a faster decision, submit a fully developed claim, where you gather all the evidence and submit it with your claim.

If you aren't filing your disability compensation claim online, you can do it by mail.

You can print VA Form 21-526EZ, fill it out and send it to:

Department of Veterans Affairs
Claims Intake Center
PO Box 4444
Janesville, WI, 53547-4444

You can also bring your application in person to the VA regional office nearest you.

Finally, you can work with a trained professional, called an accredited representative, to get help filing a disability compensation claim.

If you're going to file for disability compensation using a paper form, you might want to submit an intent to file form first. This gives you time to gather evidence while avoiding a later potential start date, which is known as an effective date. If you notify the VA of an intent to file, you may be able to get retroactive benefits.

You don't need to notify the VA of an intent to file if you file for disability compensation online, because the effective date is automatically set when you start filling out your form online, before you submit it.

Evidence Needed for a Disability Claim

When you file a disability claim, the VA will review all available evidence to determine if you qualify. The VA is looking for evidence showing:
- A current physical or mental disability, and
- An event, injury or illness that happened while you were in the military to cause the disability

You'll need to submit or give the VA permission to gather your DD214 or other separation documents, your service treatment records, and any medical evidence related to your illness or injury such as medical test results or X-rays.

The evidence you need to support a claim can vary depending on the type of claim your filing.

- Original claim: This is your first claim you file for disability benefits. You'll need evidence of a current physical or mental disability from a medical professional or layperson, and an event, injury or disease that happened during active-duty service. You'll also need to show evidence of a link between your current disability and the event, injury or disease that happened during your service. If there's no evidence, but you have a chronic illness occurring within a year after discharge, an illness caused by time spent as a POW, or an illness caused by contact with hazardous materials, VA may conclude a link.
- Increased claim: These claims are for more compensation for a disability that has already been determined service-connected and has gotten worse. For this you'll need to submit current evidence from a medical professional or layperson that

shows the disability has gotten worse.

- New claim for added benefits or other benefit requests related to an existing service-connected disability: For this type of claim you have to submit evidence of a current mental or physical disability from a medical professional or layperson, and an event injury or disease that happened during your active-duty service, and a link between your current disability and the event, injury or disease.
- Supplemental claim: These claims provide new evidence to support a disability claim that was denied.

VA Pension Benefits

Occasionally, a veteran may be entitled to both VA Compensation benefits and VA Pension benefits. The VA is prohibited from paying both benefits concurrently. In the event a veteran is entitled to both, the VA will typically pay the higher of the monetary amount of the two.

Selected Reserve and National Guard

Since September 11, 2001, numerous members of the Armed Forces Reserves have been called to active duty. Some of these individuals had already filed claims for VA compensation, based on earlier periods of active service. In September 2004, the VA General Counsel issued a precedent opinion that discussed the effect of a return to active service on a pending disability compensation claim.

In general, a veteran's return to active duty does not affect his or her claim for VA benefits and does not alter either the veteran's right or the VA's duty to develop and adjudicate the claim. If the veteran is temporarily unable to report for a medical examination or take some other required action because of his or her return to active duty, the VA must defer processing the claim until the veteran can take the required action. The VA cannot deny a claim because a veteran is temporarily unavailable due to a return to active duty.

A veteran is not entitled to receive both active duty pay and VA disability compensation for the same period of time. However, the higher monetary benefit is usually paid. If a veteran with a pending claim dies on active duty before the claim is decided, an eligible survivor may be entitled to any accrued benefits payable.

Military Retired Pay

Historically, veterans were not permitted to receive **full** military retirement pay and VA compensation benefits at the same time. Veterans who were entitled to both had to either select one of the benefits or waive the amount of retirement pay which equaled the amount of VA disability compensation to which he or she was entitled. This issue was commonly known as "concurrent receipt."

Because there was often a tax advantage to receive VA disability compensation, which is tax-free, rather than military retirement pay, most disabled retirees chose to have a $1 reduction in their retired pay for each $1 of VA disability compensation they received. Since this type of rule against "concurrent receipt" does not apply to any other group of federal or state retirees, many individuals and service organizations felt this was unfair discrimination against disabled military retirees.

This has been a very hot topic for the last several years, and Congress and the President have taken several steps toward eliminating the bar to concurrent receipt of full military retired pay and full disability compensation.

Details of CRSC Payments

As a result of the current disability process, a retiree can have both a DOD and a VA disability rating and these ratings will not necessarily be the same percentage. The percentage determined by DOD is used to determine fitness for duty and may result in the medical separation or disability retirement of the service member. The VA rating, on the other hand, was designed to reflect the average loss of earning power. CRSC is not subject to taxation. Individuals must apply for CRSC.

For detailed information, individuals should contact the following:

Army
DEPARTMENT OF THE ARMY
U.S. Army Physical Disabilities Agency/ Combat-Related Special Compensation (CRSC) 200 Stovall Street
Alexandria, VA 22332-0470
(866) 281-3254

Navy and Marine Corps
Department of Navy Naval Council of Personnel Boards Combat-Related Special Compensation Branch
720 Kennon Street S.E., Suite 309
Washington Navy Yard, DC
20374-5023
(877) 366-2772

Air Force
United States Air Force Personnel Center Disability Division (CRSC)
550 C Street West, Suite 6
Randolph AFB TX 78150-4708
(800) 525-0102

Details of Concurrent Receipt

As veterans, military retirees can apply to the VA for disability compensation. A retiree may (1) apply for VA compensation any time after leaving the service and (2) have his or her degree of disability changed by the VA as the result of a later medical reevaluation, as noted above. Many retirees seek benefits from the VA years after retirement for a condition that may have been incurred during military service, but that does not manifest itself until many years later. Typical examples include hearing loss, some cardiovascular problems, and conditions related to exposure to Agent Orange. Until 2004, the law required that military retired pay be reduced dollar-for-dollar by the amount of any VA disability compensation received. This procedure was generally referred to as an "offset." If, for example, a military retiree who received $1,500 a month in retired pay and was rated by the VA as 70% disabled (and therefore entitled to approximately $1,000 per month in disability compensation), the offset would operate to pay $500 monthly in retired pay and the $1,000 in disability compensation.

What If a Retiree Is Eligible for Both?

Retirees eligible for both programs will be able to make an election between the two programs, depending on which one is more advantageous. Because the CRSC program provides full payment immediately, versus the 10—year phase-in for concurrent receipt, the election can be changed each year. (This recognizes that a retiree who is 100% disabled, but only 60% of that is due to combat-related conditions, may find it advantageous to elect full CRSC payments for a few years until the concurrent receipt payment rises to a level that exceeds the CRSC payment. Because CRSC payments are tax-free, and non-disability retired pay is not, this could also figure into a retiree's election decision.

Disability Compensation Effective Dates

When the VA decides to pay a disability, benefit based on a claim, there is an effective dates assigned to the claim. The effective date is the day the veteran can start to get their disability benefits, and it varies with the type of benefit they're applying for, and the nature of their claim.

How Does the VA Decide Effective Dates?

The VA bases effective dates on the situation.

Direct Service Connection

The effective date for a disability that was caused or made worse by military service is whichever of the following comes later:

- The date the VA gets your claim, or
- The date you first got your illness or injury, also known as the date your entitlement arose

If the VA gets your claim within a year of the day you left active service, the effective date can be as early as the day following separation.

Presumptive Service Connection

In most cases, if VA believes your disability is related to your military service, known as a presumptive service connection, and they get your claim within a year of your separation from active service, then the effective date is the date you first got your illness or injury. If the VA gets your claim more than a year after your separation from active service, the effective date is the date the VA got your claim or when you first got your illness or injury— whichever is later.

Reopened Claims

The effective date for a reopened claim is the date the VA gets the claim to reopen, or the date you first got your illness or injury, whichever is later.

Liberalizing Law Change

If there's a change in law or VA regulation that allows the VA to pay disability compensation, the effective date could be assigned in one of these ways:

- If the VA gets your claim within a year of a law or regulation changing, the effective date may be the date the law or regulation changed.
- If the VA reviews your claim or you request a review, more than one year after the law or regulation changed, the effective date may be up to one year before the date the VA got your request or the date the VA decides to pay benefits on your claim.

Dependency and Indemnity Compensation (DIC)

For claims based on a veteran's death in service, the effective date is the first day of the month in which the veteran died or was presumed to have died. This is true only if the VA gets the claim within a year of the date of the report of the veteran's actual or presumed death. Otherwise, the effective date is the date the VA gets the claim. If the veteran's death happened after service and the VA gets the claim within a year of their death, the effective date is the first day of the month in which the veteran died. If the death happened after service and the VA gets the claim more than one year after the veteran's death, the effective date is the date the VA gets the claim.

Error In a Previous Decision

If the VA finds a clear and unmistakable error in a prior decision, the effective date of the new decision will be the date from which benefits would have been paid if there hadn't been an error in the prior decision.

Difference of Opinion

A decision that's based on a difference of opinion will have an effective date of the original decision, had it been favorable.

Increases in the Disability

The VA dates back increases in the disability rating to the earliest date when a veteran can show there was an increase in disability. This is only if the VA gets the new claim request within a year from that date. Otherwise, the effective date is the date the VA gets the claim.

Disability or Death Due to a Hospital Stay
If the VA gets a claim within a year after the date the veteran suffered an injury, or their existing injury got worse, the effective date is the date the injury happened or when it began to get worse. If the VA gets a claim within a year of the date of a veteran's death, the effective date is the first day of the month in which the veteran died. If the VA gets a claim more than a year after a veteran suffered an injury, their injury got worse, or they died, the effective date is the date the VA gets the claim.

Disabilities That Appear Within One-Year After Discharge
A veteran may be able to get disability benefits if they have signs of an illness like hypertension, arthritis, diabetes, or peptic ulcers that started within a year after they were discharged from military service. If the symptoms appear within a year after discharge, even if they weren't there while the veteran was serving, they VA concludes they're related to service.

You may be eligible for benefits if you have an illness that's at least 10% disabling and appears within one year after discharge, and you meet two requirements. First, the illness has to be listed in Title 38, Code of Federal Regulations, 3.309(a), and you didn't receive a dishonorable discharge. If you have an illness listed in Title 38, Code of Federal Regulations, 3.309(a), you don't have to show the problem started during or got worse because of your military service. These are presumptive diseases.

The following diseases are covered even if they appeared more than a year after you separated from service:
- Hansen's disease can appear within three years after discharge
- Tuberculosis can appear within three years after discharge
- Multiple sclerosis can appear within seven years after discharge
- Amyotrophic lateral sclerosis (ALS), also known as Lou Gehrig's can appear any time after discharge.

Detailed Rates of Disability Compensation
Rates Effective December 1, 2022

Dependent Status	Disability Rating									
	10%	20%	30%	40%	50%	60%	70%	80%	90%	100%
Veteran Alone	$165.92	$327.99	$508.05	$731.86	$1041.82	$1319.65	$1663.06	$1933.15	$2172.39	$3621.95
Veteran & Spouse	$165.92	$327.99	$568.05	$811.86	$1141.82	$1440.65	$1804.06	$2094.15	$2353.39	$3823.89
Veteran, Spouse &1 Child	$165.92	$327.99	$612.05	$870.86	$1215.82	$1528.65	$1907.06	$2212.15	$2486.39	$3971.78
Veteran, No Spouse & 1 Child	$165.92	$327.99	$548.05	$785.86	$1108.82	$1400.65	$1757.06	$2041.15	$2293.39	$3757.00
	10%	20%	30%	40%	50%	60%	70%	80%	90%	100%
Veteran & Spouse & No Children and 1 parent	$165.92	$327.99	$616.05	$875.86	$1222.82	$1537.65	$1917.06	$2223.15	$2498.39	$3985.96
Veteran & Spouse & 1 Child & 1 parent	$165.92	$327.99	$660.05	$934.86	$1296.82	$1625.65	$2020.06	$2341.15	$2631.39	$4133.85
Veteran & Spouse & No Child & 2 Parents	$165.92	$327.99	$664.05	$939.86	$1303.82	$1634.65	$2030.06	$2353.15	$2643.39	$4148.03
Veteran & Spouse & 1 Child & 2 Parents	$165.92	$327.99	$708.05	$998.86	$1377.82	$1722.65	$2133.06	$2470.15	$2776.39	$4295.92
Veteran & No Spouse & No Children & 1 Parent	$165.92	$327.99	$556.05	$795.86	$1122.82	$1416.65	$1776.06	$2062.15	$2317.39	$3784.02
Veteran & No Spouse & 1 Child & 1 Parent	$165.92	$327.99	$596.05	$849.86	$1189.82	$1497.65	$1870.06	$2170.15	$2483.39	$3919.07
Veteran & No Spouse & No Children & 2 Parents	$165.92	$327.99	$604.05	$859.86	$1203.82	$1513.65	$1889.06	$2191.15	$2462.39	$3946.09
Veteran & No Spouse & 1 Child & 2 Parents	$165.92	$327.99	$644.05	$913.86	$1270.82	$1594.65	$1983.06	$2299.15	$2583.39	$4081.14

Benefits Rates for Service-Connected Disability Compensation

VA compensation and pension benefits cost of living allowance (COLA) is paid based on the Social Security Administration (SSA) COLA. By statute, compensation COLA may not be more than the SSA COLA; and pension COLA is equal to the SSA COLA.

This year SSA increased COLA by 8.7%		
Basic Rates of Disability Compensation **Effective December 1, 2022**		
	Disability Rating	Monthly Benefit
(a)	10%	$165.92
(b)	20%	$327.99
(c)	30%	$508.05
(d)	40%	$731.86
(e)	50%	$1041.82
(f)	60%	$1319.65
(g)	70%	$1663.06
(h)	80%	$1933.15
(i)	90%	$2172.39
(j)	100%	$3621.95

Additional Amounts Payable for Spouse Requiring Aid & Attendance Rates Effective December 1, 2022								
Disability Rating	30%	40%	50%	60%	70%	80%	90%	100%
Monthly Benefit	$56	$74	$93	$111	$130	$148	$167	$185.21

Additional Amount Payable for Each Additional Child Under 18 Rates Effective December 1, 2022								
Disability Rating	30%	40%	50%	60%	70%	80%	90%	100%
Monthly Benefit	$30	$40	$50	$60	$70	$80	$90	$100.34

Additional Amount Payable for Each Additional Child Over Age 18 Attending School Rates Effective December 1, 2022								
Disability Rating	30%	40%	50%	60%	70%	80%	90%	100%
Monthly Benefit	$97	$129	$162	$194	$226	$259	$291	$324.12

Notes: Rates for Children over age 18 attending school are shown separately in the above chart. All other entries in the above charts reflect rates for children under age 18, or helpless.

All references in the preceding charts to parents refer to parents who have been determined to be dependent by the Secretary of Veterans Affairs.

Higher Statutory Awards for Certain Multiple Disabilities

VA special monthly compensation or SMC is a higher rate of compensation paid to veterans, spouses and surviving spouses, and parents with certain needs or disabilities.

Special monthly compensation rate payment variations—effective December 1, 2022

Levels K and Q are special rates called SMC rate variations. VA may add level K to your basic SMC rate.

SMC Letter Designation	Monthly payment in U.S. $	How this payment variation works
SMC-K	$128.62	If you qualify for SMC-K, VA adds this rate to the basic disability compensation rate for any disability rating from 0-100%. VA may also add this rate to all SMC basic rate except SMC-O, SMC-Q and SMC-R. You may receive 1 to 3 SMC-K

		awards in addition to basic and SMC rates.
SMC-Q	$67.00	This is a protected rate the VA hasn't awarded since August 19, 1968. If a veteran is awarded an SMC-Q designation, it pays this rate in place of the basic disability compensation rate.

Special monthly compensation rates for veterans without children—effective December 1, 2022

Levels L through O cover specific disabilities and situations. Level R may apply if you need help daily from another person for basic needs like dressing, bathing and eating. Level S may apply if you aren't able to leave the house because of service-connected disabilities.

SMC-L through SMC-N Rates

Start with the Basic SMC rates table and find the dependent status in the left column best describing you. Then find your SMC letter designation in the top row. Your monthly basic rate is where your dependent status and SMC letter meet. If you have more than one child or your spouse receives Aid and Attendance benefits, also check the Added Amounts table, and then add them to your amount from the basic SMC rates table.

Basic SMC Rates

Dependent Status	SMC-L in U.S. $	SMC-L ½ in U.S.$	SMC-M in U.S.$	SMC-M ½ in U.S.$	SMC-N in U.S.$
Veteran alone	4506.84	4739.83	4973.76	5315.51	5658.02
With Spouse	4708.78	4941.77	5175.70	5517.45	5859.96
With spouse and 1 parent	4870.85	5103.84	5337.77	5679.52	6022.03
With spouse and 2 parents	5032.92	5265.91	5499.84	5841.59	6184.10
With 1 parent	4668.91	4901.90	5135.83	5477.58	5820.09
With 2 parents	4830.98	5063.97	5297.90	5639.65	5982.16

Added Amounts

Dependent Status	SMC-L in U.S.$	SMC-L ½ in U.S.$	SMC-M in U.S.$	SMC-M ½ in U.S.$	SMC-N in U.S.$
Spouse receiving Aid and Attendance	185.21	185.21	185.21	185.21	185.21

SMC-N ½ through SMC-S Basic SMC Rates

Dependent Status	SMC-N ½ in U.S.$	SMC-O/P in U.S. $	SMC-R.1 in U.S.$	SMC-R.2/T in U.S.$	SMC-S in U.S.$
Veteran Alone	5990.84	6324.26	9036.89	10,365.53	4054.12
With Spouse	6192.78	6526.20	9238.83	10,567.47	4256.06
With Spouse and 1 Parent	6354.85	6688.27	9400.90	10,729.54	4418.13
With Spouse and 2 Parents	6516.92	6850.34	9562.97	10,891.61	4580.20
With 1 Parent	6152.91	6486.33	9198.96	10,527.60	4216.19
With 2 Parents	6314.98	6648.40	9361.03	10,689.67	4378.26

Added Amounts

Dependent Status	SMC-N ½ in U.S.$	SMC-O/P in U.S.$	SMC-R.1 in U.S.$	SMC-R.2/T in U.S.$	SMC-S in U.S.$
Spouse Receiving Aid and Attendance	185.21	185.21	185.21	185.21	185.21

Special monthly compensation rates for veterans with dependents including children—effective December 1. 2022

Levels L through O cover specific disabilities and situations.

Level R may apply if you need help from another person for basic needs.

Level S may apply if you can't leave the house because of your service-connected disabilities.

SMC-L through SMC-N Basic SMC Rates

Start with the Basic SMC rates table and find the dependent status in the left column that best describes you. Then find your SMC letter designation in the top row. Your monthly basic rate is where your dependent status and SMC letter meet. If you have more than one child or your spouse receives Aid and Attendance benefits, be sure to check the Added amounts table and add those to your amount from the Basic SMC rates table.

Dependent Status	SMC- L in U.S.$	SMC-L ½ in U.S.$	SMC-M in U.S.$	SMC-M ½ in U.S.$	SMC-N in U.S.$
Veteran with 1 Child	4641.89	4874.88	5108.81	5450.56	5793.07
With 1 Child and Spouse	4856.67	5089.66	5323.59	5665.34	6007.85
With 1 Child, Spouse and 1 Parent	5018.74	5251.73	5485.66	5827.41	6169.92
With 1 Child, Spouse and 2 Parents	5180.81	5413.80	5647.73	5989.48	6331.99
With 1 Child and 1 Parent	4803.96	5036.95	5270.88	5612.63	5955.14
With 1 Child and 2 Parents	4966.03	5199.02	5432.95	5774.70	6117.21

Added Amounts

Dependent Status	SMC-L in U.S.$	SMC-L ½ in U.S.$	SMC-M in U.S.$	SMC-M ½ in U.S.$	SMC-N in U.S.$
Each additional child under age 18	100.34	100.34	100.34	100.34	100.34
Each additional child over age 18 in a qualifying school program	324.12	324.12	324.12	324.12	324.12
Spouse receiving Aid and Attendance	185.21	185.21	185.21	185.21	185.21

SMC-N ½ through SMC-S

Dependent Status	SMC-N ½ in U.S.$	SMC-O/P in U.S.$	SMC-R.1 in U.S.$	SMC-R.2/T in U.S.$	SMC-S in U.S.$
Veteran with 1 Child	6125.89	6459.31	9171.94	10,500.58	4189.17
With 1 Child and Spouse	6340.67	6674.09	9386.72	10,715.36	4403.95
With 1 Child, Spouse and 1 Parent	6502.74	6836.16	9548.79	10,877.43	4566.02
With 1 Child, Spouse and 2 Parents	6664.81	6998.23	9710.86	11,039.50	4728.09
With 1 Child and 1 Parent	6287.96	6621.38	9334.01	10,662.65	4351.24
With 1 Child and 2 Parents	6450.03	6783.45	9496.08	10,824.72	4513.31

Added Amounts

Dependent Status	SMC-N ½ in U.S.$	SMC-O/P in U.S.$	SMC-R.1 in U.S.$	SMC-R.2/T in U.S.$	SMC-S in U.S.$
Each additional child under age 18	100.34	100.34	100.34	100.34	100.34
Each additional child over age 18 in a qualifying school program	324.12	324.12	324.12	324.12	324.12
Spouse receiving Aid and Attendance	185.21	185.21	185.21	185.21	185.21

How VA Assigns SMC Levels L Through O

The SMC levels are assigned based on very specific situations and combinations of situations, including:

- The amputation of one or more limbs or extremities
- The loss of use of one or more limbs or extremities
- The physical loss of one or both eyes

- The loss of sight or total blindness in one or both eyes
- Being permanently bedridden
- Needing daily help with basic needs also called Aid and Attendance

You may receive an SMC-L designation if any of these situations are true:
- You've had both feet amputated, or
- You've had one foot amputated and have lost the use of the other foot, or
- You've had one hand and one foot amputated, or
- You've had one foot amputated and have lost the use of one hand, or
- You've had one hand amputated and have lost the use of one foot, or
- You've lost the use of both feet, or
- You've lost the use of one hand and one foot, or
- You've lost sign in both eyes, or
- You're permanently bedridden, or
- You need help with daily basic needs

You may receive an SMC-L ½ designation if any of these situations are true for you:
- You've had one foot and the other knee amputated, or
- You've had one foot amputated, and have lost the use of the other knee, or
- You've had one foot and one elbow amputated, or
- You've had one foot amputated and have lost the use of one elbow, or
- You've had one knee and one hand amputated, or
- You've had one knee amputated and have lost the use of one hand
- Or you've lost the use of one foot and have had the other knee amputated, or
- You've lost the use of one foot and had one elbow amputated, or
- You've lost the use of one foot and one elbow, or
- You've lost the use of one knee and have had one hand amputated, or
- You've lost the use of one knee and one hand, or
- You have blindness in one eye and total blindness in the other eye with only the ability to perceive light, or
- You have blindness in both eyes and have lost the use of one foot—rated as less than 50% disabling

You may receive an SMC-M designation if any of these are true for you:
- You've had both hands amputated, or
- You've had one hand amputated and have lost the use of the other hand, or
- You've had both knees amputated, or
- You've had one elbow and one knee amputated, or
- You've had one foot amputated and have lost the use of one arm at the shoulder, or
- You've had one foot amputated and have had one leg amputated so close to the hip that you can't wear a prosthesis, or
- You've had one foot amputated and have had one arm amputated so close to the shoulder that you can't wear a prosthesis, or
- You've had one hand amputated and have had one leg amputated so close to the hip that you can't wear a prosthesis, or
- You've lost the use of both hands, or
- You've lost the use of both knees, or
- You've lost the use of one elbow and one knee, or
- You've lost the use of one foot and the use of one arm at the shoulder, or
- You've lost the use of one foot, and you have had one leg amputated so close to the hip that you can't wear a prosthesis, or
- You've lost the use of one foot, and have had one arm amputated so close to the

shoulder that you can't wear a prosthesis, or
- You've lost the use of one hand and have had one leg amputated so close to the hip that you can't wear a prosthesis

Or you have blindness in one eye, and:
- You've physically lost the other eye, or
- You have total blindness without the ability to perceive light in the other eye, or
- You have total blindness in the other eye with only the ability to perceive light and have total deafness in one ear, or
- You have total blindness in the other eye with only the ability to perceive light and have lost the use of one foot rated as less than 50% disabling, or
- You have blindness in both eyes considered total blindness, with only the ability to perceive light, or
- You have blindness in both eyes that requires you to have daily help with basic needs

Or you have blindness in both eyes, and:
- You have deafness in both rated as 30% or more disabling, or
- You've had one hand amputated, or
- You've lost the use of one hand, or
- You've had one foot amputated, or
- You've lost the use of one foot—rated as 50% or more disabling

You may receive an SMC-M ½ designation if any of these situations are true for you:
- You've had one knee amputated, and have had one leg amputated so close to the hip that you can't wear a prosthesis, or
- You've had one knee amputated, and have had one arm amputated so close to the shoulder that you can't wear a prosthesis, or
- You've had one elbow amputated, and have had one leg amputated so close to the hip that you can't wear a prosthesis, or
- You've had one hand and one elbow amputated, or
- You've had one hand amputated, and have lost the use of one elbow, or
- You've lost the use of one knee, and have had one leg amputated so close to the hip that you can't wear a prosthesis, or
- You've lost the use of one knee, and have had one arm amputated so close to the shoulder that you can't wear a prosthesis, or
- You've lost the use of one elbow, and have had one leg amputated so close to the hip that you can't wear a prosthesis, or
- You've lost of use of one hand, and have had one elbow amputated, or
- You've lost the use of one hand and of one elbow

Or you have total blindness with only the ability to perceive light:
- In one eye, and have physically lost the other eye, or
- In one eye, and have total blindness without the ability to perceive light in the other eye, or
- In both eyes, and have lost the use of one foot (rated as less than 50% disabling)

Or you have blindness in one eye, and:
- You've physically lost the other eye, and have total deafness in one ear, or
- You have total blindness without the ability to perceive light in the other eye, and have total deafness in one ear, or
- You have total blindness with only the ability to perceive light in the other eye, and have deafness in both ears (rated as 30% or more disabling), or
- You have total blindness with only the ability to perceive light in the other

49

eye, and have had one foot amputated, or
- You have total blindness with only the ability to perceive light in the other eye, and have lost the use of one foot (rated as 50% or more disabling), or
- You have total blindness with only the ability to perceive light in the other eye, and have had one hand amputated, or
- You have total blindness with the only ability to perceive light in the other eye, and have lost the use of one hand, or
- You have blindness in both eyes and total deafness in one ear, or
- You have blindness in both eyes that requires you to have daily help with basic needs (like eating, bathing, and dressing), and have lost the use of one foot (rated as less than 50% disabling)

You may receive an SMC-N designation if any of these situations are true for you:
- You've had both elbows amputated, or
- You've had both legs amputated so close to the hip that you can't wear a prosthesis, or
- You've had one arm and one leg amputated so close to the shoulder and hip that you can't wear a prosthesis on either, or
- You've had one hand amputated, and one arm amputated so close to the shoulder that you can't wear a prosthesis, or
- You've lost the use of both elbows, or
- You've lost the use of one hand, and have had one arm amputated so close to the shoulder that you can't wear a prosthesis, or
- You've physically lost both eyes, or
- You have total blindness without the ability to perceive light

Or you have total blindness with only the ability to perceive light in one eye, and:
- You've physically lost the other eye, and have deafness in both ears (rated as 10% or 20% disabling), or
- You have total blindness without the ability to perceive light in the other eye, and have deafness in both ears (rated as 10% or 20% disabling), or
- You've physically lost the other eye, and have lost the use of one foot (rated as less than 50% disabling), or
- You have total blindness without the ability to perceive light in the other eye, and have lost the use of one foot (rated as less than 50% disabling)

Or you have total blindness with only the ability to perceive light in both eyes, and:
- You have deafness in both ears (rated as 30% or more disabling), or
- You've had one hand amputated, or
- You've lost the use of one foot, or
- You've lost the use of one hand, or
- You've had one foot amputated, or
- You've lost the use of one foot (rated as 50% or more disabling)

Or you have blindness in one eye, and:
- You've physically lost the other eye, and have deafness in both ears (rated as 30% or more disabling), or
- You have total blindness without the ability to perceive light in the other eye, and have deafness in both ears (rated as 30% or more disabling)

Or you have blindness in both eyes that requires you to have daily help with basic needs and:
- You have deafness in both ears (rated as 30% or more disabling), or
- You've had one hand amputated, or
- You've lost the use of one hand, or

- You've had one foot amputated, or
- You've lost the use of one foot (rated as 50% disabling)

You may receive an SMC-N ½ designation if any of these situations are true for you:
- You've had one elbow amputated, and have had one arm amputated so close to the shoulder that you can't wear a prosthesis, or
- You've lost the use of one elbow, and have had one arm amputated so close to the shoulder that you can't wear a prosthesis, or
- You've physically lost both eyes, and have lost the use of one foot (rated as less than 50% disabling), or
- You have total blindness without the ability to see light, and have lost the use of one foot (rated as less than 50%)

Or you have total blindness with only the ability to perceive light in one eye, have physically lost the other eye, and:
- You have deafness in both ears (rated as 30% or more disabling), or
- You've had one foot amputated, or
- You've lost the use of one foot (rated as 50% or more disabling), or
- You've had one hand amputated, or
- You've lost the use of one hand

Or you have total blindness with only the ability to perceive light in one eye and total blindness without the ability to perceive light in the other eye, and:
- You have deafness in both ears (rated as 30% or more disabling), or
- You've had one foot amputated, or
- You've lost the use of one foot (rated as 50% or more disabling), or
- You've had one hand amputated, or
- You've lost the use of one hand

You may receive SMC-O designation if any of these situations are true for you:
- You've had both arms amputated so close to the shoulder that you can't wear a prosthesis, or
- You have complete paralysis of both legs that's resulted in being unable to control your bladder or bowels, or
- You have hearing loss in both ears (with at least one ear's deafness caused by military service) that's rated as 60% or more disabling, and you have blindness in both eyes, or
- You have hearing loss in both ears (with at least one ear's deafness caused by military service) that's rated as 40% or more disabling, and you have blindness in both eyes with only the ability to perceive light, or
- You have total deafness in one ear as well as blindness in both eyes with only the ability to perceive light

Or you have total blindness without the ability to see light, and:
- You have deafness in both ears (rated as 30% or more disabling), or
- You've had one foot amputated, or
- You've lost the use of one foot (rated as 50% or more disabling), or
- You've had one hand amputated, or
- You've lost the use of one hand

Or you have physically lost both eyes, and:
- You have deafness in both ears (rated as 30% or more disabling), or
- You've had one foot amputated, or
- You've lost the use of one foot (rated as 50% or more disabling), or
- You've had one hand amputated, or

- You've lost the use of one hand

Adjustment to Individual VA Awards

There will be no adjustment of VA awards. Special Compensation paid under 10 USC 1413 is provided under chapter 71, title 10, USC, "Computation of Retired Pay." However, it is NOT RETIRED PAY. It is to be paid from funds appropriated for pay and allowances of the recipient member's branch of service. Eligible retirees in receipt of VA disability compensation may receive this special compensation in addition to their VA disability compensation.

Presumptions

Presumption of Sound Condition

Every veteran will be assumed to have been in sound medical condition when examined, accepted, and enrolled for service, except any defects, infirmities, or disorders noted at the time of the examination, acceptance, and enrollment, or if there is clear and unmistakable evidence showing that the injury or disease did exist before acceptance and enrollment, and the injury or disease was not aggravated by such service.

Presumptions of Service-Connection Relating to Certain Chronic Diseases and Disabilities

In the case of any veteran who served for 90 days or more during a period of war, any of the following shall be considered to have been incurred in, or aggravated by such service, notwithstanding there is no record of evidence of such disease during the period of service:

A chronic disease (detailed below), becoming manifest to a degree of 10% or more within one year from the date of separation from such service.

A tropical disease (detailed below), and the resultant disorders or disease originating because of therapy, administered in connection with such diseases, or as a preventative thereof, becoming manifest to a degree of 10% or more within one year from the date of separation from such service. Additionally, if it is shown to exist at a time when standard and accepted treatises indicate that the incubation period thereof commenced during active service, it shall be deemed to have incurred during such service.

- Active tuberculosis disease developing a 10% degree of disability or more within three years from the date of separation from such service.
- Multiple sclerosis developing a 10% degree of disability or more within seven years from the date of separation from such service.
- Hansen's disease developing a 10% degree of disability or more within three years from the date of separation from such service.

Chronic Diseases

- Amyotrophic lateral sclerosis (ALS)
- Anemia, primary
- Arteriosclerosis
- Arthritis
- Atrophy, progressive muscular
- Brain hemorrhage
- Brain thrombosis
- Bronchiectasis
- Calculi of the kidney, bladder, or gallbladder
- Cardiovascular-renal disease, including hypertension
- Cirrhosis of the liver
- Coccidioidomycosis
- Diabetes mellitus
- Encephalitis lethargica residuals

- Endocarditis
- Endocrinopathies
- Epilepsies
- Hansen's disease
- Hodgkin's disease
- Leukemia
- Lupus erythematosus, systemic
- Myasthenia gravis
- Myelitis Myocarditis
- Nephritis
- Organic diseases of the nervous system
- Osteitis deformans (Paget's disease)
- Osteomalacia
- Palsy, bulbar
- Paralysis agitans
- Psychoses
- Purpura idiopathic, hemorrhagic
- Raynaud's disease
- Sarcoidosis
- Scleroderma
- Sclerosis, amyotrophic lateral Sclerosis,
- multiple Syringomyelia
- Thromboangiitis obliterans (Buerger's disease)
- Tuberculosis, active
- Tumors, malignant, or of the brain or spinal cord or peripheral nerves
- Ulcers, peptic (gastric or duodenal)
- Other chronic diseases the Secretary of Veterans Affairs may add to this list.

Tropical Diseases

- Amebiasis
- Blackwater fever
- Cholera
- Dracontiasis
- Dysentery
- Filariasis
- Hansen's disease
- Leishmaniasis, including kala-azar
- Loiasis
- Onchocerciasis
- Oroya fever
- Pinta
- Plague Schistosomiasis
- Yaws
- Yellow fever
- Other tropical diseases the Secretary of Veterans Affairs may add to this list.

Presumptions of Service-Connection Relating to Certain Diseases and Disabilities for Former Prisoners of War

In the case of any veteran who is a former prisoner of war, and who was detained or interned for not less than thirty days, any of the following which became manifest to a degree of 10% or more after active military, naval or air service, shall be considered to have been incurred in or aggravated by such service, notwithstanding that there is no record of such disease during the period of service:

In the case of any veteran who is a former prisoner of war, and **who was detained or interned for not less than thirty days,** any of the following which became manifest to a degree of 10% or more after active military, naval or air service, shall be considered to have been incurred in or aggravated by such service, notwithstanding that there is no record of such disease during the period of service:

- Avitaminosis
- Beriberi (including beriberi heart disease, which includes ischemic heart disease-coronary artery disease-for former POWs who suffered during captivity from edema-swelling of the legs or feet- also known as "wet" beriberi)
- Chronic dysentery
- Helminthiasis
- Malnutrition (including optic atrophy associated with malnutrition)
- Pellagra
- Any other nutritional deficiency
- Peripheral neuropathy, except where directly related to infectious causes
- Irritable bowel syndrome
- Peptic ulcer disease
- Cirrhosis of the liver (This condition was added as part of Public Law 108- 183, The Veterans Benefits Act of 2003.)
- Atherosclerotic heart disease or hypertensive vascular disease (including hypertensive heart disease) and their complications (including myocardial infarction, congestive heart failure, and arrhythmia). These conditions were added as part of Public Law 109-233.
- Stroke and its complications; this condition was added as part of Public Law 109-233.
- The Veterans Benefits Improvement Act added osteoporosis to the list of disabilities presumed to be service-connected (and therefore compensable through VA disability compensation) in the case of veterans who are former prisoners of war, if the Secretary determines that such veteran has post-traumatic stress disorder (PTSD).

In the case of any veteran who is a former prisoner of war, and **who was detained or interned for any period of time,** any of the following which became manifest to a degree of 10% or more after active military, naval or air service, shall be considered to have been incurred in or aggravated by such service, notwithstanding that there is no record of such disease during the period of service:

- Psychosis
- Any of the anxiety states
- Dysthymic disorder (or depressive neurosis)
- Organic residuals of frostbite, if the VA determines the veteran was interned in climatic conditions consistent with the occurrence of frostbite.
- Post-traumatic osteoarthritis

Presumptions Relating to Certain Diseases Associated with Exposure to Radiation

VA may pay compensation for radiogenic diseases under two programs specific to radiation-exposed veterans and their survivors:

Please see Chapter 26: Toxic Exposures for more details.

Lou Gehrig's Disease Amyotrophic Lateral Sclerosis (ALS)

ALS is a rapidly progressive, totally debilitating, and irreversible motor neuron disease that results in muscle weakness leading to a wide range of serious disabilities, including impaired mobility. VA adapted its rules, so veterans with service-connected ALS no longer have to file multiple claims with VA for increased benefits as their condition progresses.

Before the new Specially Adapted Housing (SAH) regulatory change, many Veterans and Servicemembers who were rated by VA for service-connected ALS, but who did not yet have symptoms debilitating enough to affect their mobility to the degree required for SAH grant eligibility, were unable to begin the process of modifying their homes to accommodate their often rapidly progressing conditions. Veterans with Lou Gehrig's or ALS are eligible for monthly VA disability compensation benefits because of a presumption of service connection established in 2008. VA amended its disability rating scale in January 2012 to assign a 100% disability evaluation for any veteran with service-connected ALS. In 2014 the VA announced military personnel and veterans with service-connected ALS would also be presumed medically eligible for grants to adapt their homes, making them eligible for the maximum grant amount.

Peacetime Disability Compensation
Basic Entitlement

A veteran may be entitled to VA disability compensation for any medical condition or injury that was incurred in or aggravated by his or her military service during any period other than a period of war. The veteran must have been discharged or released under conditions other than dishonorable from the period of service in which the injury or disease was incurred or aggravated.

No compensation shall be paid if the disability is a result of the person's own willful misconduct or abuse of alcohol or drugs.

Presumption of Sound Condition

Every person employed in the active military, naval, or air service during any period other than a period of war, for six months or more, will be assumed to have been in sound medical condition when examined, accepted, and enrolled for service, except any defects, infirmities, or disorders noted at the time of the examination, acceptance, and enrollment, or if there is clear and unmistakable evidence showing that the injury or disease did exist before acceptance and enrollment, and the injury or disease was not aggravated by such service.

Presumptions Relating to Certain Diseases

If a veteran who served in the active military, naval, or air service after December 31, 1946, during any period other than a period of war, for six months or more, contracts any of the following, it shall be considered to have been incurred in, or aggravated by such service, notwithstanding there is no record of evidence of such disease during the period of service:

- A chronic disease, becoming manifest to a degree of 10% or more within one year from the date of separation from such service.
- Tropical disease and the resultant disorders or disease originating because of therapy, administered in connection with such diseases, or as a preventative thereof, becoming manifest to a degree of 10% or more within one year from the date of separation from such service. Additionally, if it is shown to exist at a time when standard and accepted treatises indicate that the incubation period thereof commenced during active service, it shall be deemed to have incurred during such service.
- Active tuberculosis disease developing a 10% degree of disability or more within three years from the date of separation from such service.

- Multiple sclerosis developing a 10% degree of disability or more within seven years from the date of separation from such service.
- Hansen's disease developing a 10% degree of disability or more within three years from the date of separation from such service.

In the case of any veteran who served for 90 days or more during a period of war, service-connection will not be granted in any case where the disease or disorder is shown by clear and unmistakable evidence to have had its inception before or after active military, naval, or air service.

Presumptions Rebuttal

If there is affirmative evidence to the contrary, or evidence to establish that an intercurrent injury or disease which is a recognized cause of any of the diseases or disabilities mentioned in the above sections, has been suffered between the date of separation from service and the onset of any such diseases or disabilities, or if the disability is due to the veteran's own willful misconduct, payment of compensation shall not be made.

Rates of Peacetime Disability Compensation

The compensation payable shall be the same as the compensation payable for Wartime Disability Compensation. Please refer to the charts presented earlier in this chapter.

Additional Compensation for Dependents

Any veteran entitled to peacetime disability compensation, and whose disability is rated as 30% or greater, will be entitled to additional monthly compensation for dependents in the same amounts payable for Wartime Disability Compensation. Please refer to the charts presented earlier in this chapter.

Adjudication

Adjudication means a judicial decision made by the Veterans Administration in claims filed within their jurisdiction. There is an Adjudication Division in each regional office, under the direction of an adjudication officer, who is responsible for the preparation of claims.

Upon the receipt of an original application in the Adjudication Division, it will be referred to the Authorization Unit for review and development in accordance with established procedures. All reasonable assistance will be extended a claimant in the prosecution of his or her claim, and all sources from which information may be elicited will be thoroughly developed before the submission of the case to the rating board. Every legitimate assistance will be rendered a claimant in obtaining any benefit to which he or she is entitled, and the veteran will be given every opportunity to substantiate his or her claim. Information and advice to claimants will be complete and will be given in words that the average person can understand.

VA personnel must always give to claimants and other properly interested and recognized individuals courteous and satisfactory service, which is essential to good public relations. It is incumbent upon the claimant to establish his or her case following the law. This rule, however, should not be highly technical and rigid in its application. The general policy is to give the claimant every opportunity to substantiate the claim, to extend all reasonable assistance in its prosecution, and to develop all sources from which information may be obtained. Information and advice to claimants will be complete and expressed, so far as possible, in plain language, which can be easily read and understood by persons not familiar with the subject matter.

Benefits for Persons Disabled by Treatment or Vocational Rehabilitation

Compensation shall be awarded for an additional qualifying disability or a qualifying death of a veteran in the same manner as if such additional disability or death were service connected, provided:

The disability or death was caused by hospital care, medical or surgical treatment, or examination was furnished under any law administered by the VA, either by a VA employee, on a VA facility, and the proximate cause of the disability or death was:

- Carelessness, negligence, lack of proper skill, error in judgment, or similar instance of fault on the part of the VA in furnishing the hospital care, medical treatment, surgical treatment, or examination; or
- An event not reasonably foreseeable.
- The disability or death was proximately caused by the provision of training and rehabilitation services by the VA (including a service-provider used by the VA) as part of an approved rehabilitation program.

Effective December 1, 1962, if an individual is awarded a judgment against the United States in a civil action brought pursuant to Section 1346(b) of Title 28, or enters into a settlement or compromise under Section 2672 or 2677 of Title 28 by reason of a disability or death treated pursuant to this section as if it were service-connected, then no benefits shall be paid to such individual for any month beginning after the date such judgment, settlement, or compromise on account of such disability or death becomes final until the aggregate amount of benefits which would be paid out for this subsection equals the total amount included in such judgment, settlement, or compromise.

LGBT Veterans

The Obama administration announced a decision to extend veteran benefits to same sex married couples as a result of the Supreme Court's decision to strike down the Defense of Marriage Act. Veterans' spouses are now eligible to collect benefits, regardless of sexual orientation.

LGBT Veterans are eligible for the same VA benefits as any other veteran and will be treated in a welcoming environment. Comprehensive health services are available to LGBT veterans, including primary care, specialty care, mental health care, residential treatment, and reproductive health care services. VA provides management of acute and chronic illnesses, preventive care, contraceptive and gynecology services, menopause management, and cancer screenings.

Transgender veterans will be treated based upon their self- identified gender, including room assignments in residential and inpatient settings. Eligible transgender Veterans can receive cross-sex hormone therapy, gender dysphoria counseling, preoperative evaluations, as well as post-operative and long-term care following sex reassignment surgeries.

Aggravation

A preexisting injury or disease will be considered to have been aggravated by active military, naval, or air service, if there is an increase in disability during such service unless there is a specific finding that the increase in disability is due to the natural progress of the disease.

Consideration to be Accorded Time, Place and Circumstances of Service

Consideration shall be given to the places, types, and circumstances of each veteran's service. The VA will consider the veteran's service record, the official history of each organization in which such veteran served, such veteran's medical records, and all pertinent medical and lay evidence. The provisions of Public Law 98-542 – Section 5 of the Veterans' Dioxin and Radiation Exposure Compensation Standards Act shall also be applied.

In the case of any veteran who engaged in combat with the enemy in active service with a military, naval, or air organization of the United States during a period of war, campaign, or expedition, the Secretary shall accept as sufficient proof of service- connection of any disease or injury alleged to have been incurred in or aggravated by such service, if consistent with the circumstances, conditions, or hardships of such service. This provision will apply even if there is no official record of such incurrence or aggravation in such service. Every reasonable doubt in such instances will be resolved in favor of the veteran.

The service-connection of such injury or disease may be rebutted only by clear and convincing evidence to the contrary. The reasons for granting or denying service-connection in each case shall be recorded in full.

Disappearance

If a veteran who is receiving disability compensation disappears, the VA may pay the compensation otherwise payable to the veteran to such veteran's spouse, children, and parents. Payments made to such spouse, child, or parent should not exceed the amounts payable to each if the veteran died from a service-connected disability.

Combination of Certain Ratings

If the VA finds that a veteran has multiple disabilities, they use a Combined Ratings Table to calculate combined disability rating. Disability ratings are not additive, meaning that if a veteran has one disability rated 60% and a second disability 20%, the combined rating is not 80%. This is because subsequent disability ratings are applied to an already disabled veteran, so the 20% disability is applied to a veteran who is already 60% disabled. The following outlines how the VA combines ratings for more than one disability:

The disabilities are first arranged in the exact order of their severity, beginning with the greatest disability combined with use of the Combined Ratings Table. The degree of one disability will be read in the left column of the table, and the degree of the other in the top row, whoever is appropriate. The figures appearing in the space where the column and row intersect will represent the combined value of the two. This combined value is rounded to the nearest 10%. If there are more than two disabilities, the combined value for the first two will be found as previously described for two disabilities. The exact combined value— without yet rounding—is combined with the degree of the third disability. The process continues for subsequent disabilities, and the final number is rounded to the nearest 10%.

Example of Combining Two Disabilities: If a veteran has a 50 percent disability and a 30 percent disability, the combined value will be found to be 65 percent, but the 65 percent must be converted to 70 percent to represent the final degree of disability. Similarly, with a disability of 40 percent, and another disability of 20 percent, the combined value is found to be 52 percent, but the 52 percent must be converted to the nearest degree divisible by 10, which is 50 percent.

Example of Combining Three Disabilities: If there are three disabilities ratable at 60 percent, 40 percent, and 20 percent, respectively, the combined value for the first two will be found opposite 60 and under 40 and is 76 percent. This 76 will be found in the left column, then the 20 rating in the top row. The intersection of these two ratings is 81. Thus, the final rating will be rounded to 80%.

Tax Exemption

Compensation and pension may not be assigned to anyone and are exempt from taxation (including income tax). No one can attach, levy, or seize a compensation or pension check either before or after receipt. Property purchased with money received from the government is not protected.

Disabled veterans may be eligible to claim a federal tax refund based on an increase in the veteran's percentage of disability from the Department of Veterans Affairs (which may include a retroactive determination) or the combat-disabled veteran applying for, and being granted, Combat-Related Special Compensation, after an award for Concurrent Retirement and Disability.

To do so, the disabled veteran will need to file the amended return, Form 1040X, Amended U.S. Individual Income Tax Return, to correct a previously filed Form 1040, 1040A, or 1040EZ. An amended return cannot be e-filed. It must be filed as a paper return. Disabled veterans should include all documents from the Department of Veterans Affairs and any information received from Defense Finance and Accounting Services explaining proper tax treatment for the current year.

Please note: It is only in the year of the Department of Veterans Affairs reassessment of disability percentage (including any impacted retroactive year) or the year that the CRSC is initially granted or adjusted that the veteran may need to file amended returns.

Under normal circumstances, the Form 1099-R issued to the veteran by Defense Finance and Accounting Services correctly reflects the taxable portion of compensation received. No amended returns would be required since it has already been adjusted for any non-taxable awards.

If needed, veterans should seek assistance from a competent tax professional before filing amended returns based on a disability determination. Refund claims based on an incorrect interpretation of the tax law could subject the veteran to interest and/or penalty charges.

Protection of Service Connection

Service connection for any disability or death granted under this title which has been in force for ten or more years shall not be severed on or after January 1, 1962, unless it is shown that the original grant of service connection was based on fraud, or it is clearly shown from military records that the person concerned did not have the requisite service or character of discharge. The mentioned period shall be computed from the date determined by the VA as the date on which the status commenced for rating purposes.

Preservation of Ratings

Public Law 88-445, approved August 19, 1964, effective the same date, amends Section 110, Title 38, U.S. Code as follows:
The law provides that a disability that has been continuously rated at or above a given percentage for 20 years or longer for service connection compensation under laws administered by the VA shall not after that be rated at any lesser percentage except upon showing that the rating was based on fraud.

Special Consideration for Certain Cases of Loss of Paired Organs or Extremities

If a veteran has suffered any of the following, the VA shall assign and pay to the veteran the applicable rate of compensation, as if the combination of disabilities were the result of a service-connected disability:

- Blindness in one eye as a result of a service-connected disability, and blindness in the other eye as a result of a non-service-connected disability not the result of the veteran's own willful misconduct; or

- The loss or loss of use of one kidney as a result of a service-connected disability, and involvement of the other kidney as a result of a non-service-connected disability not the result of the veteran's own willful misconduct; or
- Total deafness in one ear as a result of a service-connected disability, and total deafness in the other ear as the result of non-service-connected disability not the result of the veteran's own willful misconduct; or
- The loss or loss of use of one hand or one foot as a result of a service-connected disability and the loss or loss of use of the other hand or foot as a result of non-service-connected disability not the result of the veteran's own willful misconduct; or
- Permanent service-connected disability of one lung rated 50% or more disabling, in combination with a non-service-connected disability of the other lung that is not the result of the veteran's own willful misconduct.

If a veteran described above receives any money or property of value according to an award in a judicial proceeding based upon, or a settlement or compromise of, any cause of action for damages for the non-service-connected disability, the increase in the rate of compensation otherwise payable shall not be paid for any month following a month in which any such money or property is received until the total of the amount of such increase that would otherwise have been payable equals the total of the amount of any such money received, and the fair market value of any such property received.

Payment of Disability Compensation in Disability Severance Cases

The deduction of disability severance pay from disability compensation, as required by Section 1212(c) of Title 10, shall be made at a monthly rate, not more than the rate of compensation to which the former member would be entitled based on the degree of such former member's disability, as determined on the initial Department rating.

Trial Work Periods and Vocational Rehabilitation for Certain Veterans with Total Disability Ratings

The disability rating of a qualified veteran who begins to engage in a substantially gainful occupation after January 31, 1985, may not be reduced based on the veteran having secured and followed a substantially gainful occupation unless the veteran maintains such an occupation for 12 consecutive months.

("Qualified Veteran" means a veteran who has a service-connected disability or disabilities, not rated as total, but who has been awarded a rating of total of disability because of inability to secure or follow a substantially gainful occupation as a result of such disability or disabilities.)

Counseling services, placement, and post-placement services shall be available to each qualified veteran, whether or not the veteran is participating in a vocational rehabilitation program.

Asbestos Exposure

Veterans who were exposed to asbestos while in service and developed a disease related to asbestos exposure may receive service-connected compensation benefits.

Please see Chapter 26: Toxic Exposure for more information.

Naturalization

Veterans who served before September 11, 2001, are eligible to file for naturalization based on their U.S. military service. An applicant who served three years in the U.S. military and is a lawful permanent resident is excused from any specific period of required residence, period of residence in any specific place, or physical presence within the United States if the application for naturalization is filed while the applicant is still serving in the military or within six months of honorable discharge. Applicants who file for naturalization more than six months after termination of three years of U.S. military service may count any periods of honorable service as residence and physical presence in the United States.

Aliens and non-citizen nationals with honorable service in the U.S. armed forces during specified periods of hostilities may be naturalized without having to comply with the general requirements for naturalization. This is the only section of the Immigration and Nationality Act, as amended, which allows persons who have not been lawfully admitted for permanent residence to apply for naturalization.

Any person who has served honorably during qualifying time may apply at any time in his or her life if, at the time of enlistment, reenlistment, extension of enlistment or induction, such person shall have been in the United States, the Canal Zone, American Samoa or Swain's Island, or, on or after November 18, 1997, aboard a public vessel owned or operated by the United States for non-commercial service, whether or not lawful admittance to the United States for permanent residence has been granted. Certain applicants who have served in the U.S. Armed Forces are eligible to file for naturalization based on current or prior U.S. military service. Such applicants should file the N-400 Military Naturalization Packet.

How To Avoid Overpayments

A VA overpayment is when a veteran receives more VA benefits than he or she is entitled to and therefore must pay the money back to the VA. This may happen if a veteran is delayed in submitting paperwork or forgets to update records. When discovered those funds are owed to the VA and may lead to a deduction in future monthly benefit amounts until the debt is repaid. You can make changes if anything happens that could impact your benefits by calling 1-800-827-1000.

Common overpayment situations include:
- A veteran in the Reserves may be called up for active duty and still receiving benefits from VA
- A veteran receiving education benefits doesn't complete the course requirement
- A veteran has a change in marital status and doesn't notify VA
- The veteran doesn't report a school-age child remarries
- A dependent passes away and VA isn't notified
- A veteran receives care at a VA medical facility and doesn't pay a required co-pay
- A veteran or beneficiary is incarcerated by still receives benefits during that time
- Veterans or beneficiaries receiving an income-based pension don't report a change in income
- The Vocational Rehabilitation program purchased a service or tools for a veteran who leaves the program without a valid reason

What to Remember
✓ VA disability compensation is a tax-free, monthly payment to veterans.
✓ A veteran may qualify for VA disability benefits for physical conditions like a chronic illness or injury, and mental health conditions.
✓ Rates of compensation aren't automatically adjusted for inflation and can only be increased if Congress passes legislation specifically enabling an increase
✓ The illness or injury could have developed before, during or after service.
✓ If you plan to file for disability using a paper form, you may want to submit an intent to file form first so you have time to gather evidence, and you may then get retroactive payments.

CHAPTER 5

SPECIAL CLAIMS & MISCELLEANOUS BENEFITS

If a veteran has a disability not listed as being linked to military service, they may still be able to get disability compensation or other benefits. If a veteran has a disability that the VA has concluded is because of an illness or injury caused or made worse by active-duty service, they may be able to get special compensation to help with disabilities, like the ones listed below.

Commissary Access

According to the Department of Defense, starting January 1, 2020, all service-connected Veterans, Purple Heart recipients, former prisoners of war (POW), and individuals approved and designated as the primary family caregivers of eligible Veterans under the Department of Veterans Affairs Program of Comprehensive Assistance for Family Caregivers (PCAFC) can use commissaries, exchanges, and morale, welfare, and recreation (MWR) retail facilities, in-person and online.

Who Is Eligible?

Eligible people include veterans who are Purple Heart recipients, former prisoners of war, and veterans with 0-90% service-connected disability ratings. On January 1, 2020, individuals approved and designated as the primary family caregiver of an eligible veteran under the PCAFC are eligible for these privileges.

Required Credentials
Veterans

- On January 1, 2020, veterans eligible solely under this act who are eligible to obtain a Veteran Health Identification Card must use this credential for in-person installation and privilege access. The card must display the Veteran's eligibility status (i.e., PURPLE HEART, FORMER POW, or SERVICE CONNECTED).
- Veterans eligible solely under this act who are not enrolled in or are not eligible to enroll in VA health care, or who are enrolled in VA health care, but do not possess a Veteran Health Identification Card will not have access to DoD and Coast Guard installations for in-person commissary, exchange, and MWR retail privileges, but will have full access to online exchanges and American Forces Travel.
- Medal of Honor recipients and Veterans with 100% service-connected disability ratings are eligible for DoD credentials under DoD policy.

Caregivers

Eligible caregivers will receive an eligibility letter from VA's Office of Community Care. If you are a primary family caregiver under the PCAFC and lose your eligibility letter, please call 1-877-733-7927 to request a replacement. Please allow two weeks for processing.

2023 VA Special Benefit Allowance Rates

Benefit	Monthly Payment	Effective Date
Automobile Allowance Money to help you buy a specially equipped vehicle if your service-connected disability prevents you from driving.	$24,115.12 paid once	October 1, 2022
Clothing Allowance Money to help you replace clothes damaged by a medicine or prosthetic or orthopedic device related to your service-connected disability	$968.52 paid once or once each year	December 1, 2022
Medal of Honor Pension Added compensation if you received the Medal of Honor	$1619.34 paid once each month	December 1, 2022

Automobile Assistance Program

The VA offers an automobile assistance program for eligible veterans, or eligible members of the Armed Forces serving on active duty who are suffering from a disability as described below if such disability is the result of an injury incurred or disease contracted in or aggravated in the line of duty in the active military, naval, or air service. Following are highlights of the VA program:

Financial Assistance

Qualified veterans may receive a **one-time** payment from VA of up to **$24,115.12** to be used toward the purchase of an automobile or other conveyance. The law requires that this amount be adjusted annually by the percentage increase in the Consumer Price Index.

This is available beginning October 1, 2022. You must have VA approval before purchasing an automobile or adaptive equipment

Eligibility Requirements for Receipt of One-Time Payment

Veterans or service members who are entitled to Disability Compensation under Chapter 11 of Title 38 due to one of the following service-connected losses:

- The loss or permanent loss of use of one or both feet; or
- The loss or permanent loss of use of one or both hands; or
- The permanent impairment of vision of both eyes of the following status:
- Central visual acuity of 20/200 or less in the better eye, with corrective glasses, or central visual acuity of more than 20/200 if there is a field defect in which the peripheral field has contracted to such an extent that the widest diameter of visual field subtends an angular distance no greater than twenty degrees in the better eye.

Adaptive Equipment

In addition to the one-time payment described above, VA will also pay for installation of adaptive equipment deemed necessary to ensure that the eligible veteran will be able to safely operate the vehicle and to satisfy the applicable state standards of licensure.

VA will also repair, replace, or reinstall adaptive equipment determined necessary for the operation of a vehicle acquired under this program, or for the operation of a vehicle an eligible veteran may previously or subsequently have acquired.

Eligibility Requirements for Receipt of Adaptive Equipment

- The loss or permanent loss of use of one or both feet; or
- The loss or permanent loss of use of one or both hands; or
- The permanent impairment of vision of both eyes of the following status:
  - o Central visual acuity of 20/200 or less in the better eye, with corrective glasses, or central visual acuity of more than 20/200 if there is a field defect in which the peripheral field has contracted to such an extent that the widest diameter of visual field subtends an angular distance no greater than twenty degrees in the better eye; or
  - o Ankylosis (immobility) of one or both knees; or
  - o Ankylosis (immobility) of one or both hips.

Adaptive Equipment Available for Installation

The term adaptive equipment generally means any equipment which must be part of or added to a vehicle manufactured for sale to the general public to make it safe for use by the claimant and to assist him or her in meeting the applicable standards of licensure of the proper licensing authority.

Following is a partial list of adaptive equipment available under this program:

- Power steering;
- Power brakes;
- Power window lifts;
- Power seats;
- Special equipment necessary to assist the eligible person into and out of the automobile or other conveyance;
- Air-conditioning equipment, if such equipment is necessary to the health and safety of the veteran and the safety of others, regardless of whether the automobile or other conveyance is to be operated by the eligible person or is to be operated for such person by another person;
- Any modification of the size of the interior space of the automobile or other conveyance if necessary, for the disabled person to enter or operate the vehicle;
- Other equipment, not described above, if determined necessary by the Chief Medical Director or designee in an individual case.

Eligible veterans are not entitled to adaptive equipment for more than two vehicles at any one time during any four years. (In the event an adapted vehicle is no longer available for use by the eligible veteran due to circumstances beyond his or her control, loss due to fire, theft, accident, etc., an exception to this four-year provision may be approved.

Specially Adapted Housing Grant

There are three types of grants that may be payable:

Specially Adapted Housing Grants (SAH)
Special Housing Adaptations Grants (SHA)
Temporary Residence Adaptation Grants (TRA)
Specially Adapted Housing Grant (SAH)

The VA may approve a grant of not more than 50% of the cost of building, buying, or remodeling adapted homes or paying indebtedness on homes previously acquired, up to a maximum of **$109,986.**

Public law 110-289 directed that the maximum grant amount will be adjusted annually. The amount will either increase or stay the same each year – it cannot decrease. To qualify for this grant, veterans must be entitled to compensation for permanent and total service-connected disability due to:

- Loss or loss of use of both lower extremities, which prevents movement without the aid of braces, crutches, canes, or a wheelchair; or

Disability which includes:

- Blindness in both eyes, having only light perception; **with**
- Loss or loss of use of one lower extremity; or
- Loss or loss of use of one lower extremity together with:
- Residuals of organic disease or injury; **or**
- The loss or loss of use of one upper extremity, which so affects the functions of balance or propulsion as to preclude locomotion without using braces, canes, crutches, or a wheelchair.
- Loss of, or loss of use, of both upper extremities such as to preclude the use of the arms at or above the elbows.
- A severe burn injury (as so determined.)

Eligibility for Active-Duty Members
Public Law 108-183, The Veterans Benefits Act of 2003, which became law on December 16, 2003, extended eligibility for the $109,986 grant to members of the Armed Forces serving on active duty who are suffering from a disability as described above if such disability is the result of an injury incurred or disease contracted in or aggravated in the line of duty in the active military, naval, or air service.

Public Law 110-289, the Economic and Housing Recovery Act of 2008, dated July 30, 2008, extended eligibility for active duty servicemembers to the same extent, and made them subject to the same limitations, as veterans.

An eligible veteran or active-duty service member can use his or her benefit up to three times, as long as the aggregate amount of assistance does not exceed the maximum amounts allowable.

This benefit extends to previous grant recipients; however, they cannot obtain a subsequent grant(s) to pay for adaptations made before June 15, 2006, or to reduce an existing mortgage principal balance for properties acquired before June 15, 2006.

Special Home Adaptation Grant (SHA)
VA may approve a grant for the actual cost, up to a maximum of $22,036 for adaptations to a veteran's residence that are determined by VA to be reasonably necessary. Public Law 110-289 directed that the maximum grant amount will be adjusted annually, based on a cost-of-construction index. The amount will either increase or stay the same each year- it cannot decrease.

The grant may also be used to assist veterans in acquiring a residence that has already been adapted with special features for the veteran's disability.

To qualify for this grant, veterans must be entitled to compensation for permanent and total service-connected disability due to:

- Blindness in both eyes, with 5/200 visual acuity or less; or
- Anatomical loss or loss of use of both hands

Temporary Residence Adaptation Grant (TRA)
Eligible veterans and servicemembers who are temporarily residing in a home owned by a family member may also receive a Temporary Residence Adaptation (TRA) grant to help the

veteran or servicemember adapt the family member's home to meet his or her special needs. While the SAH and SHA grants require ownership and title to a house, in creating the TRA, Congress recognized the need to allow veterans and active-duty members who may not yet own homes to have access to the adaptive housing grant program.

VA may provide a TEMPORARY RESIDENCE ADAPTATION GRANT of up to $44,299.

VA may provide a TEMPORARY RESIDENCE ADAPTATION GRANT of up to $7,910.

Specially Adapted Housing Assistive Technology Grant Program

VA through its Specially Adapted Housing Assistive Technology (SAHAT) Grant Program is authorized to award grants of up to $200,000 per fiscal year to persons or entities to encourage the development of specially adapted housing assistive technologies. The FY2020 grant solicitation period is to be determined but may open mid-December and close late February.

Program Purpose

There are many emerging technologies that could improve home adaptions or enhance a veteran's or servicemember's ability to live independently, such as voice-recognition and voice-command operations, living environment controls, and adaptive feeding equipment. VA has defined "new assistive technology" as an advancement that could aid or enhance the ability of a veteran or servicemember to live in an adapted home.

Please note: SAHAT funding does not support the construction or modification of residential dwellings for accessibility. Veterans and servicemembers interested in adapting their home for barrier-free living are encouraged to review this factsheet to identify Home Adaptation programs offered by VA.

Amount of Funding

For FY2020, VA is authorized to award up to $1 million in grant funds for the SAHAT Grant Program. The number of grants VA will fund will be based on application quality and funding availability. The maximum grant award is limited to $200,000 per grant recipient.

How to Apply

Interested applicants must register and apply through www.grants.gov during the application period stated on the Notice of Funding Availability (NOFA). Applications will not be accepted outside of the application period and each application must meet the requirements provided in the NOFA and at www.grants.gov. To locate the funding opportunity on www.grants.gov, click here or search using the following terms: Opportunity Title: Specially Adapted Housing Assistive Technology Grant, or Opportunity Number (FY2020): VA-SAHAT-20-05, or Catalog of Federal Domestic Assistance (CFDA) Number: 64.051

The SAHAT Grant Program transfers grants funds through the Department of Health and Human Services Payment Management System (PMS). SAHAT grant recipients will need to register with the PMS in order to access SAHAT grant funds. For information on the PMS, please refer to: https://pms.psc.gov/ For more information on the SAHAT Grant Program, please write to SAHAT.VBAVACO@va.gov.

Supplemental Financing—Loan Guaranty

Veterans who have available loan guaranty entitlement may also obtain a guaranteed loan or possibly a direct loan from VA to supplement the grant to acquire a specially adapted home.

Clothing Allowance

In 2011, a final regulation was published in the Federal Register, expanding the eligibility criteria for veterans with multiple prosthetic and orthopedic devices or those who use prescription medications for service-related skin conditions.

The new regulation provides the criteria for more than one annual clothing allowance in situations where distinct garments are affected. Payment for more than one clothing allowance for eligible veterans began in 2012.

The VA shall pay a clothing allowance of $968.52 per year (rate effective December 1, 2022, and payable beginning August 1, 2023) to each veteran who:

- Because of a service-connected disability, wears or uses a prosthetic or orthopedic appliance (including a wheelchair) which the VA determines tends to wear out or tear the clothing of the veteran; or
- Uses medication prescribed by a physician for a skin condition that is due to a service-connected disability, and which the VA determines causes irreparable damage to the veteran's outer garments.

Medal of Honor Pension

The law further states that VA shall pay, in a lump sum, to each person who is in receipt of the Medal of Honor Pension, an amount equal to the total amount of special pension that the person would have received during the period beginning on the 1st day of the 1st month beginning after the date of the act for which the person was awarded the Medal of Honor, and ending on the last day of the month preceding the month in which the person's special pension commenced.

For each month of a period referred to in the previous paragraph, the amount of special pension payable shall be determined using the rate of special pension that was in effect for such month.

The Secretary of the Department of the Army, the Department of the Navy, the Department of the Air Force, or the Department of Transportation will determine the eligibility of applicants to be entered on the Medal of Honor Roll and will deliver to the Secretary of the Department of Veterans Affairs a certified copy of each certificate issued in which the right of the person named in the certificate to the special pension is set forth. The special pension will be authorized based on such certification.

The special pension will be paid in addition to all other payments under laws of the United States. However, a person awarded more than one Medal of Honor may not receive more than one special pension.

The Medal of Honor Pension rate was changed 12/01/2022 to $1,619.34

Hearing Aids

To receive hearing aids through the VA, you must first register at the health administration/enrollment section of the VA Medical Center of your choice. You will need to bring a copy of your DD214, a driver's license and health insurance information if available.

You can do this in person at any VA medical center or clinic, or online by filling out the form 10-10EZ. You can also mail the completed Form 10-10EZ to the medical center of your choice.

Once registered you can schedule an appointment at the Audiology and Speech Pathology Clinic for a hearing evaluation. The audiologist then makes a clinical determination on the need for hearing aids. If hearing aids are recommended and fit, hearing aids, repairs and future batteries will be available at no charge to you if you remain VA healthcare eligible.

Ordering Batteries

If you are authorized to receive batteries from the VA, you can request them in a few different ways.

You can request them by mail using VA Form 2346, Request for Batteries and Accessories. You should have received the card/envelope with your most recent battery order. You can then complete the form and mail it to:

VA Denver Acquisition and Logistics Center
PO Box 25166
Denver, CO 80225-0166

You can also order batteries over the phone by calling the Denver Acquisition and Logistics center at 303-273-6200.

A third way to order new batteries is to use the eBenefits website. You have to apply for a Premium Account on eBenefits. Once you're logged in as a Premium User, you can go to the category "Manage Health." From there, navigate to the subcategory which is "Hearing Aid Batteries and Prosthetic Socks."

Once you order your batteries, they should arrive within 7 to 10 days.

Recovery From Surgery or an Immobilizing Disability (Convalescence)

If a veteran is recovering from a surgery or disability related to their military service that left them unable to move, they may be able to get a temporary 100% disability rating, and disability compensation or benefits.

You could be eligible for disability benefits if you had surgery or another treatment at a VA hospital, approved hospital, or an outpatient center for a disability related to your military service (service-connected disability).

If you had surgery, both of the following most be true:
- The surgery required a recovery time of at least one month or reports show that the surgery or treatment was for a service-connected disability, and
- The surgery resulted in severe injuries, like:
  - o Surgical wounds that haven't totally healed
  - o Stumps of recent amputations
  - o Being unable to move due to being put in splints or casts to help with healing, known as therapeutic immobilizations
  - o Being unable to leave your house, known as house confinement
  - o Being required to use crutches or a wheelchair

If you didn't have surgery, you must have had one or more major joints immobilized by a cast.

You may be eligible for health care, added compensation while you recover, and a temporary 100% disability rating.

You'll need to file a claim for disability compensation.

Recovery time from either a surgery or joint immobilization by a cast without surgery requires a temporary 100% disability rating for a service-connected disability. The temporary 100% rating may continue for 1 to 3 months, depending on your specific case. You may be able to get an extension for up to 3 more months if your case is severe.

Increased Disability Rating for Time in a Hospital

If a veteran spent time in a VA hospital or a VA-approved hospital for a disability that's service-connected, they may be able to get added disability compensation or benefits with a temporary 100% disability rating for the time spent in the hospital.

A veteran may be eligible for disability benefits if they meet one of the following requirements:

- The veteran spent more than 21 days in a VA hospital or other approved hospital for a service-connected disability, or
- The veteran was under hospital observation for more than 21 days at the VA's expense for a service-connected disability

If you qualify, you'll need to file a claim for disability compensation. If you weren't in a VA hospital when you file your claim, you have to provide the VA with a hospital discharge summary showing the length and cause of the hospital stay.

VA Temporary Disability (Prestabilization Rating)

If a veteran recently ended their military service and they have a service-connected disability, they might be able to get temporary disability compensation or benefits right away. If a veteran qualifies for these benefits, they get a prestabilization rating. The rating may be 50% or 100%, depending on the severity of the disability. The prestabilization rating continues for one year after discharge from active service.

Both of the following must be true:

- The veteran must have a severe service-connected disability that's unstable, meaning an illness or injury that will change or hasn't yet been fully treated, and
- Your disability is expected to continue for an unknown amount of time.

You'll need to file a claim for disability compensation. When you file, you have to so show the VA you have a severe, service-connected disability that's unstable and is expected to continue for an unknown amount of time. The information will be part of your Service Treatment Record.

VA Title 38 U.S.C. 1151 Claims

If a veteran suffered an added disability or an existing injury or disease got worse while they were getting VA medical care or taking part in a VA program designed to help the veteran find, get or keep a job, they may be able to get compensation.

At least one of the three must have led directly to an added disability or to the injury or disease getting worse:

- VA carelessness or negligence, or
- VA medical or surgical treatment, or
- A VA health exam, or
- A VA vocational rehabilitation course under 38 U.S.C. Chapter 31, or
- VA compensated work therapy under 38 U.S.C. 1718

It's necessary to file a claim for disability compensation. When a veteran files, they have to show the added disability happened because of VA medical care or a VA medical program designed to help them find, get or keep a job. The VA will award compensation payments in the same way it would if the disability was related to military service (a service-connected disability).

Individual Unemployability

If you can't work because of a disability related to your service in the military, you may qualify for Individual Unemployability, meaning you may be able to get disability compensation or benefits at the same level as a veteran with a 100% disability rating.

You have to meet both of the following:

- You have at least one service-connected disability rated at 60% or more disabling, or two or more service-connected disabilities with at least one rated at 40% or more disabling and a combined rating of 70% or more, and

- You can't hold down a steady job that supports you financially, known as substantially gainful employment because of your service-connected disability. Odd jobs, known as marginal employment, don't count.

In some cases, for example, if you often have to be in the hospital, you could qualify at a lower disability rating.

You'll need to file a claim for disability compensation. When you file, you'll have to provide evidence showing your disability prevents you from holding down a steady job. The VA will also review your work and education history.

When you file a disability claim you also need to file A Veteran's Application for Increased Compensation Based on Unemployability (VA Form 21-8940) and A Request for Employment Information in Connection with Claim for Disability Benefits (VA Form 21-4192).

What to Remember
✓ An automobile allowance is money to help a veteran buy a specially equipped vehicle if their service-connected disability prevents them from driving.
✓ Housing grants are available to help adapt a veteran's residence.
✓ The clothing allowance is money to help you replace clothes damaged by a medicine or prosthetic or orthopedic device related to your service-connected disability.
✓ If you can't work because of your service-connected disability you may be able to get increased disability payments, known as individual unemployability.
✓ If you're recovering from surgery or an immobilizing disability that leaves you unable to move, you might be able to get temporary disability payments or other benefits.

CHAPTER 6

VETERANS PENSION

The Veterans Pension program provides monthly payments to wartime veterans who meet certain age or disability requirements, and who have income and net worth within certain limits.

You may be eligible if you meet both of these requirements:
- You didn't receive a dishonorable discharge, and
- Your yearly family income and net worth meet certain limits set by Congress. Your net worth includes all personal property you own except your house, car and most home furnishings, minus any debt you owe. Your net worth includes the net worth of your spouse.

At least one of the following must be true about your service:
- You started on active duty before September 8, 1980, and you served at least 90 days on activity duty with at least one day during wartime, or
- You started on active duty as an enlisted person after September 7, 1980, and served at least 24 months or the full period for which you were called to active with some exceptions with at least one day during wartime, or
- You were an officer and started on active duty after October 16, 1981, and you hadn't previously served on active duty for at least 24 months.

At least one of the following must be true:
- You're at least 65 years old, or
- You have a permanent and total disability, or
- You're a patient in a nursing home for long-term care because of a disability, or
- You're getting Social Security Disability Insurance or Supplemental Security income.

Under current law, the VA recognizes the following wartime periods to decide eligibility for VA pension benefits:
- Mexican Border period (May 9, 1916, to April 5, 1917, for Veterans who served in Mexico, on its borders, or in adjacent waters)
- World War I (April 6, 1917, to November 11, 1918)
- World War II (December 7, 1941, to December 31, 1946)
- Korean conflict (June 27, 1950, to January 31, 1955)
- Vietnam War era (November 1, 1955, to May 7, 1975, for Veterans who served in the Republic of Vietnam during that period. August 5, 1964, to May 7, 1975, for Veterans who served outside the Republic of Vietnam.)
- Gulf War (August 2, 1990, through a future date to be set by law or presidential proclamation)

To apply for tax-free VA pension benefits as a veteran, you'll need this information:
- Social Security number or VA file number
- Military history

- Your financial information
- The financial information of your dependents
- Work history
- Bank account direct deposit information
- Medical information

You can apply online, or by mail. If you're applying by mail, you fill out an Application for Pension (VA Form 21P-527EZ). You mail the completed form to the pension management center at:

Department of Veterans Affairs
Pension Intake Center
PO Box5365
Janesville, WI 53547-5365

You can also bring your application to a VA regional office, or you can work with an accredited representative to help when you're applying for VA pension benefits.

VA Aid and Attendance Benefits and Housebound Allowance

VA Aid and Attendance or Housebound benefits provide monthly payments added to the amount of a monthly VA pension for qualified veterans and survivors. You may be eligible if you get a VA pension, and you meet at least one of these requirements:

- You need another person to help you with daily activities like dressing, feeding or bathing, or
- You have to stay in bed or spend a large portion of the day in bed because of illness, or
- You are a patient in a nursing home due to the loss of mental or physical abilities related to a disability, or
- Your eyesight is limited (even with contact lenses or glasses you have only 5/200 or less in both eyes, or concentric contraction of the visual field to 5 degrees or less).

You may be eligible for housebound allowance if you get a VA pension and spend most of your time in your home because of a permanent disability.

You can't get Aid and Attendance and Housebound benefits at the same time.

You can send a complete VA Form 21-2680 (Examination for Housebound Status or Permanent Need for Regular Aid and Attendance) to the PMC for your state. Your doctor should fill out the examination information section.

You can include other evidence like a doctor's report, details about what you normally do during the day, or details that show what kind of illness or disability affects your ability to do things on your own.

If a veteran is in a nursing home, they'll need to fill out a Request for Nursing Home Information in Connection with Claim for Aid and Attendance (VA Form 21-0779).

VA Pension Rates for Veterans

If you qualify for pension benefits, the VA bases the payment amount on the difference in your countable income and a limit set by Congress, called the Maximum Annual Pension Rate or MAPR.

Your countable income is how much you earn. This includes your Social Security benefits, investment and retirement payments, and income your dependents receive. Some expenses, like non-reimbursable medical expenses, may reduce your countable income.

Your MAPR amount is the maximum amount of pension payable. MAPR is based on how many dependents you have, if you're married to another veteran who qualifies for a pension, and if your disabilities qualify you for Housebound or Aid and Attendance benefits. MAPRs are adjusted every year for cost-of-living increases.

Example: You're a qualified veteran with a dependent, non-veteran spouse and no children. You also qualify for Aid and Attendance benefits based on your disabilities. You and your spouse have a combined yearly income of $10,000.
Your MAPR amount = $31,714
Your yearly income = $10,000
Your VA pension = $21,714 for the year (or $1,809 paid each month)

Net Worth Limit
From December 1, 2022, to November 30, 2023, the net worth limit to be eligible for Veterans Pension benefits is $150,538.

On October 18, 2018, the VA changed the way it assesses net worth to make pension entitlement rules clearer. Net worth includes the veteran and their spouse's assets and annual income.

If the veteran's child's net worth is more than the net worth limit, they aren't considered as a dependent when determining pension.

Assets
Assets include the fair market value of all real and personal property, minus the amount of any mortgages. Real property means land and buildings the veteran may own. Personal property assets can include investments like stocks and bonds, furniture and boats.

Assets don't include your primary residence where you live most of the time, your car, or basic home items like appliances that you wouldn't take with you if you moved to a new house.

Annual Income
For purposes of pensions, annual income is defined as the money earned in a year from a job or from retirement or annuity payments. It includes salary or hourly pay, bonuses, commissions, overtime and tips. The VA will subtract certain expenses from a veteran's annual income when assessing net worth. These are known as applicable deductible expenses, and they include educational expenses and medical expenses you aren't reimbursed for.

3-Year Look-Back Period
When VA receives a pension claim, they review the terms and conditions of any assets the veteran may have transferred in the three years before filing the claim. If a veteran transfers asset for less than fair market value during the look-back period, and those assets would have pushed the veteran's net worth above the limit for a VA pension, they may be subject to a penalty period of up to five years. The veteran isn't eligible for benefits during this time.

This policy took effect on October 18, 2018. If a veteran filed a claim before this date, the look-back period doesn't apply.

A penalty period is the length of time when a veteran isn't eligible for pension benefits because the transferred assets for less than fair market value during the look-back period. The penalty period rate is $2,642.

2023 VA Pension Rates for Veterans
Date of Cost-of-Living Increase: December 1, 2022
Increase Factor: 8.7%
Standard Medicare Deduction: Actual amount will be determined by SSA based on individual income

For veterans with no dependents

If You Have No Dependents and...	Your Maximum Annual Pension Rate (MAPR) Amount is
You don't qualify for Housebound or Aid and Attendance benefits	$16,037
You qualify for Housebound benefits	$19,598
You qualify for Aid and Attendance Benefits	$26,752

Note: If you have medical expenses, you may deduct only the amount that's above 5% of your MAPR amount ($801 for a veteran with no spouse or child).

For veterans with at least 1 dependent child or spouse

If You Have 1 Dependent and...	Your MAPR Amount Is
You don't qualify for Housebound or Aid and Attendance benefits	$21,001
You qualify for Housebound benefits	$24,562
You qualify for Aid and Attendance benefits	$31,714

Notes:
- If you have more than one dependent, add $2,743 to your MAPR amount for each additional dependent.
- If you have a child who works, you may exclude their wages up to $13,850.
- If you have medical expenses, you may deduct only the amount that's above 5% of your MAPR amount ($1,050 for a veteran with one dependent).

For 2 veterans who are married to each other

If You're 2 Veterans Who Are Married to Each Other and...	Your MAPR Amount Is
Neither of you qualifies for Housebound or Aid and Attendance benefits	$21,001
One of you qualifies for Housebound benefits	$24,562
Both of you qualify for Housebound benefits	$28,121
One of you qualifies for Aid and Attendance benefits	$31,714
One of you qualifies for Housebound benefits and one of you qualifies for Aid and Attendance benefits	$35,266
Both of you qualify for Aid and Attendance benefits	$42,433

Notes:
- If you have more than one dependent, add $2,743 to your MAPR amount for each additional child

75

- If you have a child who works, you may exclude their wages up to $13,850
- If you have medical expenses, you may deduct only the amount that's above 5% of your MAPR amount ($1,050 for a veteran with 1 dependent)

2023 VA Protected Pension Rates

A veteran may be eligible for protected rates if they began receiving VA disability pension payments before December 31, 1978, and they haven't elected to change to the current, improved pension programs. That means a veteran will continue to receive payments at the rates under the old program as well as a cost-of-living increase.

To qualify for protected rates, your yearly income for 2022 must be at or below a certain amount. This is the income limit. Income is any money you earn in a year, including your salary, investment and retirement payments and income from dependents.

Some expenses, like non-reimbursable medical expenses may work to reduce your countable income. VA bases a veteran's income limit on the specific pension benefits they're eligible to receive, and whether they have eligible dependents, and their yearly income. Eligible dependents may include a veteran's spouse, and the VA recognizes same-sex and common-law marriages. Dependents may also include biological, step or adopted children who are unmarried and meet at least one of the following requirements:

- The child is unmarried and under 18 years old, or
- The child is unmarried and between 18 and 23 years old and enrolled in qualified school full time, or
- The child is unmarried and was seriously disabled before age 18 and can't care for themselves

Section 306 Disability Pension Rates

The non-service-connected pension program was available from July 1, 1960, through December 31, 1978.

Section 306 Disability Pension Yearly Income Limits for Veterans without Dependents

Pension Benefit	2022 Yearly Income Limit (Effective December 1, 2022)
Basic monthly payment (veteran only, no dependent spouse or children)	Your yearly income must be $18,240 or less to continue receiving this benefit.
Special Aid and Attendance allowance, if your income is more than $18,240	Your yearly income must be $18,897 or less to continue receiving this benefit
Hospital reduction rate for Special Aid and Attendance, if you're hospitalized on or after January 1, 2022	Your yearly income must be $18,897 or less to continue receiving this benefit.

Note: the hospital reduction rate is a reduced rate of Special Aid and Attendance the VA will pay if you're hospitalized and meet certain requirements.

Section 306 Disability Pension Yearly Income Limits for Veterans with Dependents

Pension Benefit	2022 Yearly Income Limit
Basic monthly payment for a veteran with a spouse or one or more dependent children	Your yearly income must be $24,418 or less to continue receiving this benefit
Special Aid and Attendance allowance, if your income is more than $24,518	Your yearly income must be $25,172 or less to continue receiving this benefit
Hospital reduction rates, if you're hospitalized on or after January 1, 2022	Your yearly income must be $25,172 or less

Note: If you're married, the VA will include some of your spouse's income when determining if your yearly income is at or below the income limit. The current Section 306 disability pension spouse income exclusion limit is $5,826. The VA won't include the first $5,826 of a veteran's spouse's income but will include any amount above that unless the veteran can prove they don't have access to the income or that including it would cause financial hardship.

Old Law Disability Pension Rates

Effective December 1, 2022. This non-service-connected pension program was available before July 1, 1960. The below income limits include an 8.7% cost-of-living increase for the year.

Veteran Status	2022 Yearly Income Limit
Veteran alone (no spouse or dependent children)	The veteran's yearly income must be $15,973 or less to continue receiving this benefit.
Veteran with a spouse or one or more dependent children	The veteran's yearly income must be $23,020 or less to continue receiving this benefit.

Old Law Disability Pension Monthly Payments

Pension Benefit	Monthly Payment
Basic veteran pension	$66.15
Pension for veteran with 10 years of service who is at least 65 years old	$78.75
Aid and Attendance (if entitled)	$135.45
Housebound allowance (if entitled)	$100.00

Combination of Ratings
The VA shall provide that, for the purpose of determining whether or not a veteran is permanently and totally disabled, ratings for service-connected disabilities may be combined with ratings for non-service-connected disabilities. Where a veteran is found to be entitled to a Non-Service-Connected Disability Pension and is also entitled to Service-Connected Disability Compensation, the VA shall pay the veteran the greater benefit.

Vocational Training for Certain Pension Recipients
In the case of a veteran who is awarded a Non-Service-Connected Disability Pension, the VA shall base on information on file with the VA, make a preliminary finding whether such veteran, with the assistance of a vocational training program, has a good potential for achieving employment. If such potential is found to exist, the VA shall solicit from the veteran an application for vocational training. If the veteran after that applies for such training, the VA shall provide the veteran with an evaluation, which may include a personal interview, to determine whether the achievement of a vocational goal is reasonably feasible.

If the VA, based on the evaluation, determines that the achievement of a vocational goal by a veteran is reasonably feasible, the veteran shall be offered and may elect to pursue a vocational training program.

If the veteran elects to pursue such a program, the program shall be designed in consultation with the veteran to meet the veteran's individual needs and shall be outlined in an individualized written plan of vocational rehabilitation.
A vocational training program under this section:
- May not exceed 24 months unless, based on a determination by the VA that an

extension is necessary for the veteran to achieve a vocational goal identified in the written plan formulated for the veteran, the VA grants an extension for a period not to exceed 24 months.

- May not include the provision of any loan or subsistence allowance or any adaptive automobile equipment.
- May include a program of education at an institution of higher learning, only in a case in which the Secretary of the VA determines that the program involved is predominantly vocational in content.
- When a veteran completes a vocational training program, the VA may provide the veteran with counseling, placement, and post-placement services for a period not to exceed 18 months.

A veteran may not begin pursuit of a vocational training program under this chapter after the later of:

- December 31, 1995; or
- The end of a reasonable period of time, as determined by the VA, following either the evaluation of the veteran or the award of pension to the veteran.

In the case of a veteran who has been determined to have a permanent and total non-service-connected disability and who, not later than one year after the date the veteran's eligibility for counseling under this chapter expires, secures employment within the scope of a vocational goal identified in the veteran's individualized written plan of vocational rehabilitation (or in a related field which requires reasonably developed skills, and the use of some or all of the training or services furnished the veteran under such plan), the evaluation of the veteran as having a permanent and total disability may not be terminated because of the veteran's capacity to engage in such employment until the veteran first maintains such employment for a period of not less than 12 consecutive months.

Protection of Healthcare Eligibility

In the case of a veteran whose entitlement to pension is terminated after January 31, 1985, because of income from work or training, the veteran shall retain for a period of three years, beginning on the date of such termination, all eligibility for care and services that the veteran would have had if the veteran's entitlement to pension had not been terminated.

Disappearance

When a veteran receiving a non-service-connected disability pension from the VA disappears, the VA may pay the pension otherwise payable to such veteran's spouse and children. Payments made to a spouse or child shall not exceed the amount to which each would be entitled if the veteran died of a non-service-connected disability.

What to Remember
✓ Non-service-connected disability pension provides support to wartime veterans with limited income. The application is Form 21P-527EZ.
✓ Amounts payable under the pension program depend on the type and amount of income a veteran and their family members receive
✓ Monthly payments are made to bring a veteran's total income including other Social Security and retirement income to an established support level
✓ You must have not received a dishonorable discharge to be eligible.
✓ Your income and net worth must meet certain limits set by Congress. Net worth includes all personal property except your house and car, minus any debt you owe. Net worth includes that of your spouse.

CHAPTER 7

SURVIVOR AND DEPENDENT COMPENSATION (DIC)

If you're the surviving spouse, child or parent of a service member who died in the line of duty, or a veteran who died from a service-related injury or illness, you may be able to get a tax-free monetary benefit. The benefit is called VA Dependency and Indemnity Compensation or VA DIC.

2022 Update—Information for Survivors with PACT Act-Related Claims:

If you think you're eligible for VA DIC under the PACT Act, you can submit a new claim. If the VA denied your claim in the past, and it thinks you may be eligible now, they may try to contact you. They may be able to reevaluate your claim, but you don't have to wait for them to contact you before you reapply.

If you're a surviving family member of a veteran, you may be eligible for these benefits:

- A monthly VA DIC payment if you're the surviving spouse, dependent child or parent of a veteran who died from a service-connected disability
- A one-time accrued benefits payment if you're the surviving spouse, dependent child or dependent parent of a veteran who the VA owed unpaid benefits at the time of their death
- A survivor's pension if you're the surviving spouse or child of a veteran with wartime service

Definitions

Veteran: in this chapter, the term includes a person who died in the active military, naval, or air service.

Social Security Increase: in this chapter, the term means the percentage by which benefit amounts payable under Title II of the Social Security Act (42 U.S.C. 401 et seq.) are increased for any fiscal year because of a determination under section 215(i) of such Act (42 U.S.C. 415(i)).

Permanently Housebound: for the purposes of this chapter, the requirement of "permanently housebound" will be considered to have been met when the individual is substantially confined to such individual's house (ward or clinical areas, if institutionalized) or immediate premises due to a disability or disabilities, which it is reasonably certain will remain throughout such individual's lifetime.

Eligibility
Surviving Spouse
You may be eligible for VA benefits or compensation as a surviving spouse if you meet one of these requirements:
- You live with the veteran or service member without a break until their death, or
- If you're separated, you weren't at fault for the separation

One of the following must also be true:
- You married the veteran or service member within 15 years of their discharge from the period of military service during which the qualifying illness or injury started or got worse, or
- You were married to the veteran or service member for at least 1 year, or
- You had a child with the veteran or servicemember

If you're remarried, you can continue to receive compensation if one of the following is true:
- You remarried on or after December 16, 2003, and you were 57 years of age or older at the time you remarried, or
- You remarried on or after January 5, 2021, and you were 55 years of age or older at the time you remarried

You'll need to provide evidence with your claim to show one of these descriptions is true for the veteran or servicemember. Evidence may include documents like medical test results, military service records and doctor's reports.
- The servicemember died while on active duty, active duty for training, or inactive duty or training, or
- The veteran died from a service-connected illness or injury, or
- The veteran didn't die from a service-connected illness or injury but was eligible to receive VA compensation for a service-connected disability rated as totally disabling for a certain period of time

If the veteran's eligibility was due to a rating of totally disabling, they must have had this rating:
- For at least 10 years before their death, or
- Since their release from active duty and for at least 5 years immediately before their death, or
- For at least 1 year before their death if they were a former prisoner of war who died after September 30, 1999

Note: Totally disabling means the veteran's injury made it impossible for them to work.

As a Surviving Child
You may be eligible for benefits or compensation if all of the following are true as a surviving child:
- You aren't married, and
- You aren't included on the surviving spouse's compensation, and
- You're under the age of 18 or under the age of 23 if attending school.

Note: If you were adopted out of the veteran's or servicemember's family but meet the other criteria for eligibility you would still qualify for compensation.

You'll need to provide evidence with your claim. You'll have to provide evidence showing that at least one of the following is true:
- The servicemember died while on active duty, active duty for training or inactive duty training, or

- The veteran died from a service-connected illness or injury, or
- The veteran didn't die from a service-connected illness or injury but was eligible to receive VA compensation for a service-connected disability that was rated as totally disabling for a certain period of time.

If the veteran's eligibility was due to a service-connected disability rated as totally disabling, they must have had this rating:

- For at least 10 years before their death, or
- Since their release from active duty and for at least 5 years immediately before their death, or
- For at least 1 year before their death if they were a former prisoner of war who died after September 30, 1999.

Note: Totally disabling means the veteran's injuries make it impossible for them to work.

As a Surviving Parent
You may be eligible for VA benefits or compensation if you meet these requirements. **Both of these must be true:**

- You're the biological, adoptive or foster parent of the veteran or service member, and
- Your income is below a certain amount

Note: The VA defines a foster parent as someone who served in the role of a parent to the veteran or servicemember before their last entry into active service.

You'll have to provide evidence to show that certain descriptions are true for the veteran or service member. You'll have to provide evidence showing at least one of these is true:

- The servicemember died from an injury or illness while on active duty or in the line of duty while on active duty for training, or
- The servicemember died from an injury or certain illnesses in the line of duty while on inactive training, or
- The veteran died from a service-connected illness or injury.

Applying for Compensation

If you're the surviving spouse or child of a service member who died on active duty, a military casualty assistance officer will help you complete an Application for DIC, Death Pension, and/or Accrued Benefits by a Surviving Spouse or Child (VA Form 21P-534a). The officer then helps you mail the form to the correct VA regional office.

If you're the surviving spouse of a child or veteran, you fill out an Application for DIC, Death Pension, and/or Accrued Benefits (VA Form 21P-534EZ).

If you're a surviving parent, you fill out an Application for Dependency and Indemnity Compensation by Parents (VA Form 21P-535).

You can work with an accredited representative to apply for this benefit, or you can use the QuickSubmit tool, available through AccessVA. This allows you to upload your form online. You can also go to a VA regional office and get help from an employee, or mail your form to this address:

Department of Veterans Affairs
Pension Intake Center
PO Box 5365
Janesville, WI 53547-5365

You might want to submit an intent to file form before applying for DIC benefits. This can give you time to gather evidence, and avoid a later start date, which is known as an effective date. If you notify the VA of your intent to file, you might be eligible to receive retroactive payments.

Amount of DIC Payments to Surviving Spouses

Surviving spouses of veterans who died after January 1, 1993, receive a basic monthly rate of $1,562.74 (effective December 1, 2022)

Surviving spouses entitled to DIC based on the veteran's death before January 1, 1993, receive the greater of:

- The basic monthly rate of **$1,562.74**
- An amount based on the veteran's pay grade. (See following sections for Pay Grade tables and Determination of Pay Grade.)

There are additional DIC payments for dependent children. (Refer to the following charts.)

Additional Allowances for Surviving Spouses

Add **$331.84** to the basic monthly rate if, at the time of the veteran's death, the veteran was in receipt of or entitled to receive compensation for a service-connected disability rated totally disabling (including a rating based on individual unemployability) for a continuous period of at least eight years immediately preceding death, AND the surviving spouse was married to the veteran for those same eight years.

Add **$387.15** per child to the basic monthly rate for each dependent child under age 18. If the surviving spouse is entitled to Aid & Attendance, add **$387.15** to the basic monthly rate. If the surviving spouse is Permanently Housebound, add **$181.37** to the basic monthly rate.

Surviving Spouse DIC Rates If Veteran's Death was Prior to January 1, 1993			
	Monthly Rate	Pay Grade	Monthly Rate
E-1*:	$1562.74	W-4**:	$1868.86
E-2*:	$1562.74	O-1**:	$1650.00
E-3*(see footnote#1):	$1562.74	O-2**:	$1706.81
E-4*:	$1562.74	O-3**:	$1823.83
E-5*:	$1562.74	O-4:	$1933.15
E-6*:	$1562.74	O-5:	$2127.39
E-7**:	$1616.75	O-6:	$2398.79
E-8**:	$1706.81	O-7:	$2589.11
E-9**(see footnote#2):	$1780.11 or $1921.60	O-8:	$2843.81
W-1**:	$1650.22	O-9:	$3041.87
W-2**:	$1715.80	O-10 (see footnote#3):	$3336.41 or 3580.80
W-3**:	$1765.86		

Footnotes to Table:

* Add \$331.84 if veteran rated totally disabled eight continuous years before death and surviving spouse was married to veteran those same 8years.

A surviving spouse of an Aviation Cadet or other services not covered by this table is paid the DIC rate for enlisted E-3 under 34. \$1921.60 is for a veteran who served as Sergeant Major of the Army or Marine Corps, Senior Enlisted Advisor of the Navy, Chief Master Sergeant of the Air Force, or Master Chief Petty Officer of the CoastGuard. \$3580.80 is for a veteran who served as Chairman of the Joint Chiefs of Staff, Chief of Staff of the Army or Air Force, Chief of Naval Operations, or Commandant of the Marine Corps.

Determination of Pay Grade

With respect to a veteran who died in the active military, naval, or air service, such veteran's pay grade shall be determined as of the date of such veteran's death, or as of the date of promotion after death, while in a missing status.

With respect to a veteran who did not die in the active military, naval, or air service, such veteran's pay grade shall be determined as of:

- The time of such veteran's last discharge or release from active duty under conditions other than dishonorable; or
- The time of such veteran's discharge or release from any period of active duty for training or inactive duty training, if such veteran's death results from a service-connected disability incurred during such period, and if such veteran was not thereafter discharged or released under conditions other than dishonorable from active duty.
- If a veteran has satisfactorily served on active duty for a period of six months or more in a pay grade higher than that specified in the previous paragraphs of this section, and any subsequent discharge or release from active duty was under conditions other than dishonorable, the higher pay grade shall be used if it will result in greater monthly payments to such veteran's surviving spouse under this chapter. The determination as to whether an individual has served satisfactorily for the required period in a higher pay grade shall be made by the Secretary of the department in which such higher pay grade was held

DIC Payments to Children

Whenever there is no surviving spouse of a deceased veteran entitled to DIC, DIC shall be paid in equal shares to the children of the deceased veteran at the following monthly rates (effective December 1, 2021):

Number of Children	Total Payable (to be divided in equal shares):
1	\$659.83
2	\$949.21
3	\$1283.63
4	\$1474.01
5	\$1709.39
6	\$1944.77
7	\$2180.15
8	\$2415.53

If DIC is payable monthly to a person as a surviving spouse, and there is a child (of such person's deceased spouse) who has attained the age of 18, and who, while under such age, became permanently incapable of self-support, DIC shall be paid monthly to each such child, concurrently with the payment of DIC to the surviving spouse, in the amount of **\$659.83**

If DIC is payable monthly to a person as a surviving spouse, and there is a child (of such person's deceased spouse) who has attained the age of 18 and who, while under the age of 23, is pursuing a course of instruction at a VA-approved educational institution, DIC shall be paid monthly to each such child, concurrently with the payment of DIC to the surviving spouse, in the amount of **$387.15.**

DIC Payments for Parents

Parents whose child died in-service or from a service-connected disability may be entitled to DIC if they are in financial need. Parents may be biological, step, adopted, or in loco parentis. The monthly payment for parents of deceased veterans depends upon their income.

The following 3 charts outline the monthly rates payable, **effective December 1, 2022,** under various conditions.

Chart #1 Sole Surviving Parent Unremarried or Remarried Living with Spouse			
Income Not Over:	Monthly Rate:	Income Not Over:	Monthly Rate:
$ 800	$774	$ 2,500	$638
900	$766	2,600	$630
1,000	$758	2,700	$622
1,100	$750	2,800	$614
1,200	$742	2,900	$606
1,300	$734	3,000	$598
1,400	$726	3,100	$590
1,500	$718	3,200	$582
1,600	$710	3,300	$574
1,700	$702	3,400	$566
1,800	$694	3,500	$558
1,900	$686	3,600	$550
2,000	$678	3,700	$542
2,100	$670	3,800	$534
2,200	$662	3,900	$526
2,300	$654	4,000	$518
2,400	$646	4,100	$510

Income Not Over:	Monthly Rate:	Income Not Over:	Monthly Rate:
$4,200	$502	$6,300	$334
4,300	$494	6,400	$326
4,400	$486	6,500	$318
4,500	$478	6,600	$310
4,600	$470	6,700	$302
4,700	$462	6,800	$294
4,800	$454	6,900	$286
4,900	$446	7,000	$278
5,000	$438	7,100	$270

5,100	$430	7,200	$262
5,200	$422	7,300	$254
5,300	$414	7,400	$246
5,400	$406	7,500	$238
5,500	$398	7,600	$230
5,600	$390	7,700	$222
5,700	$382	7,800	$214
5,800	$374	7,900	$206
5,900	$366		
6,000	$358	If A&A add:	$420
6,100	$350	If living w/ spouse: $10,413 to $24,518	
6,200	$342	**If not living w/ spouse: $10,413 to $18,240	

Chart#2- One of Two Parents Not Living with Spouse			
Income Not Over:	**Monthly Rate:**	**Income Not Over:**	**Monthly Rate:**
$800	$561	$3,600	$337
900	$553	3,700	$329
1,000	$545	3,800	$321
1,100	$547	3,900	$313
1,200	$529	4,000	$305
1300	$521	4,100	$297
1,400	$513	4,200	$289
1,500	$505	4,300	$281
1,600	$497	4,400	$273
1,700	$489	4,500	$265
1,800	$481	4,600	$257
1,900	$473	4,700	$249
2,000	$465	4,800	$241
2,100	$457	4,900	$233
2,200	$449	5,000	$225
2,300	$441	5,100	$217
2,400	$433	5,200	$209
2,500	$425	5,300	$201
2,600	$417	5,400	$193
2,700	$409	5,500	$185
2,800	$401	5,600	$177
2,900	$393	5,700	$169

Income Not Over:	Monthly Rate:	Income Not Over:	Monthly Rate:
3,000	$385	5,800	$161
3,100	$377	5,900	$153
3,200	$369	6,000	$145
3,300	$361	$7,750 to 18,240	$5.08
3,400	$353		
3,500	$345	If A&A add:	$420

Chart #3 1 of 2 Parents Living with Spouse or Other Parent			
Income Not Over:	Monthly Rate:	Income Not Over:	Monthly Rate:
$1,000	$529	$4,000	$289
1,100	$521	4,100	$281
1,200	$513	4,200	$273
1,300	$505	4,300	$265
1,400	$497	4,400	$257
1,500	$489	4,500	$249
1,600	$481	4,600	$241
1,700	$473	4,700	$233
1,800	$465	4,800	$225
1,900	$457	4,900	$217
2,000	$449	5,000	$209
2,100	$441	5,100	$201
2,200	$433	5,200	$193
2,300	$425	5,300	$185
2,400	$417	5,400	$177
2,500	$409	5,500	$169
2,600	$401	5,600	$161
2,700	$393	5,700	$153
2,800	$385	5,800	$145
2,900	$377	5,900	$137

3,000	$369	6,000	$129
3,100	$361	6,100	$121
3,200	$353	6,200	$113
3,300	$345	6,300	$105
3,400	$337	6,400	$97
3,500	$329	6,500	$89
3,600	$321	6,600	$81
3,700	$313	6,800	$65
3,800	$305	6,900	$57
3,900	$297	If A&A add:	$420

Miscellaneous Information Regarding Income Limitations for Parents

The VA may require, as a condition of granting or continuing DIC to a parent that such parent, other than one who has attained 72 years of age and has been paid DIC during two consecutive calendar years, file for a calendar year with the VA, a report showing the total income which such parent expects to receive in that year, and the total income which such parent received in the preceding year. The parent or parents shall notify the VA whenever there is a material change in annual income.

In determining income under this section, all payments of any kind, or from any source shall be included except:
- Payments of a death gratuity.
- Donations from public or private relief or welfare organizations.
- Payments under this chapter (DIC), Chapter 11 of Title 38, United States Code (Disability Compensation), and Chapter 15 of Title 38, United States Code (Non-Service-Connected Disability/Death Pension);
- Payments under policies of servicemembers group life insurance, United States Government life insurance, or national service life insurance, and payments of servicemen's indemnity.
- 10% of the number of payments to an individual under public or private retirement, annuity, endowment, or similar plans or programs.

Amounts equal to amounts paid by a parent of a deceased veteran for:
- A deceased spouse's just debts.
- The expenses of the spouse's last illness, to the extent such expenses are not reimbursed under Chapter 51 of Title 38 of the United States Code.
- The expenses of the spouse's burial to the extent that such expenses are not reimbursed under Chapter 23 or Chapter 51 of Title 38 of the United States Code.
- Reimbursements of any kind for any casualty loss (as defined in regulations which the VA shall prescribe), but the amount excluded under this clause may not exceed the greater of the fair market value or the reasonable replacement value of the property involved at the time immediately preceding the loss.

Amounts equal to amounts paid by a parent of a deceased veteran for:
- The expenses of the veteran's last illness, and expenses of such veteran's burial, to the extent that such expenses are not reimbursed under Chapter 23 of Title 38 of the United States Code.
- Profit realized from the disposition of real or personal property other than during business.
- Payments received for discharge of jury duty or obligatory civic duties.

- Payments of annuities elected under Subchapter I of Chapter 73 of Title 10.

Where a fraction of a dollar is involved, annual income shall be fixed at the next lower dollar. The VA may provide by regulation for the exclusion from income under this section of amounts paid by a parent for unusual medical expense

What to Remember
✓   If you're the surviving spouse, child or parent of a service member who died in the line of duty or the survivor of a veteran who died from a service-related injury or illness, you may be eligible for a tax-free monetary benefit called Dependency and Indemnity Compensation (DIC).
✓   If the VA denied a Blue Water Navy veteran's service-connected disability claim in the past, survivors may be eligible for DIC benefits under the Blue Water Navy Vietnam Veterans Act of 2019.
✓   To be eligible you must be an eligible survivor of a veteran who died from a disease or injury incurred or aggravated while on active duty or active duty for training, or in the line of duty while on inactive duty or from a disability compensable by the VA
✓   Eligible surviving spouses and dependent children should apply for DIC by completing VA Form 21P-534EZ
✓   A copy of DD Form 214 and a copy of the veteran's death certificate should be included.

CHAPTER 8

SURVIVOR BENEFITS: PENSIONS

If you qualify for VA survivors' pensions as a surviving spouse or dependent child, the VA bases your payment amount on the difference between your countable income and a limit that congress sets. The limit Congress sets is called the Maximum Annual Pension Rate or MAPR.

Your **countable income** is how much you earn including your salary, investment and retirement payments. Countable income also includes any income you may have from your dependents. Some expenses, like non-reimbursable medical expenses may reduce your countable income. Non-reimbursable medical expenses are paid medical expenses not covered by your insurance provider.

Your **MAPR amount** is the maximum amount of pension payable to a veteran, surviving spouse or child. Your MAPR is based on how many dependents you have and whether you qualify for Housebound or Aid and Attendance benefits. MAPRs are adjusted each year for cost-of-living increases.

As an example, if you're a qualified surviving spouse with one dependent child, and you also qualify for Aid and Attendance and have a yearly income of $10,000, your MAPR amount is $18,867. Your yearly income is $10,000. Your VA pension is $8,867 for the year, or $739 paid each month.

Net Worth Limit to Be Eligible for Survivors Pension Benefits

From December 1, 2022, to November 30, 2023, the net worth limits to be eligible for Survivors Pension benefit is $150,538.

On October 18, 2018, the VA changed the way they assess net worth to make pension entitlement rules clearer. Net worth includes assets and annual income. When you apply for Survivors Pension benefits, you'll need to report all your assets and income.

If your child's net worth is more than the net worth limit, VA doesn't consider them to be a dependent when determining pension.

Assets

Assets include the fair market value of all your real and personal property, minus the amount of any mortgages you may have. Real property means any land and buildings you may own. Your personal property assets can include any of the following items:

- Investments like stocks and bonds
- Furniture
- Boats

Assets don't include your primary residence, your car or basic home items like appliances.

Annual Income
Annual income is the money earned in a year from a job or from retirement or annuity payments. Annual income includes salary or hourly pay, bonuses and commissions. It also includes overtime and tips.

VA subtracts certain expenses from your annual income when assessing net worth. These are called applicable deductible expenses. They include medical expenses you aren't reimbursed for and educational expenses.

Three-Year Look-Back Period for Asset Transfers
When VA receives a pension claim, it reviews the terms and conditions of any assets the survivor may have transferred in the three years before filing the claim.

If you transfer assets for less than fair market value during the look-back period, and those assets would have pushed your net worth above the limit for VA Survivors Pension you may be subject to a penalty period of up to five years. You won't be eligible for pension benefits during this time.

The new policy took effect on October 18, 2018. If you filed your claim before this date, the look back period doesn't apply.

A penalty period is a length of time when a survivor isn't eligible for pension benefits because they transferred assets for less than fair market value during the look-back period. This may apply if those transferred assets would have caused the survivor's net worth to be over the limit mentioned above, but not every asset transfer is subject to the penalty.

Maximum Annual Pension Rate (MAPR) Amount
Date of cost-of-living increase: December 1, 2022
Increase Factor: 8.7%
Standard Medicare Deduction: Actual amount will be determined by SSA, based on individual income.

For qualified surviving spouses with at least 1 dependent

If you have one dependent child and...	Your MAPR amount is:
You don't qualify for Housebound or Aid and Attendance benefits	$14,078
You qualify for Housebound benefits	$16,462
You qualify for Aid and Attendance benefits	$20,509
You qualify for Aid and Attendance benefits and you're the surviving spouse who served in the Spanish American War	$21,130

Notes:
- The Survivor Benefit Plan (SBP)/Minimum Income Annuity (MIW) limitation is $10,757.
- **If you have more than 1 child**, add $2,743 to your MAPR amount for each additional child.
- **If you have a child who works**, you may exclude their wages up to $13,850.

- **If you have medical expenses,** you may deduct only the amount that's above 5% of your MAPR amount ($703 for a surviving spouse with 1 dependent).

For qualified surviving spouses with no dependents

If you have no dependents and...	Your MAPR amount is:
You don't qualify for Housebound or Aid and Attendance benefits	$10,757
You qualify for Housebound benefits	$13,157
You qualify for Aid and Attendance benefits	$17,192
You qualify for Aid and Attendance benefits and you're the surviving spouse of a veteran who served in the Spanish-American War	$17,888

Notes:

- The Survivor Benefit Plan (SBP)/Minimum Income Annuity (MIW) limitation is $10,757

- **If you have medical expenses,** you may deduct only the amount that's above 5% of your MAPR amount ($537 for a surviving spouse with no dependent child).

- If you're a qualified surviving child, your MAPR amount is $2,743

Protected Death Pension Rates

This non-service-connected pension program was available from July 1, 1960, through December 31,1978.

Section 306 Death Pension Yearly Income Limits

Survivor Status	2022 Yearly Income Limit
Surviving spouse alone—no dependent children	Your yearly income must be $18,240 or less to continue receiving this benefit
Surviving spouse with one or more dependent children	Your yearly income must be $24,518 or less to continue receiving this benefit
Each surviving dependent child—if the veteran has no surviving spouse	Your yearly income must be $14,915 or less to continue receiving this benefit

An annuity is a fixed sum of money paid to the plan's beneficiary every year. If you're part of a Section 306 survivor benefit plan, the VA will pay you up to $10,757 this year, effective December 1, 2022. This is the minimum income widow provision. The rate includes a 8.7% cost-of-living increase.

Old Law Death Pension Rates

This non-service-connected pension program was available before July 1, 1960. The following rates are effective December 1, 2021.

Old Law Death Pension Yearly Income Limits

Survivor Status	2022 Yearly Income Limit
Surviving spouse alone—no children	Your yearly income must be $15,973 or less to continue receiving this benefit
Each surviving dependent child when the	Your yearly income must be $15,973 or less

veteran has no surviving spouse	to continue receiving this benefit
Surviving spouse with one or more children	Your yearly income must be $23,020 or less to continue receiving this benefit

An annuity is a fixed sum of money paid to the plan's beneficiary each year. If you're the beneficiary of old law death pension survivor benefit plan—also called the minimum income window provision—the VA will pay you up to $10,757 for the year.

VA Survivors Pension Eligibility

You may be eligible for this benefit if you haven't remarried after the veteran's death and if the deceased veteran didn't receive a dishonorable discharge, and their service meets at least one requirement below.

At least one of the following must be true:

- The veteran entered active duty on or before September 7, 1980, and served at least 90 days on active military service with at least one day during a covered wartime period, or
- The veteran entered active duty after September 7, 1980, and served at least 24 months or the full period for which they were called or ordered to active duty with at least one day during a covered wartime period, or
- The veteran was an officer and started on active duty after October 16, 1981 and hadn't previously served on active duty for at least 18 months.

Your yearly family income and net worth must meet certain limits set by Congress. You may be eligible for a VA Survivors Pension as the child of a deceased wartime veteran if you're unmarried and meet at least one of three criteria.

You are under the age of 18, or you're under the age of 23 attending a VA-approved school, or you're unable to care for yourself due to a disability that happened before age 18.

How To Apply for VA Survivors Pension

You can apply in one of a few ways.

- You can work with a trained professional called an accredited representative to get help filing a claim.
- You can use the direct upload tool through AccessVA to upload your form online.
- You can fill out an Application for DIC, Death Pension, and/or Accrued Benefits (VA Form 21P-534EZ).
- You may want to submit an intent to file form before you apply. This can give you time to gather the evidence you need but avoid a later potential start date. The start date is also called an effective date. When you notify VA of intent to file, you may be able to get retroactive payments.

What to Remember
✓ The Survivor's Pension Benefit is also known as the Death Pension.
✓ The Survivor's Pension Benefit is a tax-free monetary benefit.
✓ This benefit may be paid to low-income unremarried surviving spouses and/or unremarried children of a deceased veteran with wartime service. To be eligible as a child you have to be under the age of 18, or you're under the age of 23 and attending a VA-approved school, or you're unable to care for yourself because of a disability that occurred before age 18.
✓ The pension is based on your yearly family income and has to be less than the amount set by Congress to qualify.
✓ To apply, complete VA Form 21P-534EZ.

CHAPTER 9

SURVIVOR BENEFITS: SURVIVOR BENEFIT PLAN (SBP) & SSIA

Survivor Benefit Plan Overview

The Survivor Benefit Plan (SBP) lets retirees make sure after their death there's a continuous lifetime annuity for their dependents. The annuity is based on a percentage of retired paid and is called SBP. The annuity is paid to an eligible beneficiary, and it pays eligible survivors an inflation-adjusted monthly income.

A military retiree pays premiums for SBP coverage when they retire. Premiums are paid from gross retired pay and don't count as income. There are less tax and out-of-pocket costs for SBP as a result. The premiums are partially funded by the government.

The costs of operating the program are taken on by the government, so the average premiums are usually significantly below what it would cost to get a conventional insurance policy. SBP can be a good option for most retirees, but government contributions are based on assumptions in average cases and may not apply to all cases.

The maximum SBP annuity for a spouse is based on 55% of the member's retired pay. If the member retires under REDUX, it's the pay the member would have received under the high-three retirement system. A smaller amount may be chosen.

Eligible children can also be SBP beneficiaries, alone or added to spouse coverage. If children are added to spouse coverage, the children receive benefits only if the spouse dies or otherwise isn't eligible to receive the annuity. Children equally divide a benefit that is 55% of the member's elected base amount.
Child coverage is somewhat inexpensive because children only get benefits while they are considered eligible dependents.

Coverage is also available for a former spouse, or if the retiree has no spouse or children, for an insurable interest, like a parent or business partner.

SBP and Estate Planning

Retired pay can be an asset. It stops when a retiree dies, but no one will know when that might be, so SBP can be a source of protection, similar to life insurance. There are differences in SBP premiums and benefits compared to insurance plans.

Like life insurance, SBP protects survivors against a loss of financial security when a retired member dies. It also protects survivors against the potential of outliving the benefit.

SBP also protects against the risk of inflation through Cost-of-Living Adjustments or COLAs.

SBP alone is not a complete estate plan, but it can be an important part of a larger plan.

Eligible Beneficiaries

When you apply for retirement, you may have been asked to complete a Data for Payment of Retired Personnel Form. On that form, you would choose a beneficiary. The types you can choose from are detailed below.

Spouse Only

The most common election for a retiree to make is for only his or her spouse to be covered based on full retired pay. Cost is calculated at a maximum of 6.5% of the elected level of coverage.

If you have an eligible spouse and you choose anything less than full coverage, the spouse's notarized signature must be obtained for the election to be valid.

Spouse coverage is designed to provide a lifetime monthly income for your surviving spouse after you diet.

The SBP annuity is determined by the base amount you elect. The base may range from a minimum of $300 up to a maximum of full retired pay. The annuity is 55% of the base amount. The base amount and the payments to the surviving spouse will generally increase at the same time and by the same percentage that COLAs are made to retired pay.

Your surviving spouse may remarry after age 55 and continue to receive SBP payments for life. If your surviving spouse remarries before age 55, SBP payments will stop. Payments may resume if the marriage later ends due to death or divorce.

Former Spouse

SBP allows election of coverage for former spouses. Costs and benefits under this option are identical to those for spouse coverage.

When former spouse coverage is elected, the current spouse must be informed. Only one SBP election may be made. If there is more than one former spouse, the member must specify who will receive coverage.

Former spouse and children coverage may also be elected. The children covered are the eligible children from the marriage of the member to the covered former spouse. The children will only receive payments if the former spouse remarries before age 55 or dies. Eligible children will divide 55% of the covered retired pay in equal shares.

Child Only

This option pays SBP to your child regardless of your marital status. Your children will get the SBP until they turn 18 or age 22, if a full-time, unmarried student. Children mentally or physically incapable of self-support remain eligible, while unmarried, for as long as the incapacitation exists.

Disabled Dependent

You can contribute your SBP payments to a Special Needs Trust (SNT) to allow a disabled dependent to continue receiving federal disability payments.

A SNT is a trust designated for beneficiaries who are disabled, either physically or mentally. It is written so the beneficiary can enjoy the use of property that is held in the trust for their benefit. At the same time, it allows the beneficiary to receive essential needs-based government benefits.

CSB/Redux Cost and Benefits

CSB/REDUX is the only retirement system that includes a re-adjustment to its retired pay amount. At age 62, retired pay is recomputed to what it would have been under High-36. Also, at age 62, a one-time COLA adjustment is made that applies the cumulative effects of High-36 COLA (CPI) to the new retirement base. Afterwards, future COLAs are again set to CPI minus 1%.

Special Survivor Indemnity Allowance (SSIA)

Surviving family members of some veterans are entitled to the Dependency and Indemnity Compensation (DIC) benefit from VA. The benefit is for survivors of those who died on active duty or were severely disabled. Military retirees can elect to provide a monthly income for their family members if they die first, which is known as the Survivor Benefit Plan.

Federal regulators made the decision that a survivor receiving both of these payments is essentially receiving two federal incomes from the same source, which is illegal. To prevent this from happening, Congress created the Special Survivor Indemnity Allowance (SSIA). The goal of the program was to help survivors continue receiving a portion of that money.

For a number of years, the DoD withheld the amount of the DIC payment from a survivor's SBT check. In 2020, a law created the SSIA, which was previously a temporary program. Now, the SSIA is phased in through 2021 and 2022 and by 2023 there won't be an SBP/DIC offset anymore.

For 2022, affected survivors receive the full DIC amount from the VA each month. DFAS will pay the SBP payment with 1/3 of the DIC payment deducted from it, as well as an SSIA payment up to $346 a month.

Key Things to Know About SBP

The following are key facts about SBP:
- After Defense Finance Accounting Service (DFAS) is notified of your death, an SBP application is mailed to your surviving spouse.
- If the administrative aspects are properly handled, the annuity should begin about 45-60 days after the death of a retired member.
- You receive 45% of gross pay if you're a military spouse and that person dies after participating in SBP at the maximum level.
- If remarriage occurs before age 55, the annuity is suspended and can be reinstated if the remarriage ends by death or divorce. If remarriage occurs at age 66 or older, the annuity continues uninterrupted for the duration of the spouse's life.
- If the military member didn't waive military retired pay for a combined civil service annuity, the surviving spouse can receive both the uniformed service SBP annuity and a civil service/FERS SBP annuity.
- An insurable interest election is an election that can be made by the unmarried retiree who might want to provide for a relative or other person who could be hurt financially if the retiree dies.
- The election you make prior to retirement is not revocable after you retire.
- You can submit a request to voluntarily discontinue participation in SBP during a one-year period beginning on the second anniversary of the date of commencement of retired pay. For the purposes of this policy, the date of commencement of retired pay is the date the retiree became entitled to receive retired pay.

What to Remember
✓ The Survivor Benefit Plan is a way for a retiree to ensure after their death, their dependents receive a continuous lifetime annuity. The annuity is based on a percentage of retired pay.
✓ The Special Survivor Indemnity Allowance (SSIA) was made permanent in 2018.
✓ Some surviving spouse of military retirees who didn't participate in the Survivor Benefit Plan and who died before September 21, 1973, are eligible for something known as SBP-MIW.

CHAPTER 10

BURIAL AND MEMORIAL BENEFITS

Veteran Burial Benefits

VA burial allowances are flat-rate monetary benefits. They help cover eligible veterans' burial and funeral costs. Generally, they are paid at the maximum amount allowed by law. A 2014 VA regulation change helped simplify the program. Eligible surviving spouses are now paid automatically. This happens upon notification of the veteran's death. There is no need to submit a claim. However, VA may grant additional benefits after receiving a claim. These include plot or interment allowance and transportation allowance.

Burial Allowance Amounts for Service-Connected Deaths

Status	Maximum Burial Allowance
If the veteran died on or after September 11, 2001	$2,000
If the veteran died before September 11, 2001	$1,500
If the veteran is buried in a VA national cemetery	VA may pay you back for some or all of the costs of moving the veteran's remains.

Burial Allowance Amounts for a Non-Service-Connected Death If a Veteran Died on or After October 1, 2022

Benefits	Rate	Date Rate Changed	Public Law
Headstone/Marker	$231	10-01-2021	PL 95-476
Non-Service-Connected Burial	$300	10-01-2022	PL 111-275 PL 111-275 PL 100-322
Plot Allowance	$893	10-01-2022	PL 111-275 PL 107-103 PL 100-322
State Cemetery Plot Allowance	$893	10-01-2022	PL 111-275 PL 107-103 PL 100-322

Note 1: The P.L. 107-103 service-connected burial rate applies in cases where death occurred on or after 9/11/01.
Note 2: The headstone/marker allowance is payable only if the veteran died between 10/18/78 and 11/1/90. The rate payable is determined by when the headstone/marker was purchased.
Note 3: For non-service connected deaths, VA will pay up to $893 toward burial and funeral

expenses (if hospitalized by VA at time of death), or $300 toward burial and funeral expenses (if not hospitalized by VA at time of death).

Pre-Need Burial Eligibility

The Department of Veterans Affairs (VA) implemented the pre-need burial eligibility determination program to assist anyone who would like to know if they are eligible for burial in a VA national cemetery. VA is promoting pre-need eligibility determinations to encourage Veterans and their eligible family members to plan in advance to use VA burial benefits that Veterans have earned through their military service. Planning in advance for a Veteran's or loved one's final resting place can eliminate unnecessary delays and reduce stress on a family at a difficult time. Veteran families will have increased confidence that their loved ones are eligible for burial in a VA national cemetery at their time of need.

Upon request, VA will make eligibility determinations for burial in a VA national cemetery in advance of need. Eligible individuals are entitled to burial in any open VA national cemetery, which includes opening/closing of the grave, a government-furnished grave liner, perpetual care of the gravesite, and a government-furnished upright headstone or flat marker or niche cover all at no cost to the family. Veterans are also eligible for a burial flag and Presidential Memorial Certificate.

Burial in a VA national cemetery is open to all members of the armed forces and Veterans who have met minimum active-duty service requirements, as applicable by law, and were discharged under conditions other than dishonorable. Members of the reserve components of the armed forces who die while on active duty under certain circumstances or who die while on training duty are also eligible for burial, as are service members and former service members who were eligible for retired pay at the time of their death. Spouses, minor children and, under certain conditions, dependent unmarried adult children are also eligible for burial even if they predecease the Veteran.

The Department of Veterans Affairs (VA) has implemented this pre-need eligibility program so that Veterans, spouses and unmarried dependent adult children may better prepare for burial in a VA national cemetery before the time of need. Interested individuals may submit VA Form 40-10007, Application for Pre-Need Determination of Eligibility for Burial in a VA National Cemetery, along with a copy of supporting documentation of military service such as a DD214, if readily available, by Toll-free fax at 1-855-840-8299; or mail to the National Cemetery Scheduling Office, P.O. Box 510543, St. Louis, MO 63151.

Authorized representatives can also apply on behalf of eligible claimants. An authorized agent or representative is an individual authorized by the claimant to make decisions on the claimant's behalf. An authorized representative first needs to be recognized by VA as an authorized representative or agent by filing a VA Form 21-22 Appointment of Veterans Service Organization As Claimant Representative or VA Form 21-22a Appointment of Attorney or Agent As Claimant Representative.

You can access the forms at www.vba.va.gov/pubs/forms/VBA-21-22-ARE.pdf and VA Form 21-22a at www.vba.va.gov/pubs/forms/VBA-21-22A-ARE.pdf. Written authorization should be included with the VA Form 40-10007, Application for Pre-Need Determination of Eligibility for Burial in a VA National Cemetery, if available. A notarized statement is not required.

VA will review pre-need burial applications and provide written notice of a determination of eligibility. VA will store the pre-need application, supporting documentation, and the decision letter to expedite burial arrangements at the time of need. We encourage you to keep the decision letter with supporting documentation with your important papers in a safe place and to discuss your burial wishes and final arrangements with your loved ones or other representatives. Submission of a pre-need burial eligibility application does not obligate the Veteran or family member to burial in a VA national cemetery.

Applicants may indicate a preference for a VA national cemetery on the application form, but a pre-need determination of eligibility does not guarantee burial in a specific VA national cemetery or a specific gravesite. VA assigns gravesites in cemeteries with available space once death has occurred, and the burial is scheduled.

At your time of need, your next-of-kin, funeral home or another representative responsible for making your final arrangements should contact the National Cemetery Scheduling Office at (800) 535-1117 to request burial. VA will locate your pre-need decision letter and validate our determination. Because laws affecting VA burial eligibility and individual circumstances may change, upon receipt of a burial request, VA will verify pre-need decisions in accordance with the laws in effect at that time including bars to receipt of burial benefits.

Burial Benefits
Beginning July 7, 2014, VA changed its monetary burial benefits regulations to simplify the program and pay eligible survivors more quickly and efficiently. These regulations will authorize VA to pay, without a written application, most eligible surviving spouses basic monetary burial benefits at the maximum amount authorized in law through automated systems rather than reimbursing them for actual costs incurred.

The new burial regulations will permit VA to pay, at a flat rate, burial, and plot or interment allowances, thereby enabling VA to automate payment of burial benefits to most eligible surviving spouses and more efficiently process other burial benefit claims. The burial allowance for a non-service-connected death is $300, and $2,000 for a death connected to military service.

Service-related Death
VA will pay up to $2,000 toward burial expenses for deaths on or after September 11, 2001, or up to $1,500 for deaths before September 11, 2001. If the Veteran is buried in a VA national cemetery, some or all of the cost of transporting the deceased may be reimbursed.

Non-service-related Death
VA will pay up to $893 toward burial and funeral expenses for deaths on or after October 1, 2021 (if hospitalized by VA at time of death), or $300 toward burial and funeral expenses (if not hospitalized by VA at time of death), and a $893 plot-interment allowance (if not buried in a national cemetery).

For deaths on or after December 1, 2001, but before October 1, 2011, VA will pay up to $300 toward burial and funeral expenses and a $300 plot-interment allowance. For deaths on or after April 1, 1988, but before October 1, 2011, VA will pay $300 toward burial and funeral expenses (for Veterans hospitalized by VA at the time of death).

An annual increase in burial and plot allowances for deaths occurring after October 1, 2011, begins in fiscal year 2013 based on the Consumer Price Index for the preceding 12- month period.

Eligibility Requirements
- You paid for a Veteran's burial or funeral, AND
- You have not been reimbursed by another government agency or some other source, such as the deceased Veteran's employer, AND
- The Veteran was discharged under conditions other than dishonorable, AND
- The Veteran died because of a service-related disability, OR
- The Veteran was receiving VA pension or compensation at the time of death,
- OR The Veteran was entitled to receive VA pension or compensation, but decided not to reduce his/her military retirement or disability pay, OR
- The Veteran died while hospitalized by VA, or while receiving care under VA

contract at a non-VA facility, OR

- The Veteran died while traveling under proper authorization and at VA expense to or from a specified place for the purpose of examination, treatment, or care, OR
- The Veteran had an original or reopened claim pending at the time of death and has been found entitled to compensation or pension from a date prior to the date of death, OR
- The Veteran died on or after October 9, 1996, while a patient at a VA-approved state nursing home.

NOTE: VA does not pay burial benefits if the deceased:

- Died during active military service,OR
- Was a member of Congress who died while holding office, OR
- Was a federal prisoner

Evidence Requirements

- Acceptable proof of death as specified in 38 CFR 3.211, AND
- Receipted bills that show that you made payment in whole or part, OR
- A statement of account, preferably on the printed billhead of the funeral director or cemetery owner.

The statement of account must show:

- The name of the deceased Veteran for whom the services and merchandise were furnished, AND
- The nature and cost of the services and merchandise, AND
- All credits, AND
- The amount of the unpaid balance, if any

You can apply online at Vets.gov or submit a paper application by completing VA Form 21P-530, Application for Burial Allowance. When completed, mail it to the Pension Management Center that serves your state. You can also work with an accredited representative or go to your local, regional benefit office.

Dignified Burial of Unclaimed Veterans Act

On December 30, 2012, the Dignified Burial of Unclaimed Veterans Act of 2012 was sent to the President for signature into law. The law contains the following provisions:

Cemetery Matters: The bill authorizes the Secretary of Veterans Affairs (VA) to furnish a casket or urn for a deceased veteran with no known living kin.

Veterans Freedom of Conscience Protection: Under the bill, in cases where family or the next of kin of a deceased veteran can be identified, the bill requires the VA to ensure that the expressed wishes of the next of kin are met with regards to memorial service, funeral or interment at a national cemetery.

Improved Communication Between Department of Veterans Affairs and Medical Examiners and Funeral Directors: The bill requires the VA to obtain information from the relevant medical examiner, funeral director, service group, or other entity regarding the steps taken to determine if the deceased has no next of kin.

Identification and Burial of Unclaimed or Abandoned Human Remains: The VA will be required to cooperate with veterans' service organizations to assist entities in possession of unclaimed or abandoned remains of veterans, with the department authorized to determine whether such remains are eligible for burial in a national cemetery.

Restoration, Operation and Maintenance of Clark Veterans Cemetery by American Battle Monuments Commission: The bill authorizes the appropriation of $5 million for site preparation and construction at the Clark Veterans Cemetery located in the Philippines. The bill directs the American Battle Monuments Commission to restore and operate the cemetery once a cooperative agreement is reached with the government of the Philippines.

Reimbursement: The Veterans Benefits Administration administers a burial benefit program designed to assist claimants in meeting the funeral and burial costs of a deceased veteran. The type and amount of benefits payable depend on the veteran's individual service record and cause of death.

Service-Connected Death: If a veteran's death is service-connected, the VA will pay a burial allowance of up to $2,000 (for deaths on or after 9-11-2001 - prior to 9-11-2001, the burial allowance was $1,500). If the veteran is buried in a VA national cemetery, some or all of the cost of moving the deceased to the national cemetery nearest the veteran's home may also be reimbursed. There is no time limit for applying for a service-connected burial allowance. The person who bore the veteran's burial expense may claim reimbursement from any VA regional office.

Unpaid Balance Due Persons Who Performed Services: If there is an unpaid balance due the person who performed burial, funeral, and transportation services, such claim as a creditor for the statutory burial allowance will be given priority over any claim for reimbursement based on use of personal funds unless there is executed by such creditor a waiver in favor of the person or persons whose personal funds were used in making the partial payment of the account. No reimbursement may be made to a State, County, or other governmental subdivision.

Veteran's Estate: The representative of a deceased person's estate might file a claim for reimbursement if estate funds were used to pay the expenses of the veteran's burial, funeral, and transportation. Accordingly, if otherwise in order, reimbursement may be made to the estate, regardless of whether such person pre-deceased the veteran or was deceased at the time the expenses were incurred or paid, or burial of the veteran actually occurred.

Death of Active-Duty Personnel: The VA does not pay a burial or plot interment allowance if a veteran dies while on active military duty. However, such veteran is entitled to certain benefits from the military. Information should be obtained from the branch of the armed forces in which the person served at the time of death.

Correction of Discharge to Honorable Conditions: Public Law 88-3 (H.R. 212) provides that where burial allowance was not payable at the time of a veteran's death because of the nature of his discharge from service, but after death, his discharge is corrected by competent authority to reflect a discharge under conditions other than dishonorable, claim may be filed within two years from the date of correction of the discharge.

Legal Execution: The execution of a veteran as a lawful punishment for a crime does not of itself preclude payment of the statutory burial allowances.

State Burial Allowance: There are some burial allowances allowed by states or counties for the burial of needy veterans, and in some instances, of their dependents. Check with the veteran's state or county of residence for specific information.

Assistance to Claimants: It is incumbent upon the claimant to establish his case in accordance with the law. This rule, however, will not be highly technical or rigid in its application. The general policy of the VA is to give the claimant every opportunity to substantiate his claim, to extend all reasonable assistance in its prosecution, and develop all sources from which information may be obtained. Assistance may also be secured from post, state, and national service officers or veterans' organizations.

Miscellaneous: If the deceased veteran is entitled to payment in full of burial expenses by the Employees' Compensation Commission, Workman's Compensation, or employer, the Veterans Administration will make no payment.

Appeal: An appeal may be made within one year from the date of the denial of a claim. The appeal must be sent to the regional VA office that denied the claim.

Burial Flags

The VA shall furnish a United States flag, at no cost, for burial or memorial purposes in connection with the death of an eligible veteran who served honorably in the U.S. Armed Forces. Public Law 105-261, added eligibility for former member of the Selected Reserve who:

- Completed at least one enlistment as a member of the Selected Reserve or, in the case of an officer, completed the period of initial obligated service as a member of the Selected Reserve; or

- Was discharged before completion of the person's initial enlistment as a member of the Selected Reserve or, in the case of an officer, period of initial obligated service as a member of the Selected Reserve, for a disability incurred or aggravated in the line of duty; or
- Who died while a member of the Selected Reserve.

Burial flags may not be furnished on behalf of deceased veterans who committed capital crimes. After the burial of the veteran, the flag shall be given to his next of kin. If no claim is made for the flag by the next of kin, it may be given, upon request, to a close friend or associate of the deceased veteran. If a flag is given to a close friend or associate of the deceased veteran, no flag shall be given to any other person on account of the death of such veteran.

When burial is in a national, state or post cemetery, a burial flag will automatically be provided. When burial is in a private cemetery, an American flag may be obtained by a service officer, an undertaker or another interested person from the nearest Veterans Administration office or most U.S. Post Offices. VA Form 27-2008, "Application for United States Flag for Burial Purposes," must be completed and submitted along with a copy of the veteran's discharge papers. Generally, the funeral director will help the next of kin with this process. The proper way to display the flag depends upon whether the casket is open or closed. VA Form 27-2008 provides the proper method for displaying and folding the flag. The burial flag is not suitable for outside display because of its size and fabric. After burial of a veteran, the flag should be folded in military style and presented to the next of kin at the cemetery. Local veteran organizations, when presenting the flag, usually say something like this: "In the name of the United States government and name of a veterans organization (such as A.L., D.A.V., V.F.W., and American Veterans of World War II, etc.), we present you this flag, in loving memory of our departed comrade."

Presenting of the flag is sometimes a difficult task when there is a disagreement in the family. However, the Veterans Administration gives the following order of preference to be followed:

- Widow or widower (even if separated but not divorced)
- Children according to age (minor child may be issued flag on an application signed by guardian)
- Father, including adopted, step and foster father
- Mother, including adopted, step and foster mother
- Brothers or sisters, including brothers and sisters of half-blood. Uncles or aunts
- Nephews or nieces
- Cousins, grandparents, etc. (but not in-laws)
- If two relatives have equal rights, the flag will be presented to the elder one.

Following a veteran's burial, the United States flag received by the next of kin may be donated to the VA for use on national holidays at VA national cemeteries. If the next of kin choose to make such a donation, it should be given or mailed to the Director of any national cemetery selected by the donor with a written request that the flag is flown at that location.

If the flag is brought into a Veterans Service Division, it will be accepted and forwarded to the cemetery chosen by the donor. A Certificate of Appreciation is presented to the donor for providing their loved ones' burial flag to a national cemetery. Please note that VA cannot provide flag holders for placement on private headstones or markers.

These flag holders may be purchased from private manufacturing companies. The law allows the VA to issue one flag for a veteran's funeral. The VA cannot replace it if it is lost, destroyed, or stolen. However, if this occurs, a local veteran's organization or another community group may be able to assist you in obtaining another flag.

Headstones and Markers in Private Cemeteries

Ordering Grave Marker Medallion

In 2012 the VA introduced a new form, **VA Form 40-1330**, which is now used as the sole form for families to order a grave marker medallion. Prior to the change, families of the deceased used a form to order the medallion that was the same as the form used to order a government headstone or grave marker. The medallion is given instead of a traditional government headstone or marker, specifically for veterans whose death occurred on or after November 1, 1990. The medallion is intended for use for those veterans whose grave is located in a private cemetery and marked with a privately purchased headstone or marker.

The VA is required to furnish an appropriate headstone or marker for the graves of eligible veterans buried in private cemeteries, whose deaths occur on or after December 27, 2001, regardless of whether the grave is already marked with a non-government marker. Headstones or markers may not be furnished on behalf of deceased veterans who committed capital crimes.

For all deaths occurring before September 11, 2001, the VA may provide a headstone or marker only for graves that are not marked with a private headstone.

Spouses and dependents buried in a private cemetery are not eligible for a government-provided headstone or marker.

Flat markers in granite, marble, and bronze, and upright headstones in granite and marble are available. The style chosen must be consistent with existing monuments at the place of burial. Niche markers are also available to mark columbaria used for cremated remains.

Government-furnished headstones and markers must be inscribed with the name of the deceased, branch of service, and the year of birth and death, in this order. Headstones and markers also may be inscribed with other items, including an authorized emblem of belief and, space permitting, additional text including military-grade, rate or rank, war service such as "World War II," complete dates of birth and death, military awards, military organizations, and civilian or veteran affiliations. To apply, and to obtain specific information on available styles, contact the cemetery where the headstone or marker is to be placed.

When burial occurs in a private cemetery, an application for a government-furnished headstone or marker must be made to VA. The government will ship the headstone or marker free of charge but will not pay for its placement. To apply, complete VA Form 40-1330 (**application form updated in 2012**), and forward it along with a copy of the veteran's military discharge documents (do not send original discharge documents, as they will not be returned) to:

Memorial Products Service (41B)
Department of Veterans Affairs
5109 Russell Road
Quantico, VA 22134-3903

Bronze Medallion

VA made available a new medallion to be affixed to an existing privately purchased headstone or marker to signify the deceased's status as a veteran.

If requested, the medallion will be furnished in lieu of a traditional government headstone or marker for veterans that died on or after November 1, 1990, and whose grave is marked with a privately purchased headstone or marker.

The medallion is currently available in three sizes, 5 inches, 3 inches, and 1 ½ inches. Each medallion will be inscribed with the word VETERAN across the top and the Branch of Service at the bottom. Appropriate affixing adhesive, instructions, and hardware will be provided with the medallion.

Important: This benefit is only applicable if the grave is marked with a privately purchased headstone or marker. In these instances, eligible veterans are entitled to either a traditional government-furnished headstone or marker, or the new medallion, but not both.

Eligibility Rules for a Government Headstone or Marker in a Private Cemetery

Veterans and Members of the Armed Forces (Army, Navy, Air Force, Marine Corps, Coast Guard):

- Any veterans or members of The Armed Forces who die while on active duty.
- Any veteran who was discharged under conditions other than dishonorable. With certain exceptions, service beginning after September 7, 1980, as an enlisted person, and service after October 16, 1981, as an officer, must be for a minimum of 24 months or the full period for which the person was called to active duty. (Examples include those serving less than 24 months in the Gulf War or Reservists that were federalized by Presidential Act.) Undesirable, bad conduct and any other type of discharge other than honorable may or may not qualify the individual for veterans' benefits, depending upon a determination made by a VA Regional Office. Cases presenting multiple discharges of varying character are also referred for adjudication to a VA Regional Office

Members of Reserve Components and Reserve Officers' Training Corps:

- Reservists and National Guard members who, at time of death, were entitled to retired pay under Chapter 1223, Title 10, United States Code, or would have been entitled, but for being under the age of 60.
  - Members of reserve components who die while hospitalized or undergoing treatment at the expense of the United States for injury or disease contracted or incurred under honorable conditions while performing active duty for training or inactive duty training or undergoing such hospitalization or treatment.
  - Members of the Reserve Officers' Training Corps of the Army, Navy, or Air Force who die under honorable conditions while attending an authorized training camp or on an authorized cruise, while performing authorized travel to or from that camp or cruise, or while hospitalized or undergoing treatment at the expense of the United States for injury or disease contracted or incurred under honorable conditions while engaged in one of those activities.
  - Members of reserve components who, during a period of active duty for training, were disabled or died from a disease or injury incurred or aggravated in the line of duty or, during a period of inactive duty training, were disabled or died from an injury incurred or aggravated in the line of duty

Commissioned Officers, National Oceanic, and Atmospheric Administration:

- A commissioned officer of the National Oceanic and Atmospheric Administration (formerly titled the Coast and Geodetic Survey and the Environmental Science Services Administration) with full-time duty on or after July 29, 1945
- A commissioned officer who served before July 29, 1945, and:
- Was assigned to an area of immediate military hazard while in time of war, or of a

Presidentially declared national emergency as determined by the Secretary of Defense; or

- Served in the Philippine Islands on December 7, 1941, and continuously in such islands thereafter; or
- Transferred to the Department of the Army or the Department of the Navy under the provisions of the Act of May 22, 1917, Public Health Service
- A Commissioned Officer of the Regular or Reserve Corps of the Public Health Service who served on full-time duty on or after July 29, 1945. If the service of the particular Public Health Service Officer falls within the meaning of active duty for training, as defined in section 101(22), title 38, United States Code, he or she must have been disabled or died from a disease or injury incurred or aggravated in the line of duty.

A Commissioned Officer of the Regular or Reserve Corps of the Public Health Service who performed full-time duty before July 29, 1945:

- In time of war;
- On detail for duty with the Army, Navy, Air Force, Marine Corps, or Coast Guard; or,
- While the Service was part of the military forces of the United States pursuant to Executive Order of the President.
- A Commissioned Officer serving on inactive duty training as defined in section 101(23), title 38, United States Code, whose death resulted from an injury incurred or aggravated in the line of duty.

World War II Merchant Mariners
United States Merchant Mariners with oceangoing service during the period of armed conflict, December 7, 1941, to December 31, 1946. Prior to the enactment of Public Law 105-368, United States Merchant Mariners with oceangoing service during the period of armed conflict of December 7, 1941, to August 15, 1945, were eligible.

With the enactment of Public Law 105-368, the service period is extended to December 31, 1946, for those dying on or after November 11, 1998. A DD-214 documenting this service may be obtained by applying Commandant (G-MVP- 6), United States Coast Guard, 2100 2nd Street, SW, Washington, DC 20593.

Notwithstanding, the Mariner's death must have occurred after the enactment of Public Law 105-368, and the interment does not violate the applicable restrictions while meeting the requirements held therein.

Persons Not Eligible for a Headstone or Marker
Disqualifying Characters of Discharge
A person whose only separation from the Armed Forces was under dishonorable conditions or whose character of service results in a bar to veterans' benefits.

- Discharge from Draft: A person who was ordered to report to an induction station but was not actually inducted into military service.
- Person Found Guilty of a Capital Crime: Eligibility for a headstone or marker is prohibited if a person is convicted of a federal capital crime and sentenced to death or life imprisonment, or is convicted of a state capital crime, and sentenced to death or life imprisonment without parole. Federal officials are authorized to deny requests for headstones or markers to persons who are shown by clear and convincing evidence to have committed a Federal or State capital crime but were not convicted of such crime because of flight to avoid prosecution or by death before trial.
- Subversive Activities: Any person convicted of subversive activities after September 1, 1959, shall have no right to burial in a national cemetery from and after the date of commission of such offense, based on periods of active military

service commencing before the date of the commission of such offense, nor shall another person be entitled to burial on account of such an individual. Eligibility will be reinstated if the President of the United States grants a pardon.

- Active or Inactive Duty for Training: A person whose only service is active duty for training or inactive duty training in the National Guard or Reserve Component, unless the individual meets the following criteria.

Reservists and National Guard members who, at time of death, were entitled to retired pay under Chapter 1223, title 10, United States Code, or would have been entitled, but for being under the age of 60. Specific categories of individuals eligible for retired pay are delineated in section 12731 of Chapter 1223, title 10, United States Code.

Members of reserve components who die while hospitalized or undergoing treatment at the expense of the United States for injury or disease contracted or incurred under honorable conditions while performing active duty for training or inactive duty training or undergoing such hospitalization or treatment.

Members of the Reserve Officers' Training Corps of the Army, Navy, or Air Force who die under honorable conditions while attending an authorized training camp or on an authorized cruise, while performing authorized travel to or from that camp or cruise, or while hospitalized or undergoing treatment at the expense of the United States for injury or disease contracted or incurred under honorable conditions while engaged in one of those activities.

Members of reserve components who, during a period of active duty for training, were disabled or died from a disease or injury incurred or aggravated in the line of duty or, during a period of inactive duty training, were disabled or died from an injury incurred or aggravated in line of duty.

Other Groups

Members of groups whose service has been determined by the Secretary of the Air Force under the provisions of Public Law 95-202 as not warranting entitlement to benefits administered by the Secretary of Veterans Affairs.

Burial in National Cemeteries

In 2012 the VA announced plans to move forward with plans to provide burial options for veterans who live in rural areas, where no national, state or tribal cemeteries are available. The plan is part of the Rural Initiative Plan, in which the VA will work to build small National Veterans Burial Grounds that will be located inside existing rurally located public or private cemeteries. These burial grounds will also be located where the unserved veteran population is 25,000 or less, within a 75-mile radius.

The VA announced plans to open eight National Veteran Burial Grounds. A National Veterans Burial Ground will consist of a small area of three to five acres, located within the existing public or private cemetery that will be managed by the VA. The VA will provide a full range of burial options and be responsible for the maintenance and operation of the burial lots. These sections will be held to the national shrine standards that are in place at VA national cemeteries.

The National Cemetery Administration honors Veterans with a final resting place and lasting memorials that commemorate their service to our nation.

Veterans and service members who meet the eligibility requirements and their eligible dependents may be buried in one of the VA's national cemeteries. The National Cemetery Administration currently includes 122 national cemeteries. The Department of the Army administers two national cemeteries, and the Department of the Interior administers 14 national cemeteries. There are also numerous state cemeteries for veterans throughout the

U.S. For a listing of all national cemeteries as well as state veterans' cemeteries refer to the listing at the end of this chapter.

Effective under Public Law 112-154, reservations for more than one space at Arlington National Cemetery will no longer be accepted, and reservations will not be honored until the time of death unless there is a case of extraordinary circumstances.

Honoring America's Veterans and Caring for Camp Lejeune Families Act of 2012

This law, signed into effect by then-President Obama, provided sweeping legislation for veterans, including limitations on protestors at military funerals. The law prohibits protestors at military funerals from protesting two hours before or after a military funeral, and the must hold their protests at least 300 feet from the site of the funeral.

Respect for America's Fallen Heroes Act

On May 29, 2006, then-President Bush signed the Respect for America's Fallen Heroes Act into law. This law does the following:
Prohibits a demonstration on the property of a cemetery under the control of the National Cemetery Administration or the property of Arlington National Cemetery unless the demonstration has been approved by the cemetery superintendent or the director of the property on which the cemetery is located.

Prohibits, with respect to the above cemeteries, a demonstration during the period beginning 60 minutes before and ending 60 minutes after a funeral, memorial service, or ceremony is held, which:

- Takes place within 150 feet of a road, pathway, or other route of ingress to or egress from such cemetery property; or
- Is within 300 feet of such cemetery and impedes access to or egress from such cemetery.

The term "Demonstration" includes the following:

- Any picketing or similar conduct.
- Any oration, speech, use of sound amplification equipment or device, or similar conduct that is not part of a funeral, memorial service, or ceremony.
- The display of any placard, banner, flag, or similar device, unless the display is part of a funeral, memorial service, or ceremony.
- The distribution of any handbill, pamphlet, leaflet, or other written or printed matter other than what is distributed as part of a funeral, memorial service, or ceremony. Amends the federal criminal code to provide criminal penalties for violations of such prohibitions.
- Express the sense of Congress that each state should enact similar legislation to restrict demonstrations near any military funeral.

Eligibility

The VA national cemetery directors have the primary responsibility for verifying eligibility for burial in VA national cemeteries. A dependent's eligibility for burial is based upon the eligibility of the veteran. To establish a veteran's eligibility, a copy of the official military discharge document bearing an official seal, or a DD 214 is usually sufficient. The document must show that release from service was under conditions other than dishonorable. A determination of eligibility is usually made in response to a request for burial in a VA national cemetery.

The cemeteries administered by the National Cemetery Administration and the Department of the Interior use the eligibility requirements that follow. The Department of the Interior can be contacted at:

Department of the Interior National Park Service
1849 C Street, N.W.
Washington, D.C. 20240
(202) 208-4621

The Department of the Army should be contacted directly for inquiries concerning eligibility for interment in either of the two cemeteries under its jurisdiction (call (703) 695-3250).

Eligibility requirements for burial in state veterans' cemeteries are the same, or similar, to the eligibility requirements that follow. However, some states also have residency and other more restrictive requirements. Please contact the specific state cemetery for its eligibility requirements.

The following veterans and members of the Armed Forces (Army, Navy, Air Force, Marine Corps, Coast Guard) are eligible for burial in a VA national cemetery:
- Any member of the Armed Forces of the U.S. who dies on active duty.
- Any citizen of the U.S. who, during any war in which the U.S. has been engaged, served in the Armed Forces of any U.S. ally during that war, whose last active service was terminated honorably by death or otherwise, and who was a citizen of the U.S. at the time of entry into such service and at the time of death.
- Any veteran discharged under conditions other than dishonorable and who has completed the required period of service. With certain exceptions, service beginning after September 7, 1980, as an enlisted person, and service after October 16, 1981, as an officer, must be for a minimum of 24 months or the full period for which the person was called to active duty.

Members of Reserve Components and Reserve Officers' Training Corps
- Reservists and National Guard members with 20 years of qualifying service, who are entitled to retired pay, or would be entitled, if at least 60 years of age.
- Members of reserve components who die under honorable conditions while hospitalized or undergoing treatment at the expense of the U.S. for injury or disease contracted or incurred under honorable conditions while performing active duty for training, or inactive duty training, or undergoing such hospitalization or treatment.
- Members of the Reserve Officers' Training Corps of the Army, Navy, or Air Force who die under honorable conditions while attending an authorized training camp or cruise, or while traveling to or from that camp or cruise, or while hospitalized or undergoing treatment at the expense of the U.S. for injury or disease contracted or incurred under honorable conditions while engaged in one of those activities.
- Members of reserve components who, during a period of active duty for training, were disabled or died from a disease or injury incurred or aggravated in the line of duty, or, during a period of inactive duty training, were disabled or died from an injury incurred or aggravated in the line of duty.

Commissioned Officers, National Oceanic and Atmospheric Administration
- A Commissioned Officer of the National Oceanic and Atmospheric Administration with full-time duty on or after July 29, 1945.
- A Commissioned Officer of the National Oceanic and Atmospheric Administration who served before July 29, 1945, and:
- Was assigned to an area of immediate hazard described in the Act of December 3, 1942 (56 Stat. 1038; 33 U.S.C. 855a), as amended
- Served in the Philippine Islands on December 7, 1941, and continuously in such islands thereafter.

108

Public Health Service

- A Commissioned Officer of the Regular or Reserve Corps of the Public Health Service who served on full-time duty on or after July 29, 1945. If the service of such an Officer fall within the meaning of active duty for training, he must have been disabled or died from a disease or injury incurred or aggravated in the line of duty.
- A Commissioned Officer of the Regular or Reserve Corps of the Public Health Service who performed full-time duty prior to July 29,1945:
- In the time of war; or
- On detail for duty with the Army, Navy, Air Force, Marine Corps, or Coast Guard, or While the service was part of the military forces of the U.S. under Executive Order of the President.
- A Commissioned Officer of the Regular or Reserve Corps of the Public Health Service serving on inactive duty training as defined in Section 101 (23), title 39, U.S. Code, whose death resulted from an injury incurred or aggravated in the line of duty.

World War II Merchant Mariners

U.S. Merchant Mariners with oceangoing service during the period of armed conflict, December 7, 1941, to December 31, 1946. Prior to the enactment of Public Law 105-368, United States Merchant Mariners with oceangoing service during the period of armed conflict of December 7, 1941, to August 15, 1945, were eligible. With enactment of Public Law 105-368, the service period is extended to December 31, 1946, for those dying on or after November 11, 1998. Merchant Mariners who served on blockships in support of Operation Mulberry during WWII.

The Philippine Armed Forces

Any Philippine veteran who was a citizen of the United States or an alien lawfully admitted for permanent residence in the United States at the time of their death, and resided in the United States at the time of their death; and:

Was a person who served before July 1, 1946, in the organized military forces of the Government of the Commonwealth of the Philippines, while such forces were in the service of the Armed Forces of the United States pursuant to the military order of the President dated July 26, 1941, including organized guerilla forces under commanders appointed, designated, or subsequently recognized by the Commander in Chief, Southwest Pacific Area, or other competent authority in the Army of the United States who dies on or after November 1, 2000; or

Was a person who enlisted between October 6, 1945, and June 30, 1947, with the Armed Forces of the United States with the consent of the Philippine government, pursuant to Section 14 of the Armed Forces Voluntary Recruitment Act of 1945, and who died on or after December 16, 2003.

Spouses and Dependents

The spouse or surviving spouse of an eligible veteran is eligible for interment in a national cemetery even if that veteran is not buried or memorialized in a national cemetery. In addition, the spouse or surviving spouse of a member of the Armed Forces of the United States whose remains are unavailable for burial is also eligible for burial.

The surviving spouse of an eligible veteran who had a subsequent remarriage to a non-veteran and whose death occurred on or after January 1, 2000, is eligible for burial in a national cemetery, based on his or her marriage to the eligible veteran.

The minor children of an eligible veteran. For the purpose of burial in a national cemetery, a minor child is a child who is unmarried and:

- Under 21 years of age; or

- Under 23 years of age and pursuing a full-time course of instruction at an approved educational institution.
- An unmarried adult child of an eligible veteran if the child became permanently physically or mentally disabled and incapable of self-support before reaching 21 years of age, or before reaching 23 years of age if pursuing a full-time course of instruction at an approved educational institution.
- Any other persons or classes of persons as designated by the Secretary of Veterans Affairs or the Secretary of the Air Force.

The Veterans' Benefits Act of 2010 contains a provision that allows a parent whose child gave their lives in service to our country to be buried in a national cemetery with that child when their veteran child has no living spouse or children.

The following are not eligible for burial in a VA National Cemetery:

- A former spouse of an eligible individual whose marriage to that individual has been terminated by annulment or divorce, if not otherwise eligible.
- Family members other than those specifically described above.
- A person whose only separation from the Armed Forces was under dishonorable conditions, or whose character of service results in a bar to veterans' benefits.
- A person ordered to report to an induction station, but not actually inducted into military service.
- Any person found guilty of a capital crime. Interment or memorialization in a VA cemetery or in Arlington National Cemetery is prohibited if a person is convicted of a federal capital crime and sentenced to death or life imprisonment, or is convicted of a state capital crime, and sentenced to death or life imprisonment without parole. Federal officials are authorized to deny burial in veterans' cemeteries to persons who are shown by clear and convincing evidence to have committed a federal or state capital crime but were not convicted of such crime because of flight to avoid prosecution or by death prior to trial. The Secretary is authorized to provide aid to states for the establishment, expansion and/or improvement of veterans cemeteries on the condition that the state is willing to prohibit interment or memorialization in such cemeteries of individuals convicted of federal or state capital crimes, or found by clear and convincing evidence to have committed such crimes, without having been convicted of the crimes due to flight to avoid prosecution or death prior to trial.
- Any person convicted of subversive activities after September 1, 1959, shall have no right to burial in a national cemetery from and after the date of commission of such offense, based on periods of active military service commencing before the date of the commission of such offense, nor shall another person be entitled to burial on account of such an individual. Eligibility will be reinstated if the President of the United States grants a pardon.

Requests for Gravesites in National Cemeteries

Under Public Law 112-154 the Secretary of the Army will be responsible for decisions regarding organizations wishing to place monuments honoring the service of a group or individual at Arlington. This became law in 2012.

An eligible veteran or family member may be buried in the VA national cemetery of his choice, provided space is available. A veteran may not reserve a gravesite in his name prior to his death. However, any reservations made under previous programs will be honored. The funeral director or loved one making the burial arrangements must apply for a gravesite at the time of death. No special forms are required when requesting burial in a VA national cemetery. The person making burial arrangements should contact the national cemetery in which burial is desired at the time of need.

If possible, the following information concerning the deceased should be provided when the cemetery is first contacted:
- Full name and military rank

- Branch of service.
- Social security number.
- VA claim number, if applicable.
- Date and place of birth.
- Date and place of death.
- Date of retirement or last separation from active duty; and
- Copy of any military separation documents (such as DD-214)

When a death occurs, and eligibility for interment in a national cemetery is determined, grave space is assigned by the cemetery director in the name of the veteran or family member

One gravesite is permitted for the internment of all eligible family members unless soil conditions or the number of family decedents necessitate more than one grave. There is no charge for burial in a national cemetery.

The availability of grave space varies among each National Cemetery. In many cases, if a national cemetery does not have space for a full-casket burial, it may still intern cremated remains. Full-casket gravesites occasionally become available in such cemeteries due to disinterment or cancellation of prior reservations. The cemetery director can answer such questions at the time of need.

Most national cemeteries do not typically conduct burials on weekends. However, weekend callers trying to make burial arrangements for the following week will be provided with the phone number of one of three strategically located VA cemetery offices that remain open during weekends.

Burial at Sea
The National Cemetery Administration cannot provide burial at sea. For information, contact the United States Navy Mortuary Affairs office toll-free at 866-787-0081.

Furnishing and Placement of Headstones and Markers
In addition to the gravesite, burial in a VA national cemetery also includes furnishing and placement of the headstone or marker, opening and closing of the grave, and perpetual care. Many national cemeteries also have columbaria or gravesites for cremated remains. (Some state veterans' cemeteries may charge a nominal fee for placing a government-provided headstone or marker.)

Special Notes:
For persons with 20-years' service in the National Guard or Reserves, entitlement to retired pay must be subsequent to November 1, 1990, in order to qualify for a government-provided headstone or marker. A copy of the Reserve Retirement Eligibility Benefits Letter must accompany the application. Active-duty service while in the National Guard or Reserves also establishes eligibility.

Service prior to World War I require detailed documentation to prove eligibility such as muster rolls extracts from State files, military or State organizations where served, pension or land warrants, etc.

Flat bronze, granite, or marble grave markers and upright marble or granite headstones are available to mark graves in the style consistent with existing monuments in the national cemetery. Niche markers are also available for identifying cremated remains in columbaria. The following is a brief description of each type:
- Upright Marble and Upright Granite: 42 inches long, 13 inches wide, 4 inches thick, approximately 230 pounds.
- Flat Bronze: 24 inches long, 12 inches wide, ¾ inch rise, approximately 18 pounds.

111

Anchor bolts, nuts, and washers for fastening to the base are supplied with marker. The government does not provide the base.

- Flat Granite and Flat Marble: 24 inches long, 12 inches wide, 4 inches thick, approximately 130 pounds.
- Bronze Niche: 8 ½ inches long, 5 ½ inches wide, 7/16-inch rise, approximately 3 pounds. Mounting bolts and washers are supplied with the marker.

There are also two special styles of upright marble headstones and flat markers available - one for those who served with the Union Forces during the Civil War or the Spanish-American War, and one for those who served with the Confederate forces during the Civil War.

Inscriptions

Headstones and markers are inscribed with the name of the deceased, branch of service, year of birth, and year of death. The word "Korea" may be included on government headstones and markers for the graves of those members and former members of the United States Armed Forces who served within the areas of military operations in the Korean Theater between June 27, 1950, and July 27, 1954; and for headstones and markers for active-duty decedents who lost their lives in Korea or adjacent waters as a result of hostile action subsequent to the 1953 Armistice

The word "Vietnam" may be included on government headstones and markers for the graves of those members and former members of the United States Armed Forces who died in Vietnam or whose death was attributable to service in Vietnam, and on the headstones and markers of all decedents who were on active duty on or after August 5, 1964. The words "Lebanon" or "Grenada" may be included on government headstones and markers for those killed as a result of those military actions.

The words "Panama" and "Persian Gulf" may be included on government headstones and markers for those killed as a result of those military actions
If desired, the following inscriptions can also be made (space permitting):

Military grade
Military rank
Military rate
Identification of war service
Months and days of birth and death
An authorized emblem of religious belief (see the following list of available emblems)
Military awards (documentation of award must be provided)
Military organizations
Civilian or veteran affiliations

With the VA's approval, terms of endearment that meet acceptable standards of good taste may also be added. Most optional inscriptions are placed as the last lines of the inscription. No other graphics are permitted on government-provided headstones and markers, and inscriptions will be in English text only. Civilian titles such as "Doctor or Reverend" are not permitted on the name line of government-provided headstones and markers.

Available Emblems of Belief for Placement on Government Headstones and Markers:

Christian Cross
Buddhist (Wheel of Righteousness)
Hebrew (Star of David)
Presbyterian Cross
Russian Orthodox Cross
Lutheran Cross

Episcopal Cross
Unitarian Church (Flaming Chalice)
United Methodist Church
Aaronic Order
Church Mormon (Angel Moroni)
Native American Church of North America
Serbian Orthodox
Greek Cross
Bahai (9-Pointed Star)
Atheist
Muslim (Crescent and Star)
Hindu
Konko-Kyo
Faith Community of Christ
Sufism Reoriented Tenrikyo Church
Seicho-No-Ie
Church of World Messianity (Izunome)
United Church of Religious Science
Christian Reformed Church
United Moravian Church
Eckankar
Christian Church United Church of Christ
Christian & Missionary Alliance
Humanist Emblem of Spirit
Presbyterian Church (USA)
Izumo Taishakyo Mission of Hawaii
Soka Gakkai International - USA Sikh (Khanda)
Christian Scientist (Cross and Crown)
Muslim (Islamic 5-Pointed Star)

Ordering a Headstone or Marker

For burial of a veteran in a national, state veteran, or military post cemetery, the cemetery will order the headstone or marker and can give loved one's information on style, inscription, and shipment. Shipment and placement of the headstone or marker is provided at no cost.

When burial is in a private cemetery, VA Form 40-1330, "Application for Standard Government Headstone or Marker" must be submitted by the next of kin or a representative, such as funeral director, cemetery official or veterans' counselor, along with the veteran's military discharge documents, to request a Government-provided headstone or marker. Do not send original documents, as they will not be returned.

Headstones and Markers for Families and Children

The VA will issue a headstone or marker for an eligible spouse or child buried in a national, state veteran or military post cemetery. However, the VA cannot issue a headstone or marker for a spouse or child buried in a private cemetery.

The applicant can, however, request to reserve inscription space below the veteran's inscription, so that the non-veteran's commemorative information can be inscribed locally, at private expense, when the non-veteran is buried. The applicant may also choose to have his/her name and date of birth added at Government expense when the headstone or marker is ordered. The date of death may then be added, at private expense, at the time of his/her death.

Commemoration of Unidentified Remains

Many national cemeteries have areas suitable to commemorate veterans whose remains were not recovered or identified, were buried at sea, or are otherwise unavailable for interment. In such instances, the VA will provide a memorial headstone or marker to be placed in that section of the cemetery. The words "In Memory of" must precede the authorized inscription.

Checking Order Status

To check the status of a previously ordered headstone or marker for placement in a private cemetery, applicants may call the VA's Applicant Assistance Line at (800) 697- 6947. The line is open weekdays, from 8:00 A.M. to 5:00 P.M., Eastern Standard Time

Replacement of Headstones and Markers

The government will replace a previously furnished headstone or marker if it becomes badly damaged, is vandalized, is stolen, becomes badly deteriorated, the inscription becomes illegible, is different from that ordered by the applicant, becomes damaged in transit, or the inscription is incorrect.

Government headstones or markers in private cemeteries damaged by cemetery personnel will not be replaced at government expense.

If a marble or granite headstone or marker is permanently removed from a grave, it must be destroyed. Bronze markers must be returned to the contractor.

Military Honors

The rendering of Military Funeral Honors is a way to show the nation's deep gratitude to those who have faithfully defended our country in times of war and peace. It is the final demonstration a grateful nation can provide to the veterans' families.

The following service members are eligible for Military Funeral Honors:

- Military members on active duty or in the Selected Reserve at time of death.
- Former military members who served on active duty and departed under conditions other than dishonorable.
- Former military members who completed at least one term of enlistment or period of initial obligated service in the Selected Reserve, and departed under conditions other than dishonorable.
- Former military members discharged from the Selected Reserve due to a disability incurred or aggravated in the line of duty.

All eligible veterans receive basic Military Funeral Honors if requested by the deceased veteran's family. As provided by law, an honor guard detail for the burial of an eligible veteran shall consist of not less than two members of the Armed Forces. One member of the detail shall be a representative of the parent Service of the deceased veteran. The honor detail will, at a minimum, perform a ceremony that includes the folding and presenting of the American flag to the next of kin, and the playing of Taps. Taps will be played by a bugler, if available, or by electronic recording.

The ceremonial bugle consists of a small cone-shaped device inserted deep into the bell of a bugle that plays an exceptionally high-quality rendition of Taps that is virtually indistinguishable from a live bugler. The ceremonial bugle will be offered to families as an alternative to the pre-recorded Taps played on a stereo but will not be used as a substitute for a live bugler when one is available. Live buglers will continue to play at veterans' funerals whenever available.

Funeral Directors have the responsibility of assisting loved ones in requesting military honors. A toll-free telephone number has been established for funeral directors requesting military honors. That number is 1-800-535-1117.

The Department of Defense (DOD) has provided registered funeral home directors with a military funeral honors kit and information on how to contact the appropriate military organization to perform the honors ceremony.

Questions about the Military Funeral Honors program should be sent to
Deputy Assistant Secretary of Defense
(Military Community and Family Policy)
4000 Defense Pentagon, Room 5A726
Washington DC 20380

Preparations Which Can Be Made Prior to Death

It is suggested that veterans and their families prepare in advance by discussing cemetery options, collecting the veteran's military information (including discharge papers), and by contacting the cemetery where the veteran wishes to be buried. If burial will be in a private cemetery and a government headstone or marker will be requested, VA Form 40-1330 can be completed in advance and placed with the veteran's military discharge papers for use at the time of need.

Burial Location Assistance

The VA National Cemetery Administration can provide limited burial location assistance to family members and close friends of descendants thought to be buried in a VA national cemetery.

The National Cemetery Administration will research its records to determine if the decedent is buried in one of VA's national cemeteries. A request can include a maximum of ten specific names to locate. The National Cemetery Administration does not have information on persons buried in cemeteries other than its national cemeteries. Its records do not contain any personal, military or family information – only information regarding whether or not an individual is buried in a VA national cemetery, and if so, where can be provided

The "Gravesite Locator" tool is available online at the VA's website. If the online tool is not available or helpful, you may submit a written request. No form is required to request this information, and no fee is charged. The following information should be provided.

- Full name, including any alternate spellings, of descendent.
- Date and place of birth.
- Date and place of death.
- State from which the individual entered active duty.
- Military service branch.
- Mailing address and phone number of individuals requesting the information.

Allow approximately 4 weeks for a reply.

Requests should be sent to:
U.S. Department of Veterans Affairs
National Cemetery Administration (41C1)
Burial Location Request
810 Vermont Avenue, N.W.
Washington, D.C. 20420

U.S. Military Cemeteries and Monuments Overseas

The American Battle Monuments Commission (ABMC) a small, independent agency of the government's executive branch, maintains 24 American military cemeteries, and 27 memorials, monuments, or markers. (The VA is not responsible for maintaining cemeteries and monuments honoring deceased veterans buried on foreign soil.)

ABMC Services

The ABMC can provide interested parties with:

Name, location, and information on cemeteries and memorials.
Plot, row and grave number or memorialization location of Honored War Dead.
Best in-country routes and modes of travel to cemeteries or memorials.
Information on accommodations near cemeteries or memorials.
Escort service for relatives to grave and memorial sites within the cemeteries.
Letters authorizing fee-free passports for members of the immediate family traveling overseas to visit a grave or memorialization site.
Black and white photographs of headstones and Tablets of the Missing on which the names of dead or missing are engraved.
Arrangements for floral decorations placed at graves and memorialization sites.
An Honor Roll Certificate containing data on a Korean War casualty suitable for framing.
Polaroid color photographs of donated floral decorations in place.

The Andrews Project

The commission also provides friends and relatives of those interred in its cemeteries or memorialized on its Tablets of the Missing with color lithographs of the cemetery or memorial on which is mounted a photograph of the headstone or commemorative inscription. The Andrews Project, named in honor of its sponsor, the late Congressman George W. Andrews, is ABMC's most popular service.

For further information, contact the ABMC at:
American Battle Monuments Commission
Arlington Court House Plaza II
2300 Clarendon Blvd., Suite 500
Arlington, VA 22201
(703) 696-6897

Passports to Visit Overseas Cemeteries

Family members who wish to visit overseas graves and memorial sites of World War I and World War II veterans are eligible for "no-fee" passports. Family members eligible for the "no-fee" passports include surviving spouses, parents, children, sisters, brothers and guardians of the deceased veteran buried or commemorated in American military cemeteries on foreign soil. For further information contact the American Battle Monuments Commission at the address and phone number indicated in the previous section.

Presidential Memorial Certificate

Under Public Law 112-154, service members who die on active duty are eligible for Presidential Memorial Certificates. A Presidential Memorial Certificate is an engraved paper certificate that has been signed by the current president, honoring the memory of any honorably discharged deceased veteran. Presidential Memorial Certificates may be distributed to a deceased veteran's next of kin and loved ones. More than one certificate can be provided per family, and there is no time limit for applying for the certificate. Presidential Memorial Certificates may not be furnished on behalf of deceased veterans who committed capital crimes.

Automatic Presidential Memorial Certificate

In 2020, the VA will automatically present a PMC to a veteran's next of kin at the burial, when the veteran is buried in a national cemetery. If the family or close friends would like to request more than one PMC, they can apply for more copies using the Presidential Memorial Certificate Request Form (VA Form 40-0247). If the veteran is eligible for burial in a national cemetery, but is instead buried in a private cemetery, the family member or close friend can still apply to get a PMC. Fill out the Presidential Memorial Certificate Request Form (VA Form 40-0247); again, more than one PMC can be requested. To speed up the processing of the claim, submit the veteran's military discharge documents and death certificate. Don't send original documents, as they won't be returned.

Presidential Memorial Certificates(41A1C)
Department of Veterans Affairs
5109 Russell Road
Quantico, VA 22134-3903

What to Remember
✓ VA burial allowances are flat-rate monetary benefits to help cover eligible veterans' burial and funeral costs.
✓ Eligible surviving spouses are now paid automatically to simplify the program without the need to submit a claim.
✓ The VA implemented the pre-need burial eligibility determination program to help anyone who would like to know if they are eligible for burial in a VA national cemetery.
✓ VA will furnish a U.S. flag at no cost for burial or memorial purposes in connection with the death of an eligible veteran who served honorably in the Armed Forces.
✓ Some veterans are eligible to receive Military Funeral Honors.

List of National Cemeteries

Some national cemeteries can bury only cremated remains or casketed remains of eligible family members of those already buried. Contact the cemetery director for information on the availability of space

Department of Veterans Affairs National Cemeteries

ALABAMA
Alabama National Cemetery
3133 Highway 119
Montevallo, AL 35115
(205) 665-9039

Fort Mitchell National Cemetery
553 Highway 165
Ft. Mitchell, AL 36856
(334) 855-4731

Mobile National Cemetery
1202 Virginia Street
Mobile, AL 36604
(850) 453-4635
Burial Space- Closed

ALASKA
Fort Richardson National Cemetery
Building #58-512,
Davis Highway
Fort Richardson, AK 99505
(907) 384-7075

Sitka National Cemetery
803 Sawmill CreekRoad
Sitka, AK99835
(907) 384-7075

ARIZONA
National Memorial Cemetery of Arizona
23029 North Cave Creek Road
Phoenix, AZ 85024
(480) 513-3600

Prescott National Cemetery
500 Highway 89 North
Prescott, AZ 86303
(928) 717-7569
Cremation Only

ARKANSAS
Fayetteville National Cemetery
700 Government Avenue
Fayetteville, AR 72701
(479) 444-5051

Fort Smith National Cemetery
522 Garland Avenue
Fort Smith, AR 72901
(479) 783-5345

Little Rock National Cemetery
2523 Springer Blvd.
Little Rock AR 72206
(479) 783-5345
Closed

CALIFORNIA
Baker National Cemetery
30338 East Bear Mountain Blvd.
Arvin, CA 93203
(661) 867-2250

Fort Rosecrans National Cemetery
Cabrillo Memorial Dr.
San Diego, CA 92106
(619) 553-2084
Closed

Golden Gate National Cemetery
1300 Sneath Lane
San Bruno, CA 94066
(650) 589-7737
Closed

Los Angeles National Cemetery
950 South Sepulveda Boulevard
Los Angeles, CA 90049
(310) 268-4675
Closed

Miramar National Cemetery
5795 Nobel Dr.
San Diego, CA 92122
(858) 658-7360

Riverside National Cemetery
22495 Van Buren Boulevard
Riverside, CA 92518
(951) 653-5233

Sacramento Valley National Cemetery
5810 Midway Road
Dixon, CA 9562
(707) 693-2460
Closed

San Francisco National Cemetery
1 Lincoln Blvd.
Presidio of San Francisco
San Francisco, CA 94129
(650) 589-7737
Closed

San Joaquin Valley National Cemetery
32053 West McCabe Road
Santa Nella, CA 95322
(209) 854-1040

COLORADO
Fort Logan National Cemetery
4400 W. Kenyon Ave.
Denver, CO 80236
(303) 761-0117

Fort Lyon National Cemetery
15700 County Road HH
Las Animas, CO 81054
(303) 761-0117

FLORIDA
Barrancas National Cemetery
1 Cemetery Rd.
Pensacola, FL 32508
(850) 453-4108

Bay Pines National Cemetery
1000 Bay Pines Boulevard North
St. Petersburg, FL 33708
(727) 319-6479
Cremation Only

Florida National Cemetery
6502 SW 102nd Avenue
Bushnell, FL 33513
(352) 793-7740

Jacksonville National Cemetery
4083 Lannie Road
Jacksonville, FL 32218-1247
(904) 766-5222

Sarasota National Cemetery
9810 State Hwy 72
Sarasota, FL 34241
(877) 861-9840

South Florida National Cemetery
6501 S. State Road
Lake Worth, FL 33449
(561) 649-6489

St. Augustine National Cemetery
104 Marine Street
St. Augustine, FL 32084
(904) 766-5222
Closed

Tallahassee National Cemetery
5015 Apalachee Parkway
Tallahassee, FL 32311
(850) 402-8941

GEORGIA
Georgia National Cemetery
1080 Scott Hudgens Drive
Canton, GA 30114
(866) 236-8159

Marietta National Cemetery
500 Washington Avenue
Marietta, GA 30060
(866) 236-8159
Closed

HAWAII
National Memorial Cemetery of the Pacific
2177 Puowaina Drive
Honolulu, HI 96813-1729
(808) 523-3720
Cremation Only

ILLINOIS
Abraham Lincoln National Cemetery
20953 W. Hoff Road
Elwood, IL 60421
(815) 423-9958

Alton National Cemetery
600 Pearl Street
Alton, IL 62003
(314) 845-8320

Camp Butler National Cemetery 5063
Camp Butler Road Springfield, IL
92707-9722
(217) 492-4070

Danville National Cemetery
1900 East Main Street
Danville, IL 61832
(217) 554-4550

Mound City National Cemetery
141 State Highway 37
Mound City, IL 62963
(618) 748-9107

Quincy National Cemetery
36th and Maine Street
Quincy, IL 62301
(309) 782-2094
Closed

Rock Island National Cemetery
Bldg. 118
Rock Island Arsenal
Rock Island, IL 61299-7090
(309) 782-2094

Confederate Mound
Oak Woods Cemetery
1035 E. 67th St.
Chicago IL 60637
(815) 423-9958
Closed

North Alton Confederate Cemetery
635 Rozier St.
Alton, IL 62003
(314) 845-8355
Closed

Rock Island Confederate Cemetery
Rodman Avenue
Rock Island Arsenal
Rock Island, IL 61299
(309) 782-2094
Closed

INDIANA
Crown Hill National Cemetery
700 West 38th
Street Indianapolis, IN 46208
(765) 674-0284
Closed

Marion National Cemetery
1700 East 38th Street
Maion, IN 46952
(765) 674-0284

New Albany National Cemetery
1943 Ekin Avenue
New Albany, IN 47150
(502) 893-3852
Cremation Only

Crown Hill Cemetery Confederate Plot
700 West 38th St.
Indianapolis, IN 46208
(765) 674-0284
Closed

Woodlawn Monument site
North 3rd St. and 4th Avenue
Terre Haute, IN 47802
(815) 423-9958
Closed

IOWA
Keokuk National Cemetery
1701 J Street
Keokuk, IA 52632
(319) 524-1304

KANSAS
Fort Leavenworth Natl Cemetery
395 Biddle Blvd.
Fort Leavenworth, KS 66027
(913) 758-4105
Cremation Only

Fort Scott National Cemetery
900 East National Avenue
Fort Scott, KS 66701
(620) 223-2840

Leavenworth National Cemetery
150 Muncie Rd.
Leavenworth, KS 66048
(913) 758-4136

Baxter Springs City Cemetery
Baxter Springs, KS 66713
(913)758-4136
Closed

Mound City Cemetery Soldiers' Lot
Woodland Cemetery
Mound City, KS 66506
(913) 758-4105

KENTUCKY
Camp Nelson National Cemetery
6980 Danville Road
Nicholasville, KY 40356
(859) 885-5727

Cave Hill National Cemetery
701 Baxter Avenue
Louisville, KY 40204
For information please contact:
Zachary Taylor National Cemetery
(502) 893-3852
Closed

Danville National Cemetery
277 North First Street
Danville, KY 40442
(859) 885-5727

Closed

Lebanon National Cemetery
20 Highway 208
Lebanon, KY 40033
(270) 692-3390

Lexington National Cemetery
833 West Main Street
Lexington, KY 40508
(859) 885-5727
Closed

Mill Springs National Cemetery
9044 West Highway 80
Nancy, KY 42544
(859) 885-5727

Zachary Taylor National Cemetery
4701 Brownsboro Road
Louisville, KY 40207
(502) 893-3852
Closed

Evergreen Cemetery Soldiers' Lot
25 South Alexandria Pike
Southgate, KY 41071
(859) 885-5727
Closed

LOUISIANA
Alexandria National Cemetery
209 East Shamrock Street
Pineville, LA 71360
(601) 445-4891
Closed

Baton Rouge National Cemetery
220 North 19th Street
Baton Rouge, LA 70806
(225) 654-1988
Closed

Louisiana National Cemetery
303 W. Mount Pleasant Rd.
Zachary, LA 70791
(225) 654-1988

Port Hudson National Cemetery
20978 Port Hickey Road
Zachary, LA 70791
(225) 654-3767
Cremation Only

MAINE
Togus National Cemetery
VA Medical and Regional Office Center

Togus, ME 04330
(508) 563-7113
Closed

MARYLAND
Annapolis National Cemetery
800 West Street
Annapolis, MD 21401
(410) 644-9696
Closed

Baltimore National Cemetery
5501 Frederick Avenue
Baltimore, MD 21228
(410) 644-9696
Cremation Only

Loudon Park National Cemetery
3445 Frederick Avenue
Baltimore, MD 21228
(410) 644-9696

MASSACHUSETTS
Massachusetts National Cemetery
Connery Ave.
Bourne, MA 02532
(508) 563-7113

MICHIGAN
Fort Custer National Cemetery
15501 Dickman Road
Augusta, MI 49012
(269) 731-4164

Great Lakes National Cemetery
4200 Belford Road
Holly, MI 48842
(866) 348-8603

MINNESOTA
Fort Snelling National Cemetery
7601 34th Avenue
South Minneapolis, MN 55450-1199
(612) 726-1127

MISSISSIPPI
Biloxi National Cemetery
400 Veterans Avenue
Bld. 1001
Biloxi, MS 39535-4968
(228) 388-6668

Corinth National Cemetery
1551 Horton Street
Corinth, MS 38834
(901) 386-8311

Natchez National Cemetery
41 Cemetery Road
Natchez, MS 39120
(601) 445-4981

MISSOURI
Jefferson Barracks National Cemetery
2900 Sheridan Road
St. Louis, MO 63125
(314) 845-8320

Jefferson City National Cemetery
1024 East McCarty Street
Jefferson City, MO 65101
(314) 845-8320
Closed

Springfield National Cemetery
1702 East Seminole Street
Springfield, MO 65804
(417) 881-9499
Cremation Only

NEBRASKA
Fort McPherson National Cemetery
12004 South Spur 56A
Maxwell, NE 69151
(308) 582-4433

Omaha National Cemetery
14250 Schram Rd.
Omaha, NE 68138
(402) 253-3949

NEW JERSEY
Beverly National Cemetery
916 Bridgeboro Rd.
Beverly, NJ 08010
(215) 504-5610
Finn's Point National Cemetery
454 Fort Mott Road,
Pennsville, NJ 08070
(215) 504-5610
Cremation only

NEW MEXICO
Fort Bayard National Cemetery
200 Camino De Paz
Fort Bayard, NM 88036
(505) 988-6400

Santa Fe National Cemetery
501 North Guadalupe Street
Santa Fe, NM 87501
(505) 988-6400

NEW YORK
Bath National Cemetery

San Juan Avenue
Bath, NY 14810
(607) 664-4853

Calverton National Cemetery
210 Princeton Boulevard
Calverton, NY 11933-1031
(631) 727-5410

Cypress Hills National Cemetery
625 Jamaica Avenue
Brooklyn, NY 11208
(631) 454-4949
Closed

George B.H. Solomon Saratoga
National Cemetery
200 Duell Rd.
Schuylerville, NY 12871
(631) 454-4949
Cremation Only

Long Island National Cemetery
2040 Wellwood Avenue
Farmingdale, NY 11735-1211
(631) 454-4949
Cremation Only

Woodlawn National Cemetery
1825 Davis Street
Elmira, NY 14901
(607) 732-5411
Cremation Only

NORTH CAROLINA
New Bern National Cemetery
1711 National Avenue
New Bern, NC 28560
(252) 637-2912
Closed

Raleigh National Cemetery
501 Rock Quarry Road
Raleigh, NC 27610
(252) 637-2912d
Closed

Salisbury National Cemetery
501 Statesville Blvd.
Salisbury, NC 28144
(704) 636-2661

Wilmington National Cemetery
2011 Market Street
Wilmington, NC 28403
(910) 815-4877
Closed

OHIO
Dayton National Cemetery
4400 West Third St.
Dayton, OH 45428-1008
(937) 268-2221

Ohio Western Reserve National
Cemetery
10175 Rawiga Road
Seville, OH 44273
(330) 335-3069

OKLAHOMA
Fort Gibson National Cemetery
1423 Cemetery Road
Fort Gibson, OK 74434
(918) 478-2334

Fort Sill National Cemetery
2648 NE Jake Dunn Rd.
Elgin, OK 73538
(580) 492-3200

OREGON
Eagle Point National Cemetery
2763 Riley Road
Eagle Point, OR 97524
(541) 826-2511

Roseburg National Cemetery
913 NW Garden Valley Blvd.
Roseburg, OR 97471
(541) 671-3152

Willamette National Cemetery
11800 S.E. Mt. Scott Boulevard
Portland, OR 97086
(503) 273-5250

PENNSYLVANIA
Indiantown Gap National Cemetery
Indiantown Gap Rd.
Annville, PA 17003-9618
(717) 865-5254

National Cemetery of the Alleghenies
1158 Morgan Road
Bridgeville, PA 15017
(724) 746-4363

Philadelphia National Cemetery
Haines Street and Limekiln Pike
Philadelphia, PA 19138
(215) 504-5611
Closed

PUERTO RICO
Puerto Rico National Cemetery
Avenue Cementario Nacional
#50 Bayamon, PR 00960
(787) 798-8400

SOUTH CAROLINA
Beaufort National Cemetery
1601 Boundary Street
Beaufort, SC 29902
(843) 524-3925

Florence National Cemetery
803 East National Cemetery Road
Florence, SC 29501
(843) 669-8783

Fort Jackson National Cemetery
4170 Percival Rd
Columbia, SC 29229
(866) 577-5248

SOUTH DAKOTA
Black Hills National Cemetery
20901 Pleasant Valley Dr.
Sturgis, SD 57785
(605)347- 3830

Fort Meade National Cemetery
Old Stone Road
Sturgis, SD 57785
(605) 347-3830
Closed

Hot Springs National Cemetery
VA Medical Center
Hot Springs, SD 57747
(605) 347-3830
Closed

TENNESSEE
Chattanooga National Cemetery
1200 Bailey Avenue
Chattanooga, TN 37404
(423) 855-6590

Knoxville National Cemetery
939 Tyson Street, N.W.
Knoxville, TN 37917
(423) 855-6590
Cremation Only

Memphis National Cemetery
3568 Townes Avenue
Memphis, TN 38122
(901) 386-8311
Cremation Only

123

Mountain Home National Cemetery
53 Memorial Avenue
Mountain Home, TN 37684
(423) 979-3535

Nashville National Cemetery
1420 Gallatin Road S
Madison, TN 37115-4619
(615) 860-0086
Cremation Only

TEXAS
Dallas-Fort Worth National Cemetery
2000 Mountain Creek Parkway
Dallas, TX 75211
(214) 467-3374

Fort Bliss National Cemetery
5200 Fred Wilson Road
El Paso, TX 79906
(915) 564-0201

Fort Sam Houston National Cemetery
1520 Harry Wurzbach Road
San Antonio, TX 78209
(210) 820-3891

Houston National Cemetery
10410 Veterans Memorial Drive
Houston, TX 77038
(281) 447-8686

Kerrville National Cemetery
3600 Memorial Boulevard
Kerrville, TX 78028
(210) 820-3891
Closed

San Antonio National Cemetery
517 Paso Hondo Street
San Antonio, TX 78202
(210) 820-3891
Cremation Only

VIRGINIA
Alexandria National Cemetery
1450 Wilkes Street
Alexandria, VA 22314
(703) 221-2183
Cremation Only

Balls Bluff National Cemetery
Route 7
Leesburg, VA 22075
(540) 825-0027
Closed

City Point National Cemetery
10th Avenue and Davis Street
Hopewell, VA 23860
(804) 795-2031
Closed

Cold Harbor National Cemetery
6038 Cold Harbor Rd.
Mechanicsville, VA 23111
(804) 795-2031
Closed

Culpeper National Cemetery
305 U.S. Avenue
Culpeper, VA 22701
(540) 825-0027

Danville National Cemetery
721 Lee Street
Danville, VA 24541
(704) 636-2661
Cremation Only

Fort Harrison National Cemetery
8620 Varina Road
Richmond, VA 23231
(804) 795-2031
Closed

Glendale National Cemetery
8301 Willis Church Road
Richmond, VA 23231
For information please contact:
Fort Harrison National Cemetery
(804) 795-2031

Hampton National Cemetery Road at
Marshall Avenue
Hampton, VA 23667
(757) 723-7104

Hampton National Cemetery
VA Medical Center
Emancipation Drive
Hampton, VA 23667
(757) 723-7104

Quantico National Cemetery
P. O. Box 10
18424 Joplin Road (Route 619)
Triangle, VA 22172
(703) 221-2183 (local)

Richmond National Cemetery
1701 Williamsburg Road
Richmond, VA 23231
For information please contact: Fort

Harrison National Cemetery
(804) 795-2031

Seven Pines National Cemetery
400 East Williamsburg Road
Sandston, VA 2315
For information please contact: Fort
Harrison National Cemetery
(804) 795-2031

Staunton National Cemetery
901 Richmond Avenue
Staunton, VA 24401
(540) 825-0027

Winchester National Cemetery
401 National Avenue
Winchester, VA 22601
For information please contact:
Culpeper National Cemetery
(540) 825-0027

WASHINGTON
Tahoma National Cemetery
18600 Southeast 240th Street
Kent, WA 98042-4868
(425) 413-9614

WEST VIRGINIA
Grafton National Cemetery
431 Walnut Street
Grafton, WV 26354
For information please contact:
West Virginia National Cemetery
(304) 265-2044

West Virginia National Cemetery
Route 2, Box 127
Grafton, WV 26354
(304) 265-2044

WISCONSIN
Wood National Cemetery
5000 W National Ave, Bldg 1301
Milwaukee, WI 53295-4000
(414) 382-5300

**DEPARTMENT OF THE INTERIOR
NATIONAL CEMETERIES**

DISTRICT OF COLUMBIA
Battleground National Cemetery
C/O Superintendent,
Rock Creek Park
3545 Williamsburg Lane, NW
Washington, DC 20008
(202) 282-1063

GEORGIA
Andersonville National Historic Site
Route 1, Box 800
Andersonville, GA 31711
(912) 924-0343

LOUISIANA
Chalmette National Cemetery C/O Jean
Lafitte National Historical Park and
Preserve
365 Canal Street, Suite 2400
New Orleans, LA 70130
(504) 589-3882
(504) 589-4430

MARYLAND
Antietam National Battlefield
Box 158
Sharpsburg, MD 21782-0158
(301) 432-5124

MISSISSIPPI
Vicksburg National Military Park
3201 Clay Street
Vicksburg, MS 39180
(601) 636-0583

MONTANA
Little Bighorn Battle National
Monument Custer National Cemetery
P. O. Box 39
Crow Agency, MT 59022

PENNSYLVANIA
Gettysburg National Military Park
97 Taneytown Road
Gettysburg, PA 17325-2804
(717) 334-1124

TENNESSEE
Andrew Johnson National Historic Site
P. O. Box 1088
Greeneville, TN 37744
(423) 638-3551

Fort Donelson National Battlefield
P. O. Box 434
Dover, TN 37058
(615) 232-5348

Shiloh National Military Park
Route 1, Box 9
Shiloh, TN 38376-9704
(901) 689-5275

Stones River National Battlefield
3501 Old Nashville Highway
Murfreesboro, TN 37129
(615) 893-9501

VIRGINIA
Fredericksburg and Spotsylvania
County Battlefields Memorial National
Military Park
120 Chatham Lane
Fredericksburg, VA 22405
(540) 371-0802

Poplar Grove National Cemetery
Petersburg National Battlefield
1539 (Yorktown Battlefield Cemetery
Colonial National Historical Park
P. O. Box 210
Yorktown, VA 23690
(757) 898-3400

**DEPARTMENT OF THE ARMY
NATIONAL CEMETERIES**

United States Soldiers' &
Airmen's Home National
Cemetery
21 Harewood Road, NW
Washington, DC 20011
(202) 829-1829

STATE VETERANS' CEMETERIES

ALABAMA
Alabama State Veterans Memorial
Cemetery at Spanish Fort
34904 State Highway 225
Spanish Fort, AL 36577

ARIZONA
Southern Arizona Veterans Memorial
14317 Veterans Drive
Camp Navajo
PO Box 16419
Bellemont, AZ
(928) 214-3474

ARKANSAS
Arkansas Veterans Cemetery
1501 W Maryland Avenue
North Little Rock, AR 72120
(501) 683-2259

CALIFORNIA
Veterans Memorial Grove Cemetery
Veterans Home of California
Yountville, CA 94599
(707) 944-4600

Northern California Veterans Cemetery
P.O. Box 76
11800 Gas Point Road
Igo, CA 96047
(866) 777-4533

COLORADO
Colorado State Veterans Cemetery at
Homelake
3749 Sherman Avenue
Monte Vista, CO 81144
(719) 852-5118

Veterans Memorial Cemetery of
Western Colorado
2830 D Road
Grand Junction, CO 81505
(970) 263-8986

CONNECTICUT
Colonel Raymond F. Gates Memorial
Cemetery Veterans Home and Hospital
287 West Street
Rocky Hill, CT 06067
(860) 721-5838

Spring Grove Veterans Cemetery
Darien, CT C/O Veterans Home and
Hospital
287 West Street
Rocky Hill, CT 06067
(860) 721-5838

Middletown Veterans Cemetery C/O
Veterans Home and Hospital
287 West Street
Rocky Hill, CT 06067
(860) 721-5838

DELAWARE
Delaware Veterans Memorial Cemetery
2465 Chesapeake City Road
Bear, DE 19701
(302) 834-8046

Delaware Veterans Memorial
Cemetery-Sussex County
RD 5 Box 100
Millsboro, DE 19966
(302) 934-5653

GEORGIA
Georgia Veterans Memorial Cemetery
2617 Vinson Highway
Milledgeville, Georgia 31061
(478) 445-3363

Georgia Veterans Memorial Cemetery
8819 U.S. Highway 301
Glennville, Georgia 30427
(912)-654-5398

HAWAII
Director, Office of Veterans Services
459 Patterson Road
E-Wing, Room 1-A103
Honolulu, HI 96189
(808) 433-0420

Hawaii State Veterans Cemetery
45-349 Kamehameha Highway
Kaneohe, HI 96744
(808) 233-3630

East Hawaii Veterans Cemetery
No. I County of Hawaii
25 Aupuni Street
Hilo, HI 96720
(Island of Hawaii)
(808) 961-8311

East Hawaii Veterans Cemetery
-No. II
County of Hawaii
25 Aupuni Street
Hilo, HI 96720
(Island of Hawaii)
(808) 961-8311

Kauai Veterans Cemetery County of
Kauai Public Works
3021 Umi Street
Lihue, HI 96766
(Island of Kauai)
(808) 241-6670

Maui Veterans Cemetery
1295 Makawao Avenue,
Box 117
Makawao, HI 96768 (Island of Maui)
(808) 572-7272

Hoolehua Veterans
Cemetery (Molokai)
P. O. Box 526
Kauna Kakai, HI 96748
(Island of Molokai)
(808) 553-3204

Lanai Veterans Cemetery Maui County
(Island of Lanai)
PO Box 630359
Lanai City, Hawaii 96763

IDAHO
Idaho Veterans Cemetery
10101 North Horseshoe Bend Road
Boise, ID 83714
(208) 334-4796

ILLINOIS
Sunset Cemetery Illinois Veterans
Home
1707 North 12th Street
Quincy, IL 62301
(217) 222-8641

INDIANA
Indiana State Soldiers Home Cemetery
3851 North River Road
West Lafayette, IN 47906-3765
(765) 463-1502

Indiana Veterans Memorial Cemetery
1415 North Gate Road
Madison, IN 47250
(812) 273-9220

IOWA
Iowa Veterans Home and Cemetery
13th & Summit Streets
Marshalltown, IA 50158
(641) 753-4309

Iowa Veterans Cemetery
34024 Veterans Memorial Drive
Adel, Iowa 50003-3300
(515) 996-9048

KANSAS
Kansas Veterans Cemetery at Fort
Dodge
714 Sheridan, Unit #66
Fort Dodge, KS 67801
(620) 338-8775

Kansas Veterans Cemetery at
Wakeeney
P.O. Box 185
4035 13th Street
Wakeeney, KS 67672
(785) 743-5685

Kansas Veterans Cemetery at Winfield
1208 North College
Winfield, KS 67156
(620) 229-2287

KENTUCKY
Kentucky Veterans Cemetery
Central
1111 Louisville Road
Frankfort, Kentucky
(502) 564-9281

Kentucky Veterans Cemetery
North 205 Eibeck Lane
P.O. Box 467
Williamstown, Kentucky 41097
(859)-823-0720

Kentucky Veteran's Cemetery West
5817 Fort Campbell Boulevard
Hopkinsville, Kentucky 42240
(270) 707-9653

LOUISIANA
Northwest Louisiana Veterans
Cemetery
7970 Mike Clark Road
Keithville, Louisiana 71047
(318) 925-0612

MAINE
Maine Veterans Memorial
Cemetery (Closed)
Civic Center Drive
Augusta, Maine

Maine Veterans Memorial Cemetery--
Mt. Vernon Rd.
163 Mt. Vernon Road
Augusta, ME 04330
(207) 287-3481

Northern Maine Veterans Cemetery-
Caribou
37 Lombard Road
Caribou, ME 04736
(207)-492-1173

MARYLAND
Maryland State Veterans Cemeteries
Federal Building- 31 Hopkins Plaza
Baltimore, MD 21201
(410) 962-4700

Cheltenham Veterans
Cemetery
11301 Crain Highway
P. O. Box 10
Cheltenham, MD 20623
(301) 372-6398

Crownsville Veterans Cemetery
1080 Sunrise Beach Road Crownsville,
MD 21032
(410) 987-6320

Eastern Shore Veterans Cemetery
6827 East New Market Ellwood Road
Hurlock, MD 21643
(410) 943-3420

Garrison Forest Veterans Cemetery
11501 Garrison Forest Road
Owings Mills, MD 21117
(410) 363-6090

Rocky Gap Veterans Cemetery
14205 Pleasant Valley Road, NE
Flintstone, MD 21530
(301) 777-2185

MASSACHUSETTS
Massachusetts State Veterans
Cemetery (Agawam & Winchendon)
1390 Main Street
Agawam, MA 01001
(413) 821-9500

Winchendon Veterans Cemetery
111 Glenallen Street
Winchendon, MA 01475
(978) 297-9501

MICHIGAN
Grand Rapids Home for Veterans
Cemetery
3000 Monroe, NW
Grand Rapids, MI 49505
(616) 364-5400

MINNESOTA
MN State Veterans Cemetery
15550 HWY 15
Little Falls, MN 56345
(320) 616-2527

MISSOURI
St. James Missouri Veterans Home
Cemetery
620 North Jefferson
St. James, MO 65559
(573) 265-3271

Missouri Veterans Cemetery
20109 Bus. Hwy. 13
Higgensville, MO 64037
(660) 584-5252

Missouri Veterans Cemetery
Springfield, MO
5201 South Southwood Road
Springfield, MO 65804
(417) 823-3944

Missouri State Veterans Cemetery
Bloomfield
17357 Stars and Stripes Way
Bloomfield, Missouri 63825
(573) 568-3871

Missouri State Cemetery
Jacksonville
1479 County Road 1675
Jacksonville, Missouri 65260
(660) 295-4237

MONTANA
State Veterans Cemetery
Fort William H. Harrison Box 5715
Helena, MT 59604
(406) 324-3740

State Veterans Cemetery Miles City
Highway 59
Miles City, MT 59301
(406) 324-3740

Montana Veterans Home Cemetery
P. O. Box 250
Columbia Falls, MT 59912
(406) 892-3256

NEBRASKA
Nebraska Veterans Home Cemetery
Burkett Station
Grand Island, NE 68803
(308) 385-6252, Ext. 230

NEVADA
Commissioner of Veterans Affairs
1201 Terminal Way, Room 108
Reno, NV 89520
(775) 688-1155

Northern Nevada Veterans
Memorial Cemetery
14 Veterans Way
Fernley, NV 8940
(775) 575-4441

Southern Nevada Veterans Memorial
Cemetery
1900 Buchanan Boulevard
Boulder City, NV 89005
(702) 486-5920

NEW HAMPSHIRE
NH State Veterans Cemetery
110 Daniel Webster Hwy, Route 3
Boscawen, NH 03303
(603) 796-2026

NEW JERSEY
Brigadier General William C. Doyle
Veterans Memorial Cemetery
350 Provenceline Road,
Route #2
Wrightstown, NJ 08562
(609) 758-7250

(901) 543-7005 New Jersey Memorial
Home Cemetery (Closed)
524 N.W. Boulevard
Vineland, NJ 08360
(609) 696-6350

NORTH CAROLINA
Western Carolina State Veterans
Cemetery
962 Old Highway 70,
West Black Mountain, NC 28711
(828) 669-0684

Coastal Carolina State Veterans
Cemetery
P. O. Box 1486
Jacksonville, NC 28541
(910) 347-4550 or 3570

Sandhills State Veterans Cemetery
P. O. Box 39
400 Murchison Road
Spring Lake, NC 28390
(910) 436-5630 or 5635

NORTH DAKOTA
North Dakota Veterans Cemetery
1825 46th Street
Mandan, ND 58554
(701) 667-1418

OHIO
Ohio Veterans Home Cemetery
3416 Columbus Avenue
Sandusky, OH 44870
(419) 625-2454, Ext. 200

OKLAHOMA
Oklahoma Veterans Cemetery Military
Department (OKFAC)
3501 Military Circle N.E.
Oklahoma City, OK 73111-4398
(405) 228-5334

PENNSYLVANIA
Pennsylvania Soldiers and Sailors
Home Cemetery
P. O. Box 6239
560 East Third Street
Erie, PA 16512-6239
(814) 871-4531

RHODE ISLAND
Rhode Island Veterans Cemetery
301 South County
Trail Exeter, RI 02822-9712
(401) 268-3088

SOUTH CAROLINA
M.J. "Dolly" Cooper Veterans Cemetery
140 Inway Drive
Anderson, SC 29621
(864) 332-8022

SOUTH DAKOTA
SD Veterans Home Cemetery
2500 Minnekahta Avenue
Hot Springs, SD 57747
(605) 745-5127

TENNESSEE
East Tennessee State Veterans
Cemetery
5901 Lyons View Pike
Knoxville, TN 37919
(865) 594-6776

Middle Tennessee Veterans Cemetery
7931 McCrory Lane
Nashville, TN 37221
(615) 532-2238

West Tennessee Veterans Cemetery
4000 Forest Hill/Irene Road
Memphis, TN 38125

TEXAS
Central Texas State Veterans Cemetery
11463 South Highway 195
Killeen, Texas 76542
(512) 463-5977

Rio Grande Valley
State Veterans Cemetery
2520 South Inspiration Road
Mission, TX 78572
(956) 583-7227

UTAH
Utah State Veterans Cemetery
Utah Parks and Recreation

17111 South Camp Williams Road
Bluffdale, UT 84065
(801) 254-9036

VERMONT
Vermont Veterans Home War
Memorial Cemetery
325 North Street
Bennington, VT 05201
(802) 442-6353

Vermont Veterans Memorial Cemetery
120 State Street
Montpelier, VT 05602-4401
(802) 828-3379

VIRGINIA
Virginia Veterans Cemetery
10300 Pridesville Road
Amelia, VA 23002
(804) 561-1475

Albert G. Horton, Jr. Memorial
Veterans Cemetery
5310 Milner's Road
Suffolk, Virginia 23434
(757) 334-4731

WASHINGTON
Washington Soldiers Home Colony and
Cemetery
1301 Orting-Kapowsin Highway
Orting, WA 98360
(360) 893-4500

Washington Veterans Home Cemetery
PO. Box 698
Retsil, WA 98378
(360) 895-4700

WISCONSIN
Northern Wisconsin Veterans
Memorial Cemetery
N4063 Veterans Way
Spooner, WI 54801
(715) 635-5360

Wisconsin Veterans Memorial
Cemetery
Wisconsin Veterans Home
N2665 Highway QQ
King, WI 54946
(715) 258-5586

Southern Wisconsin Veterans
Memorial Cemetery
21731 Spring Street
Union Grove, WI 53182

WYOMING
Oregon Trail Veterans Cemetery
89 Cemetery Road, Box 669
Evansville, WY 82636
(262) 878-5660

TERRITORIES
Guam Veterans Cemetery
490 Chalan Palayso
Agatna Heights, Guam 96910
(671) 475-8388

CHAPTER 11

HEALTHCARE BENEFITS

If you qualify for VA health care, you'll receive coverage for the services required to stay healthy. Every veteran will have a unique benefits package. All veterans receive coverage for most care and services, but some qualify for added benefits like dental care. The full list of covered benefits depends on a veteran's priority group, the advice of their VA primary care provider, and the medical standards for treating any condition a veteran has.

Being signed up for VA health care meets the Affordable Care Act (ACA) health coverage requirement of having minimum essential health coverage.

Eligibility

You may be eligible for VA health care benefits if you served in the active military, naval or air service and didn't receive a dishonorable discharge.

If you enlisted after September 7, 1980, or entered active duty after October 16, 1981:
You must have served 24 continuous months or the full period for which you were called to active duty unless any of the below descriptions are true for you:
- You were discharged for a disability that was caused or made worse by your active-duty service, or
- You were discharged for a hardship or early out, or
- You served prior to September 7, 1980.

If you're a current or former member of the Reserves or National Guard:
You must have been called to active duty by a federal order and completed the full period for which you were called or ordered to active duty. If you had or have active-duty status for training purposes only, you don't qualify for VA health care.

If you served in certain locations and time periods during the Vietnam War era:
You're eligible for VA health care.

You may qualify for enhanced eligibility status if you meet at least one of the following requirements. Enhanced eligibility means you're in a higher priority group, making you more likely to get benefits.

At least one must be true:
- You receive financial compensation from VA for a service-connected disability
- You were discharged for a disability resulting from something that happened to you in the line of duty
- You were discharged for a disability that got worse in the line of duty
- You're a combat veteran discharged or released on or after September 11, 2001
- You get a VA pension
- You're a former prisoner of war

- You have received a Purple Heart
- You have received a Medal of Honor
- You receive or qualify for Medicaid benefits
- You served in Southwest Asia during the Gulf War between August 2, 1990, and November 11, 1998
- You served at least 30 days at Camp Lejeune between August 1, 1953, and December 31, 1987

Or you must have served in any of these locations during the Vietnam War era:

- Any U.S. or Royal Thai military base in Thailand from January 9, 1962, through June 30, 1976
- Laos from December 1, 1965, through September 30, 1969
- Cambodia at Mimot or Krek, Kampong Cham Province from April 16, 1969, through April 30, 1969
- Guam or American Samoa or in the territorial waters off Guam or American Samoa from January 9, 1962, through July 31, 1980
- Johnston Atoll or on a ship that called at Johnston Atoll from January 1, 1972, through September 30, 1977
- Republic of Vietnam from January 9, 1962, through May 7, 1975

If none of these descriptions apply to you, you may still qualify for care based on your income.

VA Priority Groups

When a veteran applies for VA health care, they assign you to 1 of 8 priority groups. The system is meant to make sure veterans who need care immediately get signed up quickly. The priority group a veteran is assigned can affect how soon they receive health care benefits, and how much, if anything, they pay toward the cost of their care.

Priority groups are based on:

- Military service history, and
- Disability rating, and
- Income level, and
- Qualification for Medicaid, and
- Other benefits being received, like a pension from VA

VA assigns veterans with service-connected disabilities the highest priority. Veterans who earn a higher income and don't have any service-connected disabilities are assigned a lower priority. If a veteran qualifies for more than one group, they're assigned to the higher one.

Priority Group 1
You may be assigned to this Priority Group if any of the following are true:

- You have a service-connected disability rated as 50% or more disabling, or
- You have a service-connected disability the VA concludes makes you unable to work (you're considered unemployable) or
- You received the Medal of Honor (MOH)

Priority Group 2
You may be assigned to Priority Group 2 if you have a service-connected disability the VA has rated as 30% or 40% disabling.

Priority Group 3
You may be assigned to Priority Group 3 if any of the following are true:

- You're a former prisoner of war (POW), or

- You received the Purple Heart medal, or
- You were discharged for a disability that was caused by or got worse because of your active-duty service, or
- You have a service-connected disability the VA has rated as 10% or 20% disabling, or
- You were award special eligibility classification under Title 38, USC 1151.

Priority Group 4
You may be assigned to Priority Group 4 if either of the following is true:
- You receive VA aid and attendance or housebound benefits, or
- You've received a VA determination of being catastrophically disabled

Priority Group 5
You may be assigned to Priority Group 5 if any of the following are true:
- You don't have a service-connected disability, or you have a non-compensable service-connected disability the VA has rated as 0% disabling, and you have an annual income level that's below the VA's adjusted income limits based on your zip code, or
- You're receiving VA pension benefits, or
- You're eligible for Medicaid programs

Priority Group 6
VA may assign you to Priority Group 6 if any of the following are true:
- You have a compensable service-connected disability that we've rated as 0% disabling, or
- You were exposed to ionizing radiation during atmospheric testing or during the occupation of Hiroshima and Nagasaki, or
- You participated in Project 112/SHAD, or
- You served in the Republic of Vietnam between January 9, 1962, and May 7, 1975, or
- You served in the Persian Gulf War between August 2, 1990, and November 11, 1998, or
- You served on active duty at Camp Lejeune for at least 30 days between August 1, 1953, and December 31, 1987

VA may also assign you to Priority Group 6 if you meet all of these requirements:
- You're currently or newly enrolled in VA health care, and
- You served in a theater of combat operations after November 11, 1998, and
- You were discharged less than 10 years ago

Note: As a returning combat veteran, you're eligible for these enhanced benefits for 10 years after your discharge. At the end of your enhanced enrollment period, VA will assign you to the highest priority group you qualify for at that time.

Priority Group 7
You may be assigned to Priority Group 7 if both is true for you:
- Your gross household income is below the geographically adjusted income limits for where you live, and
- You agree to pay copays

Priority Group 8
You may be assigned to Priority Group 8 if both are true for you:
- Your gross household income is above VA income limits and geographically adjusted income limits for where you live, and
- You agree to pay copays

If you're assigned to Priority Group 8, your eligibility for VA health care benefits will depend on the subpriority group you're placed in.

You may be eligible for VA health care benefits if VA places you in one of these subpriority groups:

Subpriority group a
All of these must be true:
- You have a non-compensable service-connected condition that VA rated as 0% disabling, and
- You enrolled in the VA health care program before January 16, 2003, and
- You have remained enrolled since that date and/or were placed in this subpriority group because your eligibility status changed

Subpriority group b
All of these must be true:
- You have a non-compensable service-connected condition that VA rated as 0% disabling, and
- You enrolled in the VA health care program on or after June 15, 2009, and
- You have income that exceeds current VA or geographical limits by 10% or less

Subpriority group c
All of these must be true:
- You don't have a service-connected condition, and
- You enrolled in the VA health care program as of January 16, 2003, and
- You have remained enrolled since that date and/or were placed in this subpriority group because your eligibility status changed

Subpriority group d
All of these must be true:
- You don't have a service-connected condition, and
- You enrolled in the VA health care program on or after June 15, 2009, and
- You have income that exceeds current VA or geographical limits by 10% or less

You're not eligible for VA health care benefits if VA places you in one of these subpriority groups:

Subpriority group e
All of these must be true:
- You have a non-compensable service-connected condition that VA rated as 0% disabling, and
- You don't meet the criteria for subpriority group a or b above

Note: You're eligible for care for your service-connected condition only.

Subpriority group g
All of these must be true:
- You don't have a service-connected condition, and
- You don't meet the criteria for subpriority group c or d

Once your enrolled in VA health care, your priority group may change if your income changes, or your service-connected disability gets worse, and VA gives you a higher disability rating.
Note: If you're currently enrolled or newly enrolled in VA health care, and you served in a

theater of combat operations after November 11, 1998, and were discharged from active duty on or after September 11, 2001, you're eligible for enhanced benefits for 10 years after discharge. During this time, VA assigns you to priority group 6.

How to Apply for VA Health Care

You'll gather the information below to fill out an Application for Health Benefits (VA Form 10-10EZ). The information you need includes:

- Social Security numbers for you, your spouse, and your qualified dependents
- Your military discharge papers (DD214 or other separation documents)
- Insurance card information for all insurance companies that cover you, including any coverage provided through a spouse or significant other. This includes Medicare, private insurance, or insurance from your employer.
- Gross household income from the previous calendar year for you, your spouse, and your dependents. This includes income from a job and any other sources. Gross household income is your income before taxes and any other deductions.
- Your deductible expenses for the past year. These include certain health care and education costs.

Note: You don't have to tell VA about your income and expenses when you apply, but if you aren't eligible based on other factors, the VA will need this information to make a decision.

After you apply for VA health care, you'll get a letter that will tell you if your application has been approved. If more than a week has passed sent you sent your application and VA hasn't contacted, you don't apply again. Instead, call 877-222-8387.

Basic Health Care Services VA Covers

VA will cover preventative services including:

- Health exams, including gender-specific exams
- Health education, including nutrition education
- Immunization against infectious diseases
- Counseling on genetic diseases that run in families

Inpatient hospital services like the following may be covered:

- Surgeries
- Medical treatments
- Kidney dialysis
- Acute care, which is short-term treatment for severe injuries or illnesses, or after surgery
- Specialized care including organ transplants, care for traumatic injuries, and intensive care for mental and physical conditions

Urgent and emergency care services may be covered. To use VA urgent care services, you'll need to have been receiving care from the VA within the past 24 months.

When someone signs up for VA health care, they become part of the nation's largest integrated health care system. If a veteran needs specialized care or services, they may be referred to other locations.

Health Care Costs

You can get free VA health care for any injury or illness that's service connected. VA also provides certain other services for free, including readjustment counseling and related mental health services, care for issues related to military sexual trauma, and a registry health exam to determine if you're at risk of health problems related to your military service.

You may qualify for additional free VA health care depending on your income, disability rating or other eligibility factors.

You may need to pay a fixed amount for some types of care, tests and medications from a VA care provider or approved community health provider. This is a copayment. Whether or not you have a copay and how much depends on your disability rating, income level, military service record and your priority group.

Services that don't require a copay, no matter your disability rating or priority group are:
- Readjustment counseling and related mental health services
- Counseling and care for issues related to military sexual trauma
- Exams to determine the risk of health problems linked to military service
- Care that may be related to combat service for veterans that served in a theater of combat operations after November 11, 1998
- VA claims exams
- Care related to a VA-rated service-connected disability
- Care for cancer of head or neck caused by nose or throat radium treatments received while in the military
- Individual or group programs to help quit smoking or lose weight
- Care that's part of a VA research project like the Million Veteran Program
- Lab tests
- EKGs or ECGs to check for heart disease or other health problems
- VA health initiatives that are open to the public

VA Health Care and Other Insurance

If you have other forms of health care coverage, such as Medicare, Medicaid, TRICARE or private insurance, you can use your VA health care benefits with these plans. You provide information to the VA about your health insurance coverage, including coverage under a spouse's plan. The VA will have to bill your private health insurance provider for care, supplies or medicine provided to treat illnesses or injuries not related to service. VA doesn't bill Medicare or Medicaid but may bill Medicare supplemental health insurance for covered services.

If your health insurance provider doesn't cover all the non-service-connected care the VA bills them for, you don't have to pay your unpaid balance. Depending on your assigned priority group, you may have a copay for non-service-connected care.

Whether or not you have insurance doesn't affect the VA health care benefits you're eligible for.

Community Care
VA provides health care for veterans from other providers in the local community outside of VA. Veterans may be eligible to receive care from a community provider when VA can't provide the needed care. The care is provided on behalf of and paid for by VA. Community care is available to veterans based on certain conditions and eligibility requirements, and in consideration of the specific circumstances and needs of the veteran. Community care has to be authorized by the VA before the veteran can receive care.

Veterans are charged a copay for nonservice-connected care.

Family Health Benefits
If someone is a spouse, surviving spouse, dependent child, or family caregiver of a veteran or service member, they may qualify for health care benefits. They might also qualify for health care benefits due to a disability related to their veteran's service.

Health care programs that might be available to family members include:

- TRICARE: This program provides comprehensive health coverage including health plans, prescription medicine, dental plans and more to family of active duty, retired or deceased servicemembers, as well as National Guard soldiers, Reservists, and Medical of Honor Recipients.
- CHAMPVA: The CHAMPVA program is available to current or surviving spouses or children of veterans with disabilities or servicemembers who died in the line of duty. If someone doesn't qualify for TRICARE, they may be eligible for insurance through CHAMPVA.
- The Program of Comprehensive Assistance for Family Caregivers: Under the PCAFC program, support, and services for family caregivers for eligible veterans are available. Services include access to health insurance, respite care, and mental health counseling.
- The Camp Lejeune Family Member Program: This program is available to people who lived at Camp Lejeune or Marine Corps Air Station with an active-duty veteran who was their spouse or parent. It's available to people who lived there for at least 30 cumulative days from August 1953 through December 1987. Eligible family members may qualify for health care benefits through VA.
- Spina Bifida Health Care Benefits Program: If someone is the biological child of a Korean or Vietnam War veteran and they've been diagnosed with spina bifida they may qualify for disability benefits, including health care benefits.
- The Children of Women Vietnam Veterans Health Care Benefits Program: People who are biological children of women Vietnam War veterans diagnosed with certain birth defects may qualify for VA health care benefits.
- Pharmacy Benefits: If someone qualifies for CHAMPVA, Spina Bifida or Children of Women Vietnam Veterans programs, they can get prescription benefits from their local pharmacy or through the VA's Meds by Mail program.

VA Travel Pay Reimbursement

VA will pay reimbursement for mileage and other travel expenses to and from approved health care appointments. Health care travel reimbursement covers regular transportation, and approved meals and lodging expenses. A claim can be filed online through the Beneficiary Travel Self Service System, accessible in the Access VA travel claim portal.

As a veteran, you may be eligible for reimbursement if you meet the following requirements:

- You're traveling for care at a VA health facility or for VA-approved care at a non-VA health facility in your community

At least one of these things must also be true:

- You have a disability rating of 30% or higher, or
- You're traveling for treatment of a service-connected condition, even if your VA disability rating is less than 30%, or
- You receive a VA pension, or
- You have an income that's below the maximum annual VA pension rate, or
- You can't afford to pay for your travel, as defined by VA guidelines, or
- You're traveling for one of these reasons: As scheduled VA claim exam also known as a compensation and pension exam, to get a service dog or for VA-approved transplant care.

The VA may pay for transportation and related lodging and meals for non-veterans if the person meets any one of these requirements:

- The person is a family caregiver under the National Caregiver Program traveling to receive caregiving training or to support the veteran's care, or

- The person is a veteran's medically required attendant traveling to support the care of the veteran, or
- The person is the veteran's transplant care donor or support person

The VA may also pay for care for an allied beneficiary when the appropriate foreign government agency authorizes their care, or for the beneficiary of another federal agency when that agency has approved their care.

Before filing a claim, keep all receipts for transportation and approved meals and lodging, and track mileage to and from appointments. The claim must be filed within 30 days of the appointment, or within 30 days of when the veteran becomes eligible for reimbursement. A new claim needs to be filed for every appointment.

Veteran Health Identification Card (VHIC)
A Veteran Health Identification Card or VHIC is a photo ID card that is used to check in at VA health care appointments.

You'll have to be enrolled to get a VHIC. You can contact your nearest VA medical Center and ask to speak to an enrollment coordinator, who will help you arrange to get your picture taken for your card.

You can also use AccessVA to request a VHIC.

Copayment Rates – Current as of January 2022
Urgent Care Copay Rates—Care for Minor Illnesses and Injuries
There's no limit to how many times you can use urgent care. To be eligible for urgent care benefits, including through the VA network of approved community providers, you must be enrolled in the VA health care system, and have received care from VA within the past 25 months. You won't have any copay for a visit when you're only getting a flu shot, no matter your priority group.

2022 Urgent Care Copay Rates

Priority Group	Copay amount for the first 3 visits in each calendar year	Copay amount for each additional visit in the same year
1 to 5	$0 (no copay)	$30
6	If related to a condition that's covered by a special authority: $0 If not related to a condition covered by a special authority: $30 each visit	$30
7 to 8	$30	$30

* Special authorities include conditions related to combat service and exposures (like Agent Orange, active duty at Camp Lejeune, ionizing radiation, Project Shipboard Hazard and Defense (SHAD/Project 112), Southwest Asia Conditions) as well as military sexual trauma, and presumptions applicable to certain Veterans with psychosis and other mental illness.

Outpatient Care Copay Rates—Primary or Specialty Care That Doesn't Require an Overnight Stay
If you have a service-connected disability rating of 10% or higher, you won't need to pay a copay for outpatient care. If you don't have a service-connected disability rating of 10% or

higher, you may need to pay a copay for outpatient care for conditions not related to your military service, at the rates listed below.

2022 Outpatient Care Copay Rates

Type of Outpatient Care	Copay amount for each visit or test
Primary care services (like a visit to your primary care doctor)	$15
Specialty care services (like a visit to a hearing specialist, eye doctor, surgeon, or cardiologist)	$50
Specialty tests (like an MRI or CT scan)	$50

Note: You won't need to pay any copays for X-rays, lab tests, or preventive tests and services like health screenings or immunizations.

Inpatient care copay rates—Care that requires you to stay one or more days in a hospital

If you have a service-connected disability rating of 10% or higher, you won't need to pay a copay for inpatient care. If you're in priority group 7 or 8, you'll either pay the full copay rate or a reduced copay rate. If you live in a high-cost area you may qualify for a reduced inpatient copay rate no matter what priority group, you're in. You can find out if you're eligible for a reduced inpatient copay rate by calling 877-222-8387.

2022 Reduced Inpatient Care Copay Rates for Priority Group 7

Length of Stay	Copay Amount
First 90 days of care during a 365-day period	$320 copay + $2 charge per day
Each additional 90 days of care during a 365-day period	$160 copay + $2 charge per day

Note: You may be in priority group 7 and qualify for these rates if you don't meet eligibility requirements for priority groups 1 through 6, but you have a gross household income below our income limits for where you live and you agree to pay copays.

2022 Full Inpatient Care Copay Rates for Priority Group 8

Length of Stay	Copay Amount
First 90 days of care during a 365-day period	$1,600 copay + $10 charge per day
Each additional 90 days of care during a 365-day period	$800 copay + $10 charge per day

Note: You may be in priority group 8 and qualify for these rates if you don't meet eligibility requirements for priority groups 1 through 6, and you have a gross household income above our income limits for where you live, agree to pay copays, and meet other specific enrollment and service-connected eligibility criteria.

Medication Copay Rates

If you're in priority group 1, you won't pay a copay for any medications. You may be in priority group 1 if VA rated your service-connected disability at 50% or more disabling, if VA determined you can't work because of your service-connected disability (called unemployable), or you've received the Medal of Honor.

If you're in priority groups 2 through 8 you'll pay a copay for medications your health care provider prescribes to treat non-service-connected conditions and over-the-counter medications, you get from a VA pharmacy. The cost for any medications you receive while staying in a VA or other approved hospital or health care facility are covered by the inpatient care copay.

The amount you'll pay for the medicines depends on the tier and the amount of medication you get. Once you've been charged $700 in medication copays within a calendar year you won't have to pay any more that year, even if you get more medication.

2022 Outpatient Medication Copay Amounts

Outpatient Medication Tier	1–30-day supply	31–60-day supply	61–90-day supply
Tier 0 Prescription and over-the-counter medicines with no copay	$0	$0	$0
Tier 1 Preferred generic prescription medications	$5	$10	$15
Tier 2 Non-preferred generic prescription medicines and some OTC medicines	$8	$16	$24
Tier 3 Brand-name prescription medicines	$11	$22	$33

Geriatric and extended copay rates

You won't need to make a copay for geriatric care or extended care for the first 21 days of care in a 12-month period. Starting on the 22nd day of care, the VA will base copay on two factors. The first is the level of care and the second is the financial information provided on the Application for Extended Care Services (VA Form 10-10EC).

2022 Geriatric and Extended Care Copay Amounts by Level of Care

Level of Care	Type of Care Included	Copay amount for each day of care
Inpatient care	• Short-term or long-term stays in a community living center (formerly called nursing homes) • Overnight respite care (in-home or onsite care designed to give family caregivers a	Up to $97

	break, available up to 30 days each calendar year) • Overnight geriatric evaluations (evaluations by a team of health care providers to help you and your family decide on a care plan)	
Outpatient care	• Adult day health care (care in your home or at a facility that provides daytime social activities, companionship, recreation, care and support) • Daily respite care • Geriatric evaluations that don't require an overnight stay	Up to $15
Domiciliary care for homeless veterans	• Short-term rehabilitation • Long-term maintenance care	Up to $5

Hardship Determination

If gross household income decreases, a veteran may be eligible for a hardship which may qualify them for copayment exemption for the remaining calendar year and enrollment in a higher priority group.

To request a hardship determination, veterans should send a letter explaining the financial hardship their copayment charges will cause them and a completed Request for Hardship Determination (**VA Form10-10HS**).

Letters and forms can be submitted in person at a local Veteran Affairs Medical Center Business Office or Health Administration Office.
They can also be sent by mail to the Business Office/Health Administration Service at a local VA Medical Center.

Medicare Prescription Benefits

Beginning January 1, 2006, Medicare prescription drug coverage (Medicare Part D) became available to everyone with Medicare Part A or B coverage. The Medicare prescription drug coverage is wholly voluntary on the part of the participant. Each individual must decide whether to participate based on his or her own circumstances.

How This Affects Veterans

Each veteran must decide whether to enroll in a Medicare Part D plan based on his or her own situation. An individual's VA prescription drug coverage will not change based on his or her decision to participate in Medicare Part D. VA prescription drug coverage is considered by Medicare to be at least as good as Medicare Part D coverage. (Therefore, it is considered to be creditable coverage.

Refer to the following section for more information regarding creditable coverage. If an individual's spouse is covered by Medicare, he or she must decide whether to enroll in a Medicare Part D plan regardless of the veteran's decision to participate.

Creditable Coverage

Most entities that currently provide prescription drug coverage to Medicare beneficiaries, including VA, must disclose whether the entity's coverage is "creditable prescription drug coverage."

Enrollment in the VA health care system is creditable coverage. This means that VA prescription drug coverage is at least as good as the Medicare Part D coverage.

Because they have creditable coverage, veterans enrolled in the VA health care program who chose not to enroll in a Medicare Part D plan before May 15, 2006 will not have to pay a higher premium on a permanent basis ("late enrollment penalty") if they enroll in a Medicare drug plan during a later enrollment period.

However, if an individual un-enrolls in VA health care or if he or she loses his or her enrollment status through no fault of his/her own (such as an enrollment decision by VA that would further restrict access to certain Priority Groups), he or she may be subject to the late enrollment penalty unless he or she enrolls in a Medicare Part D plan within 62 days of losing VA coverage.

If a veteran becomes a patient or inmate in an institution of another government agency (for example, state veterans' home, a state institution, a jail, or a corrections facility), he or she may not have creditable coverage from VA while in that institution. For further information, individuals should contact the institution where the veteran resides, the VA Health Benefits Service Center at (877) 222-VETS, or the local VA medical facility.

Covered Services—Standard Benefits

VA's medical benefits package provides the following health care services to all enrolled veterans:

Preventative Care Services
- Immunizations
- Physical Examinations (including eye and hearing examinations)
- Health Care Assessments
- Screening Tests
- Health Education Programs

Ambulatory (Outpatient) Diagnostic and Treatment Services
- Emergency outpatient care in VA facilities
- Medical
- Surgical (including reconstructive/plastic surgery as a result of disease or trauma)
- Chiropractic Care
- Bereavement Counseling
- Mental Health
- Substance Abuse

Hospital (Inpatient) Diagnostic and Treatment
- Emergency inpatient care in VA facilities
- Medical
- Surgical (including reconstructive/plastic surgery as a result of disease or trauma)
- Mental Health
- Substance Abuse

Limited Benefits
The following is a partial listing of acute care services which may have limitations and special eligibility criteria:
- Ambulance Services
- Dental Care (refer to Chapter 11 of this book for additional details.)
- Durable Medical Equipment
- Eyeglasses (see footnote below)
- Hearing Aids (see footnote below)
- Home Health Care
- Homeless Programs
- Maternity and Parturition Services—usually provided in non-VA contracted hospitals at VA expense, care is limited to the mother (costs associated with the care of newborn are not covered)

Non-VA Healthcare Services
- Orthopedic, Prosthetic, and Rehabilitative Devices
- Rehabilitative Services
- Readjustment Counseling
- Sexual Trauma Counseling

Footnote: To qualify for hearing aids or eyeglasses, the individual must have a VA service-connected disability rating of 10% or more. An individual may also qualify if he or she is a former prisoner of war, Purple Heart recipient, require this benefit for treatment of a 0% service-connected condition, or are receiving increased pension based on the need for regular aid and attendance or being permanently housebound.

General Exclusions

The following is a partial listing of general exclusions:
- Abortions and abortion counseling
- Cosmetic surgery except where determined by VA to be medically necessary for reconstructive or psychiatric care
- Gender alteration
- Health club or spa membership, even for rehabilitation
- Drugs, biological, and medical devices not approved by the Food and Drug Administration unless part of formal clinical trial under an approved research program or when prescribed under a compassionate use exemption.
- Medical care for a veteran who is either a patient or inmate in an institution of another government agency if that agency has a duty to provide the care or services.
- Services not ordered and provided by licensed/accredited professional staff
- Special private duty nursing

Emergency Care

Revised Regulations for Emergency Treatment Reimbursement

in 2018, the Department of Veterans Affairs announced through a Federal Registrar notice that there are revised regulations relating to payment or reimbursement for emergency treatment for non-service-connected conditions at non-VA facilities.

The VA announced it would begin processing claims for reimbursement of reasonable costs that were only partially paid by the veteran's other health insurance OHI. These costs may include hospital charges, professional fees, and emergency transportation such as ambulances. Effective January 9, 2018, the VA updated a portion of its regulations in response to an April 2016 U.S. Court of Appeals for Veterans Claims decision stating the VA can no longer deny reimbursement when OHI pays a portion of the treatment expenses.

VA will apply the updated regulations to claims pending with VA on or after April 8, 2016, and to new claims. By law, VA may still not reimburse veterans for the costs of copayments, cost shares and deductibles required by their OHI. VA will work directly with community providers to get additional information needed to review and process these claims. Previous claims do not have to be resubmitted unless requested by the VA.

What Is a Medical Emergency?

A medical emergency is defined as a condition of such nation that it would be expected that a delay in immediate medical attention would be life-threatening. With VA medical care, a veteran may receive emergency care at a non-VA health care facility at VA expense when a VA facility or other federal health care facility with which the VA has an agreement can't furnish economical care due to your distance from the facility, or when the VA is unable to furnish the needed emergency services.

VA Payment for Emergency Care

Since payment may be limited to the point when your condition is stable enough for you to travel to a VA facility, you need to contact the nearest VA medical facility as possible. VA may pay for your non-VA emergency care based on the following situations

If you are service connected:

The VA may pay for your non-VA emergency care for a rated service-connected disability, or for your non-service-connected condition associated with and held to be aggravating for your service- connected condition, or any condition if you are an active participant in the VA Chapter 31 Vocational Rehabilitation program, and you need treatment to make possible your entrance into a course of training or to prevent interruption of a course of training or other approved reason for any condition if you were rated as having a total disability permanent in nature resulting from your service-connected disability

VA Payment for Emergency Care of Non-Service-Connected Conditions Without Prior Authorization

VA may pay for emergency care provided in a non-VA facility for treatment of a non-service-connected condition only if all the following conditions are met:
If you are service-connected, not permanently and totally disabled, or nonservice-connected, then the VA may pay for your non-VA emergency care for treatment of a non-service-connected condition if all of the following conditions are met:

- The episode of care can't be paid for under another VA authority, and
- Based on an average knowledge of health and medicine you reasonably expected that a delay in seeking immediate medical attention for have been hazardous to life or health and
- A VA or other federal facility/provider wasn't feasibly available and
- You receive VA medical care within a 24-month period preceding the non-VA emergency care and

- The services were furnished by an emergency department, or similar facility held out to provide emergency care to the general public and
- You are financially liable to the health care provider for emergency care and
- You have no other coverage under a health plan including Medicare, Medicaid and Worker's Compensation and
- You have no contractual or legal recourse against a third party that would, in whole, extinguish your liability

FAQs About Veteran Emergency Care

What Is an Emergency?

A medical emergency is defined as an injury or illness so severe that without immediate treatment, it threatens your life or health. Your situation is an emergency if you believe your life or health is in danger. If you believe your life or health is in danger, call 911 or go to the ER immediately. You don't need to call the VA before calling an ambulance or going to the ER.

When Should I Contact the VA About an ER Visit?

You, your family, friends, or hospital staff should call the closest VACM as possible, and it's best if this happens within 72 hours of an emergency. The VA will need information about your emergency and the services being provided, and the VA can provide guidance on what charges are covered.

If the Doctor Wants Me to Be Admitted to the Hospital, Do I Need Advance Approval from the VA?

If you are being admitted because of an emergency, an advanced approval isn't required, but prompt VA notification is important.

If I Am Admitted to the Hospital as a Result of An Emergency, How Much Will VA Pay?

Depending on your VA eligibility, VA may pay all, some or none of the charges. Since payment may be limited to the point when your condition is stable enough for you to travel to a VA facility, you need to contact the nearest VA medical facility as soon as possible. An emergency is deemed to have ended at the point when a VA provider has determined that, based on sound medical judgment, you should be transferred from the non-VA facility to a VA medical center.

If you are service-connected the VA may pay for your non-VA emergency care for a rated Service-connected disability, or for your Nonservice-connected condition associated with and held to be aggravating your Service-connected condition, or any condition, if you are an active participant in the VA Chapter 31 Vocational Rehabilitation program, and you need treatment to make possible your entrance into a course of training or to prevent interruption of a course of training or other approved reason or any condition if you are rated as having a total disability permanent in nature resulting from your Service-connected disability.

If you are service-connected, not permanently and totally disabled, or non-service connected, then VA may pay for non-VA emergency care for treatment of a nonservice-connected condition if all the following conditions are met:

- The episode of care can't be paid under another VA authority and
- Based on an average knowledge of health and medicine, you reasonably expected that delay in seeking immediate medical attention would have been hazardous to your life or health and
- A VA or other federal facility or provider was not feasibly available and
- You received VA medical care within a 24-month period preceding the non-VA emergency care and
- The services were furnished by an Emergency Department or similar facility held out to provide emergency care to the general public and
- You have no other coverage under a health plan including Medicare, Medicaid

and Worker's Compensation and

- You have no contractual resource against a third party that would in whole extinguish your liability

Ambulance Transport

The Department of Veterans Affairs may provide or reimburse for land or air ambulance transport of certain eligible veterans in relation to VA care or VA-authorized community care.

VA pays for ambulance transport when the transport has been preauthorized and in certain emergency situations without preauthorization. Two criteria must be met for the VA to pay for ambulance transport.

First, the claimant must meet appropriate administrative eligibility, and a VA provider must determine medical need for ambulance transport. VA must be providing medical care or paying a community care provider for medical care in order to pay for the transport in relation to that care.

Preauthorized Ambulance Transport
Transport is arranged for eligible veterans before inpatient or outpatient care. To qualify a veteran must meet the following administrative requirements:

- Has a single or combined service-connected rating of 30 percent or more, or
- Veteran is in receipt of VA pension, or
- Previous calendar year income does not exceed maximum VA pension rate, or
- Projected income in travel year does not exceed maximum VA pension rate, or
- Projected income in travel year does not exceed maximum VA pension rate, or
- Travel is in connection with care for a service-connected disability, or
- Travel is for a Compensation and Pension exam, or
- Travel is to obtain a service dog, and
- A VA clinician must determine and document that special mode transportation is medically required

Unauthorized Ambulance Transport
Transport must be preauthorized by VA unless it is in relationship to a medical emergency. Veterans do not have to contact VA in advance of a medical emergency and are encouraged to call 911 or go to the nearest medical emergency room.

VA may pay for ambulance transport that's not preauthorized in the following situations:

- Transport from point of community emergency to a VA facility if the veteran meets administrative and eligibility criteria noted under "Preauthorized ambulance transport"
- Transport from point of community emergency to a community care facility if VA pays for the emergency care in the community care facility under the nonservice-connected or service-connected authorities detailed below
- VA is contacted within 72 hours of care at a community care facility and retroactively authorizes the community care, and the veteran meets administrative and medical travel eligibility noted under "preauthorized ambulance transport."

In order for VA to pay for unauthorized ambulance transport, the care associated with the transport must meet one of the following authorities for VA payment:
Emergent care for nonservice-connected conditions (38 United States Code (U.S.C.) 1725 ["Mill Bill"])

- Based on average knowledge of health and medicine, it is reasonably expected that a delay in seeking immediate medical attention would have been hazardous to life or health and

- The episode of care can't be paid under another VA authority and
- A VA or other federal facility/provider was not feasibly available and
- VA medical care was received within 24 months prior to the episode of emergency care and
- The services were furnished by an Emergency Department or similar facility that provides emergency care to the general public and
- Veteran is financial liable for the emergency care and
- Veteran has no other coverage under a health care plan including Medicare, Medicaid or Worker's Compensation, and
- There is no contractual or legal recourse against a third party that could, in whole, extinguish liability

Emergent Care for Service-Connect Conditions (38 U.S.C. 1728)
- Care is for a service-connected disability or
- Care is for a nonservice-connected condition associated with and aggravating a service-connected condition or
- Care is for any condition of an active participant in the VA Chapter 31 Vocational Rehabilitation program and is needed to make a possible entrance into a course of training or to prevent interruption of a course of training, or
- Care is for any condition of a veteran rated as having a total disability permanent in nature resulting from a service-connected disability and
- Based on an average knowledge of health and medicine, it is reasonably expected that delay in seeking immediate medical attention would have been hazardous to life or health and
- A VA or other federal facility/provider was not feasibly available

Retroactive Preauthorization (38 CFR 17.54)
In case of an emergency which existed at the time of treatment, VA may retroactively preauthorize the care if:
- An application for VA payment of care provided is made within 72 hours after the emergency care initiated and
- Veteran meets the eligibility criteria for community care at VA expense of 38 U.S.C. 1703

Reimbursement Considerations
- If the emergency room visit and/or admission meets eligibility for VA reimbursement, and the veteran meets beneficiary travel requirements, the ambulance will be paid from the scene of the incident to the first community care facility providing necessary care
- If a veteran arrives via ambulance but leaves the hospital before being treated by a physician, the ambulance is not guaranteed to be covered by VA regardless of eligibility
- Accepted VA payments are payments in full. Balance due billing of VA or veterans is prohibited. VA pays the authorized amount or not at all.

Documents Needed to Process Claims
In order to consider a claim for VA payment of emergency care provided and associated ambulance transport, VA needs the following documents:
- Documented request or application for VA payment of emergency transportation (usually a Health Care Financing Administration form or a bill). Unless transport is preauthorized, the application must be made within 30 days of transport.
- Ambulance trip report documenting circumstances of medical event and care provided by the ambulance service.
- Invoice from ambulance service and community care provider.
- Community care facility records of care provided to the veteran—VA will request

these from the facility.

All required documents must be received prior to payment consideration. Payment for associated ambulance transport can't occur unless VA is providing or paying for emergency care.

Appeals
If a claim does not meet VA payment criteria (is not payable,) then it's denied and both the community care provider and veteran are provided an explanation of denial and notified of the right to appeal the decision (VA Form 4107, Notification of Rights to Appeal Decision).

Veterans Transportation Program

The Veterans Transportation Service (VTS) provides safe and reliable transportation to veterans who require assistance traveling to and from VA health care facilities and authorized non-VA health care appointments. VTS also partners with service providers in local communities to serve veteran's transportation needs. Partners include Veterans Service Organizations (VSOs), local and national non-profit groups, and federal, state and local transportation services.

VA recognizes Veterans who are visually impaired, elderly or immobilized due to disease or disability, and particularly those living in remote and rural areas face challenges traveling to their VA health care appointments. Veterans Transportation Service (VTS) is working to establish Mobility Managers at each local VA facility to help Veterans meet their transportation needs.

VTS has established a network of transportation options for Veterans through joint efforts with VA's Office of Rural Health and organizations, such as Veterans Service Organizations (VSOs); community transportation providers; federal, state and local government transportation agencies; non-profits and Veterans Transportation Community Living Initiative (VTCLI) grantees. Veterans who are eligible for VA health care benefits and have a VA-authorized appointment are eligible for transportation through the VTS program based on the availability and guidelines in place at their local facility. Each local VA authorized facility has ridership guidelines based on their capabilities.

Highly Rural Transportation Grants

Highly Rural Transportation Grants (HRTG) is a grant-based program that helps veterans in highly rural areas travel to VA or VA-authorized health care facilities. This program provides grant funding to Veteran Service Organizations and State Veterans Service Agencies to provide transportation services in eligible counties. HRTGs provide transportation programs in counties with fewer than seven people per square mile. There is no cost to participate in the program for veterans who live in an area where HRTG is available.

Grantee Organization Name	Details
North Dakota Department of Veterans Affairs	**Eligible Counties**: Adams, Billings, Bottineau, Bowman, Burke, Cavalier, Divide, Dickey, Dunn, Emmons, Foster, Golden Valley, Grant, Griggs, Hettinger, Kidder, Lamoure, Logan, McHenry, McIntosh, McKenzie, McLean, Mountrail, Nelson, Oliver, Pierce, Renville, Sargent, Sheridan, Sioux, Slope, Steele, Towner Call for Transportation Services: 1-800-920-9595
North Dakota-Robert Tovsrud VFW Post 757	**Eligible Counties**: Benson, Eddie, Wells

	Call for Transportation Services: 701-438-2192
South Dakota-American Legion Stanley Post 20	**Eligible Counties:** Clark, Dewey, Hand, Hyde, Haakon, Jerauld, Jones, Kingsbury, Lyman, Miner, Potter, Sandborn, Spink, Stanley, Sully, Ziebach. Call for Transportation Services: 605-945-2360 or Toll Free at 1-877-587-5776 **Eligible Counties:** Butte, Custer, Fall River Call for Transportation Services: 605-642-6668 or Toll Free at 1-877-673-3687
Washington State Department of Veterans Affairs	Eligible County: Skamania Call for Transportation Services: 509-427-3990 Eligible County: Ferry Call for Transportation Services: 1-800-776-9026 or 509-684-2961
Nevada Department of Veterans Services	Eligible County: Elko Call for Transportation Services: 775-777-1428 Eligible County: Nye Call for Transportation Services: 775-572-VETS (8387)
California-VFW Post 8988	Eligible County: Mono, Inyo Call for Transportation Services: 760-873-7850
Texas Veterans Commission	Eligible County: Duval Call for Transportation Services: 361-279-6219 Eligible County: Hansford Call for Transportation Services: 806-659-4100 Eligible County: Kimble Call for Transportation Services: 325-396-4682 Eligible County: Jim Hogg Call for Transportation Services: 361-527-5845 Eligible County: McMullen Call for Transportation Services: 361-279-6219 Eligible County: Briscoe Call for Transportation Services: 806-823-2131 Eligible County: Cochran Call for Transportation Services: 806-266-5508 Eligible County: Kent Call for Transportation Services: 806-237-3373 Eligible County: Menard Call for Transportation Services:

	325-396-4682
Texas-American Legion Post 142	**Eligible Counties:** Baylor, Cottle, Foard, Hardeman Call for Transportation Services: 1-800-633-0852
Texas VFW Post 7207	**Eligible Counties:** Borden, Crane, Glasscock, Loving, Martin, Pecos, Reeves, Terrell Call for Transportation Services: 1-800-245-9028 Eligible counties: Brewster, Culberson, Hudspeth, Jeff Davis, Presidio Call for Transportation Services: 1-855-879-8729
Maine Veterans of Foreign Wars of the U.S.	Eligible County: Piscataquis **Call for Transportation Services:** Bangor Clinic: 207-561-3637 or Lincoln Clinic: 207-403-2012
Montana-American Legion Rocky Boy Post 67	**Eligible Counties:** Blaine, Chouteau, Hill Call for Transportation Services: 406-395-5610
Idaho-American Legion Post 12	Eligible County: Boise Call for Transportation Services: 208-392- 9934 or 9935
Oregon Department of Veterans Affairs	Eligible Counties & the number to call for Transportation Services: Baker, 541-523-6591 Gilliam, 541-384-2252 Grant, 541-575-2721 Malheur, 541-881-0000 Morrow, 541-922-6420 Sherman 541-565-3553 Wallowa, 541-426-3840 Wheeler 541-468-2859
Alaska Department of Military and Veterans Affairs	Interior Alaska Bus Southwest Fairbanks Call for Transportation Services: 1-800-770-6652 Valley People Mover **Eligible County:** Matanuska-Susitna Call for Transportation Services: 907-892-8800 Alaska Marine Highway System **Eligible Counties:** Kodiak Island, Kenai Peninsula Call for Transportation Services: 1-800-642-0066 Inter-Island Ferry **Eligible County:** Prince of Wales-Hyder Call for Transportation Services: 1-866-308-4848

In Vitro Fertilization (IVF)

The VA announced fertility regulations would be expanded to include in vitro fertilization. The new benefit makes IVF available as an option for eligible veterans with service-connected disabilities that result in infertility. The benefit also provides spouses of eligible veterans with access to assisted reproductive technologies, including IVF.

The Office of Women's Health Services (10P4W) is responsible for the contents of this VHA directive. Questions may be referred to the Director of Reproductive Health at 202-461-0373.

This Veterans Health Administration (VHA) directive establishes policy and procedures for the Department of Veterans Affairs (VA) health care systems for the evaluation and treatment of infertility as authorized under the VA medical benefits package for eligible Veterans enrolled in the VA health care system.

NOTE: Except for certain Veterans who have a service-connected disability that results in their inability to procreate without the use of assisted reproductive technology (ART), and their spouses, VA cannot perform or pay for in vitro fertilization (IVF) because it is specifically excluded from the VA medical benefits package (Title 38 Code of Federal Regulations (CFR) 17.38). VA may also provide ART and fertility counseling and treatment that is available under the medical benefits package to spouses of Veterans authorized to receive IVF. AUTHORITY: Title 38 United States Code (U.S.C.) 7301(b), Public Law 114-223 section 260, 38 CFR §§ 17.38, 17.380, and 17.412.

The Office of Women's Health Services (Reproductive Health), in collaboration with experts designated by the National Surgery Office (NSO), will be available to assist with Veterans Integrated Service Network (VISN) level consultations when the medical standards for providing services and treatment related to gamete (sperm or oocyte) cryopreservation or other infertility services are unavailable or unclear. Women's Health Services (WHS) may need to involve Spinal Cord Injury/Disorders (SCI/D) National Program Office, National Center for Ethics in Health Care, and the NSO).

Eligibility

Veterans eligible to receive health care under the medical benefits package may receive VA-covered infertility evaluation, management, and treatment. IVF is specifically excluded from the medical benefits package.

However, ART including IVF is authorized under 38 CFR 17.380 for certain Veterans who have a service-connected disability that results in the inability of the Veteran to procreate without the use of assisted reproductive technology (ART).

The spouse of a Veteran authorized to receive IVF may be provided with fertility counseling and treatment that is available under the medical benefits package, as well as ART, including IVF (38 CFR 17.412). The aforementioned ART/IVF benefit for certain Veterans and their spouse is further delineated in VHA Directive 1334, Assisted Reproductive Technology (ART) Services for the Benefit of Veterans with Service-Connected Illness or Injury Resulting in Infertility, pending publication.

Otherwise, non-Veteran partners (spouses or significant others), if applicable, are not eligible to receive infertility treatment services from VA unless they are eligible for the Civilian Health and Medical Program of the Department of Veterans Affairs (CHAMPVA), which allows VA to provide infertility services and treatment to certain family members of Veterans under 38 CFR 17.270- 278.

In cases where a non-Veteran partner is not eligible for VA-covered infertility services, the Veteran and non-Veteran partner must be informed that payment for such service to the community provider is the responsibility of the Veteran and non-Veteran partner. These requirements must be discussed with the Veteran before any treatment course is undertaken.

Exclusions

The following procedures or services are not covered VA medical benefits: (1) Gestational surrogacy treatment; (2) Costs of obtaining, transporting, and storing donor sperm and oocytes; (3) IVF procedures, except for certain Veterans who have a service-connected disability that results in the inability of the Veteran to procreate without the use of assisted reproductive technology (ART) (38 CFR 17.380). The spouse of a Veteran authorized to receive IVF may be provided with fertility counseling and treatment that is available under the medical benefits package, as well as IVF (38 CFR 17.412). (4) Costs of cryopreservation, storage, and transport of embryo(s), except for certain Veterans who have a service-connected disability that results in the inability of the Veteran to procreate without the use of ART and their non-Veteran spouses. (38 CFR 17.380 and 17.412); and (5) Infertility, evaluation, and management of non-Veteran partners except for the spouse of certain veterans who have a service-connected disability that results in the inability of the veteran to procreate without the use of assisted reproductive technology (ART), including IVF.

What to Remember
✓ You may be eligible for VA health care benefits if you served in the active military, naval or air service and didn't receive a dishonorable discharge.
✓ If you enlisted after September 7, 1980, or entered active duty after October 16, 1981, you must have served 24 continuous months for the full period for which you were called to active duty unless certain things are true for you, such as you were discharged for a disability caused or made worse by your active-duty service.
✓ You may qualify for enhanced eligibility status, meaning you're placed in a higher priority group if you meet certain requirements such as receiving compensation from VA for a service-connected disability.
✓ When you apply for VA health care, you'll be assigned 1 of 8 priority groups. Your priority group may affect how soon the VA signs you up for health care benefits.
✓ Your priority group can also affect how much you have to pay toward the cost of your care, if anything.

CHAPTER 12

NURSING HOME AND LONG-TERM CARE BENEFITS

Veterans may be able to get assisted living, residential or home health care through VA. The care is available in different settings, some of which are run by VA and others are run by state or community organizations the VA inspects and approves.

Standard Benefits

The following long-term care services are available to all enrolled veterans:

Geriatric Evaluation
A geriatric evaluation is the comprehensive assessment of a veteran's ability to care for him/herself, his/her physical health, and the social environment, which leads to a plan of care. The plan could include treatment, rehabilitation, health promotion, and social services. These evaluations are performed by inpatient Geriatric Evaluation and Management (GEM) Units, GEM clinics, geriatric primary care clinics, and other outpatient settings.

Adult Day Health Care
The adult day health care (ADHC) program is a therapeutic daycare program, providing medical and rehabilitation services to disabled veterans in a combined setting.

Respite Care
Respite care provides supportive care to veterans on a short-term basis to give the caregiver a planned period of relief from the physical and emotional demands associated with providing care. Respite care can be provided in the home or other noninstitutionalized settings.

Home Care
Skilled home care is provided by VA and contract agencies to veterans that are homebound with chronic diseases and include nursing, physical/occupational therapy, and social services.

Hospice/Palliative Care
Hospice/palliative care programs offer pain management, symptom control, and other medical services to terminally ill veterans or veterans in the late stages of the chronic disease process. Services also include respite care as well as bereavement counseling to family members.

Financial Assessment for Long-Term Care Services

For veterans who are not automatically exempt from making copayments for long-term care services, a separate financial assessment must be completed to determine whether they qualify for cost-free services or to what extent they are required to make long term care copayments. For those veterans who do not qualify for cost-free services, the financial assessment for long term care services is used to determine the copayment requirement. Unlike copayments for other VA health care services, which are based on fixed charges for all, long-term care copayment charges are individually adjusted based on each veteran's financial status.

Limited Benefits

Nursing Home Care
While some veterans qualify for indefinite nursing home care services, other veterans may qualify for a limited period of time. Among those that automatically qualify for indefinite nursing home care are veterans whose service-connected condition is clinically determined to require nursing home care and veterans with a service connected rating of 70% or more. Other veterans—with priority given to those with service-connected conditions—may be provided short-term nursing home care if space and resources are available.

The Department of Veterans Affairs (VA) provides both short-term and long-term care in nursing homes to veterans who aren't sick enough to be in the hospital but are too disabled or elderly to take care of themselves. Priority is given to veterans with service- connected disabilities.

Priority Groups
The VA is required to provide nursing home care to any veteran who
- Needs nursing home care because of a service-connected disability
- Has a combined disability rating of 70% or more, or
- Has a disability rating of at least 60% and is:
  - Deemed unemployable, or
  - Has been rated permanently and totally disabled.

Other veterans in need of nursing care will be provided services if resources are available after the above groups are taken care of.

Types of Nursing Care Available

Community Living Centers
Some VA Medical Centers have Community Living Centers (these used to be called Nursing Home Care Units or VA Nursing Homes). These centers are typically located within the VA Medical Center itself or in a separate building.

Contract Nursing Home Care
Nursing home care in public or private nursing homes is also available to some veterans. Stays in these nursing homes can be limited; however, for veterans with ratings less than 70% and for veterans who do not need care due to a service-connected disability.

State Veterans Homes
State Veterans Homes are nursing homes run by the state and approved by the VA. Sometimes the VA will pay for part of the care a veteran gets at a state veterans' home.

Eligibility for Community Living Centers (CLCS)

To receive care in a Community Living Center/VA nursing home, a veteran must:
- Be enrolled in the VA Health Care System
- Be psychiatrically and medically stable.

- Provide documentation specifying whether short or long-term care is needed, an estimation of how long the stay will be, and when discharge occurs, and
- Show priority for a stay in aCLC.

However, meeting the above criteria does not automatically ensure admission. CLCs make decisions about whether to admit a veteran based on the following factors:

- Availability of services in theCLC
- What sort of care the veteran needs,and
- Whether the CLC can competently provide the type of care the veteran needs.

Co-Pays
Veterans required to make co-pays are typically those:

- Without a service-connected disability rated at least 10%, and
- Whose income is higher than the VA's maximum annual pension rate.

How to Apply for CLC Care
Typically, a veteran's physician will submit the application requesting care in a CLC. Veterans who are not exempt from co-pays must complete VA Form 10-10EC, Application for Extended CareServices.

Eligibility for Contract Nursing Home Care
Any veteran who needs Contract Nursing Home Care for a service-connected disability or is receiving VA home health care after discharge from a VA hospital is eligible for direct admission. To be admitted, all that is required is for a VA physician or authorized private physician to determine that nursing home care is needed. Veterans rated 70%, or more service-connected should also be eligible.

Other veterans are eligible to be transferred into Contract Nursing Home Care (also called a Community Nursing Home) if the VA determines the care is needed and:

- The veteran is in a VA hospital, nursing home, domiciliary, or has been receiving VA outpatient care, or
- An active member of the Armed Forces who was in a DOD hospital needs nursing care and will be an eligible veteran upon discharge.

Time Limits for Contract Nursing Home Care
Veterans who are not in the priority groups are technically limited to six months of care, but this may be reduced to 30 to 60 days if resources are limited. Veterans in the priority groups are technically entitled to unlimited free care but again may receive shorter stays due to a lack of funding and resources to accommodate them.

Many veterans can extend their stay by relying on payments from Medicare and Medicaid.

How to Apply for Contract Nursing Home Care
Typically, an application will be made by a veterans' doctor, social worker or nurse, using VA Form 10-0415, Geriatrics and Extended Care (GEC) Referral.

Eligibility for State Veterans Homes
In some cases, the VA will help pay for a veteran's care at a State Veterans Home. The payments the VA will make are called per diem aid. A home must meet the VA standards for nursing home care to receive per diem aid. In addition, the VA will not pay more than half the cost of the veteran's care.

State homes provide hospital care, nursing home care, domiciliary care, and sometimes adult daycare. To receive per diem aid, veterans must meet VA eligibility requirements for the type of care they will receive.

States usually have their own eligibility requirements, in addition to the VA's requirements, such as residency requirements. The veterans' home will apply for VA aid for a veteran's care by submitting VA Form 10-10EZ, Application for Medical Benefits. The VA will pay per diem aid for a veteran's care indefinitely.

Basic State Home Per Diem Rates

The Basic State Home Per Diem Rates for Fiscal Year (FY) 2023 are as follows:
Adult Day Health Care: $101.32
Domiciliary: $54.89
Nursing Homes: $127.17

Reduction in Pension

Congress establishes the maximum annual Veterans Pension rates. Payments are reduced by the amount of countable income of the veteran, spouse, and dependent children. When a veteran without a spouse or a child is furnished nursing home or domiciliary care by VA, the pension is reduced to an amount not to exceed $90 per month after three calendar months of care. The reduction may be delayed if nursing-home care is being continued to provide the veteran with rehabilitation services.

Domiciliary Care

Domiciliary care provides rehabilitative and long-term, health maintenance care for veterans who require some medical care, but who do not require all the services provided in nursing homes. Domiciliary care emphasizes rehabilitation and return to the community. VA may provide domiciliary care to veterans whose annual income does not exceed the maximum annual rate of VA pension or to veterans who have no adequate means of support.

Services for Blind Veterans

Veterans with corrected central vision of 20/200 or less in both eyes, or field loss to 20 degrees or less in both eyes are considered to be blind.

Blind veterans may be eligible for many of the benefits detailed throughout this book, including, but not limited to: Disability Compensation, Health Insurance, Adaptive Equipment, and Training & Rehabilitation. In addition to these benefits, there are a number of miscellaneous benefits due veterans of all wars who were blinded as the result of their war service. Many individual states offer special programs and benefits for the blind.

Services are available at all VA medical facilities through Visual Impairment Services Team (VIST) coordinators.

The VIST Coordinator is a case manager who has major responsibility for the coordination of all services for legally blind veterans and their families. Duties include providing and/or arranging for appropriate treatment, identifying new cases of blindness, providing professional counseling, resolving problems, arranging annual healthcare reviews, and conducting education programs relating to blindness.

Blind veterans may be eligible for services at a VA medical center or for admission to a VA blind rehabilitation center or clinic. In addition, blind veterans entitled to receive disability compensation may receive VA aids for the blind, which may include:
- A total health and benefits review by a VA Visual Impairment Services team.
- Adjustment to blindness training;

- Home Improvements and Structural Alterations to homes (HISA Program);
- Specially adapted housing and adaptations;
- Low-vision aids and training in their use;
- Electronic and mechanical aids for the blind, including adaptive computers and computer-assisted devices;
- Guide dogs, including the expense of training the veteran to use the dog, and the cost of the dog's medical care;
- Talking books, tapes, and Braille literature, provided from the Library of Congress.

Guide Dogs/Service Dogs

As previously mentioned, VA may provide guide dogs to blind veterans. Additionally, Public Law 107-135 (signed by then-President Bush January 23, 2002) states that VA may provide:

- Service dogs trained for the aid of the hearing impaired to veterans who are hearing impaired, and are enrolled under Section 1705 of Title 38; and
- Service dogs trained for the aid of persons with spinal cord injury or dysfunction or other chronic impairment that substantially limits mobility to veterans with such injury, dysfunction, or impairment who are enrolled under section 1705 of Title 38.
- VA may also pay travel and incidental expenses for the veteran to travel to and from the veteran's home while becoming adjusted to the dog.

The VA operates nine blind rehabilitation centers in the United States and Puerto Rico. Rehabilitation centers offer comprehensive programs to guide individuals through a process that eventually leads to maximum adjustment to the disability, reorganization of the person's life, and return to a contributing place in the family and community. To achieve these goals, the rehabilitation centers offer a variety of skill courses to veterans, which are designed to help achieve a realistic level of independence. Services offered at rehabilitation centers include:

- Orientation and mobility;
- Living skills;
- Communication skills;
- Activities of daily living;
- Independent daily living program;
- Manual skills;
- Visual skills;
- Computer Access Training Section;
- Physical conditioning;
- Recreation;
- Adjustment to blindness;
- Group meetings.

The VA also employs Blind Rehabilitation Outpatient Specialists (BROS) in several areas, including:

- Albuquerque, NM
- Ann Arbor, MI
- Bay Pines / St. Petersburg, FL
- Baltimore, MD
- Boston, MA
- Cleveland, OH
- Dallas, TX
- Gainesville, FL
- Los Angeles, CA
- Phoenix, AZ
- Portland, OR

- San Antonio, TX
- San Juan, PR
- Seattle, WA
- West Haven, CT

Home Improvements and Structural Alterations (HISA)

The HISA program provides funding for disabled veterans to make home improvements necessary for the continuation of treatment or for disability access to the home, essential lavatory and sanitary facilities.

Disabled veterans may be eligible for HISA when it is determined medically necessary or appropriate for the effective and economical treatment of the service-connected disability.

Veterans who are rated 50% or higher are not required to prove their service-connected disability is the reason for their HISA request, although veterans rated less than 50% must demonstrate their disability is the cause of the HISA improvements.

Lifetime home improvement benefits may not exceed $6,800 (U.S.). Any costs that exceed $6,800 (U.S.) will be the veteran's responsibility. This amount is provided to veterans with a service-connected condition or a non-service-connected condition of a veteran rated 50 percent or more service-connected. Home improvement benefits of up to $2,000 may be provided to all other veterans registered in the VA healthcare system.

To apply, veterans must submit the following documentation:
- A letter from describing the physical disability and a description of the request home improvement and/or structural alteration;
- A statement from the attending physician or therapist. This statement should describe the physical condition/disability, why the home modification(s) are medically necessary or appropriate for the effective and economical treatment of the service-connected disabling condition;
- A drawing of the work to be undertaken. This drawing can be hand sketched. It should include the height, width and length dimensions. It does not have to be a formal blueprint or architectural drawing;
- A completed, signed home ownership or a rental lease form with a letter from the landlord agreeing with the modifications if required;
- A completed and signed "VA Form 10-0103", Veterans Application for Assistance;
- Documentation verifying that the provider/contractor is licensed/bonded;
- An itemized cost estimate of the proposed improvement and/or structural alteration from the provider/contractor.

Preauthorization must be obtained before beginning any alterations, otherwise, HISA benefits will be denied.

VA Caregiver Support

Please see Chapter 29 for more information about VA caregiver benefits.

What to Remember
✓ Veterans may be able to get assisted living, live-in or home health care through VA.
✓ To access the services, a veteran has to be signed up for VA healthcare. VA also has to conclude the veteran needs the specific service to help with ongoing treatment and personal care.
✓ Factors such as service-connected disability status and insurance coverage may be considered.
✓ VA may cover some of the services under standard health benefits if the veteran is signed up for VA health care. The veteran may still be responsible for some covered services.
✓ Veterans can get nursing home care in community living centers, community nursing homes or state veterans' homes.

CHAPTER 13

DENTAL BENEFITS

If you qualify for VA dental care benefits, you may be able to get some or all of your dental care through VA. Whether or not you can get VA dental care benefits for some, or all of your dental care depends on factors like your military service history, your current health, and your living situation. Based on this information, you are placed in a benefits class. You get specific benefits assigned to that class.

VA Outpatient Dental Benefits

Eligibility and benefits for veteran dental care are restricted by law and categorized into VA dental classifications or classes. If you just got out of the service, you may be entitled to a one-time course of dental care. You must apply for dental care within 180 days of your discharge, and your discharge must be under conditions under than dishonorable, from a period of active duty of 90 days or more. If not eligible for VA Dental Care, the national VA Dental Insurance Program gives enrolled veterans and CHAMPVA beneficiaries the opportunity to purchase dental insurance at a reduced cost.

The eligibility for outpatient dental care isn't the same as for most other VA medical benefits and is categorized into classes. If you are eligible for VA dental care under Class I, IIA, IIC, or IV, you are eligible for any necessary dental care to maintain or restore oral health and masticatory function, including repeat care. Other classes have time and/or service limitations.

If You:	You Are Eligible For:	Through:
Have a service-connected compensable dental disability or condition	Any needed dental care	Class I
Are a former prisoner of war	Any needed dental care	Class IIC
Have service-connected disabilities rated 100% (total) disabling or are unemployable and paid at the 100% rate due to service-connected conditions	Any needed dental care (Note: veterans paid at the 100% rate based on a temporary rating are not eligible for comprehensive outpatient dental services based on this temporary rating)	Class IV
Request dental care within 180 days of discharge or release under conditions other than dishonorable from a period of active duty of 90 days or more	One-time dental care if your DD214 certificate of discharge does not indicate a complete dental exam and all appropriate treatment had been rendered prior to discharge	Class II

Have a service-connected noncompensable (0%) dental condition/disability resulting from combat wounds or service trauma	Any dental care necessary to provide and maintain a functioning dentition. A VA Regional Office Rating Decision Letter (VA Form 10-7131) or the historical Dental Trauma Rating (VA Form 10-564-D) identifies the tooth/teeth/conditions that are trauma related	Class IIA
Have a dental condition clinically determined by VA to be associated with and aggravating a service-connected medical condition	Dental care to treat the oral conditions that are determined by a VA dental professional to have a direct and material detrimental effect to your service-connected medical condition	Class III
Are actively engaged in a Title 38, USC Chapter 31 Vocational Rehabilitation and Employment Program	Dental care to the extent determined by a VA dental professional to meet certain requirements.	Class V
Are receiving VA care or are scheduled for inpatient care and require dental care for a condition complicating a medical condition currently under treatment	Dental care to the treat the oral conditions that are determined by a VA dental professional to complicate your medical condition currently under treatment	Class VI
Are enrolled in a qualifying VA sponsored homeless residential rehabilitation program for at least 60 days	A one-time course of dental care that is determined medically necessary for certain reasons	Class IIB

VA Dental Insurance Program (VADIP)

If you can't get VA dental care benefits, you may be able to buy dental insurance at a reduced cost through the VA Dental Insurance Program (VADIP). You have to meet at least one of two requirements. The first is that you're signed up for VA health care, or you are signed up for CHAMPVA.

Insurance carriers may offer separate coverage options for dependents who aren't CHAMPVA beneficiaries.

VADIP provides coverage throughout the U.S. and its territories including Puerto Rico, Guam, the U.S. Virgin Islands, American Samoa, and the Commonwealth of the Northern Mariana Islands.

If you're not eligible for free VA dental care, VADIP can help you buy private dental insurance at a reduced cost. If you're eligible for free VA care for some of your dental needs, you can buy a VADIP plan if you want added dental insurance. If you sign up for VADIP, it doesn't affect your ability to get free dental care.

VADIP plans cover common dental procedures, which may include:
- Diagnostic services
- Preventive dental care
- Root canals and other services to manage oral health problems and restore function (called endodontic or restorative services)
- Dental surgery
- Emergency dental care

The costs of coverage depend on the insurance company and plan selected.

Based on your plan, you'll pay the full insurance premium for each individual on your plan, and any required copays when you get care.

You can choose between a Delta Dental Plan or a MetLife Plan, and you enroll online with the company. Once you enroll you can use your insurance provider's website to manage your plan and benefits.

If you participated in the VADIP pilot program, that ended in 2017, so you need to enroll again to get new coverage.

TRICARE Retiree Dental Program

RDP ended in 2018 and was replaced by the Office of Personnel Management's Federal Employees Dental and Vision Insurance Program (FEDVIP). If you were eligible for TRDP, you might be able to enroll in a FEDVIP dental plan.

You must enroll in FEDVIP dental during the 2020 Open Season or after a FEDVIP Qualifying Life Event.

FEDVIP is a voluntary, enrollee-pay-all dental and vision program available to Federal employees and annuitants and certain uniformed service members. It is sponsored by the U.S. Office of Personnel Management (OPM) and offers eligible participants a choice between ten dental and four vision carriers.

BENEFEDS is the government-authorized and OPM-sponsored enrollment portal that eligible participants use to enroll in and manage their FEDVIP coverage. BENEFEDS also manages the billing systems and customer service functions necessary for the collection of FEDVIP premiums.

FEDVIP dental plans provide comprehensive dental coverage, including in-network preventive services covered 100%; no deductibles when using in-network dentists; no waiting period for major services such as crowns, bridges, dentures, and implants; and no 12-month waiting period or age limit for orthodontic coverage under some plans.

TRICARE Selected Reserve Dental Program Benefits

Individuals with at least one year of service commitment remaining who are serving in the Army Reserve, Naval Reserve, Air Force Reserve, Marine Corps Reserve, Coast Guard Reserve, Army National Guard or Air National Guard, may be eligible to enroll in the Tricare Selected Reserve Dental Program (TSRDP).

The Department of Defense works in conjunction with Humana Military Healthcare Services to offer and administer the TRICARE Selected Reserve Dental Program.

Coverage remains available as long as an individual maintains his or her Reserve status and is shown as eligible on the DEERS record.

The information provided in this section is intended only as a brief overview. Humana Military Healthcare Services has a staff of trained Beneficiary Services Representatives who are available to answer your questions.

TRICARE Dental Program Survivor Benefit Plan

If your sponsor died while serving on active duty, you may qualify for the TRICARE Dental Program Survivor Benefit Plan.

This includes 100% payment of monthly premiums, and TRICARE pays for cost shares for covered service.

If you were using the TRICARE Dental Program when your sponsor died, you're automatically transferred to the Survivor Benefit Plan. If not, you can enroll at any time. After three years, surviving spouses lose eligibility for the TRICARE Dental Program. They can purchase the TRICARE Retiree Dental Program only if their sponsor died while on active duty for more than 30 days.

Surviving children can remain enrolled in the TRICARE Dental Plan until they lose TRICARE eligibility for other reasons.

What to Remember
✓ To be eligible for VA dental care, your discharge status must be a condition other than dishonorable unless an exception is made by a VA Regional Office.
✓ Dental care is categorized into classes.
✓ Class I is Service-Connected Dental Disability.
✓ Class IIC is Former Prisoner of War
✓ Class IV is for someone who receives service-connected disability compensation
✓ If you aren't eligible for free VA dental care, VADIP can help you buy private dental insurance at a reduced cost.
✓ If you're eligible for free VA dental care for some of your needs, you can buy a VADIP plan if you want additional dental insurance.

CHAPTER 14

CHAMPVA

If you are the spouse, surviving spouse or child of a veteran with disability or a veteran who died and you don't qualify for TRICARE, you may be eligible for health insurance through the Civilian Health and Medical Program of the Department of Veterans Affairs (CHAMPVA).

CHAMPVA Eligibility

You may be eligible for health care through CHAMPVA if you don't qualify for TRICARE and at least one of the below is true for you:

At least one of the following must be true:
- You're the spouse or child of a veteran who's been rated permanently and totally disabled for a service-connected disability by a regional VA office, or
- You're the surviving spouse or child of a veteran who died from a VA-rated service-connected disability, or
- You're the surviving spouse or child of a veteran who was at the time of the death rated permanently and totally disabled from a service-connected disability, or
- You're the surviving spouse or child of a servicemember who died in the line of duty, not due to misconduct—in most of these cases, family members qualify for TRICARE rather than CHAMPVA.

A service-connected disability is one that the VA concludes is caused or made worse by the veteran's active-duty service. A permanent disability isn't expected to improve.

A veteran who's the qualifying CHAMPVA sponsor for their family may also qualify for the VA health care program based on their own veteran status. If two spouses are both veterans who qualify as CHAMPVA sponsors for their family, they both may now qualify for CHAMPVA benefits. Each time they need care, they may choose to get it through the VA health care program or using their CHAMPVA coverage.

CHAMPVA Benefits

With CHAMPVA, you're covered for services and supplies when the VA determines they are medically necessary and were received from an authorized provider. When providers perform services within the scope of their license or certification, the VA considers them to be authorized.

Covered services under CHAMPVA include:
- Ambulance service
- Ambulatory surgery
- Durable medical equipment
- Family planning and maternity
- Hospice

- Inpatient services
- Mental health services
- Outpatient services
- Prescription medicines
- Skilled nursing care
- Transplants

How to Get CHAMPVA Benefits

To apply for CHAMPVA benefits you'll need the Application for CHAMPVA Benefits (VA Form 10-10d) and Other Health Insurance Certification (VA Form 10-7959c).

You'll also need documents related to your Medicare status. If you qualify for Medicare for any reason, you'll need to submit a copy of your Medicare card. If you're 65 years or older and don't qualify for Medicare, you'll need to send the VA documentation from the Social Security Administration confirming you don't qualify for Medicare benefits under anyone else's Social Security number.

To speed up the process, you can also send copies of optional documents. Optional documents include the page from the VBA rating decision showing your veteran is permanently and totally disabled, or the death rating if you're a survivor.

You can also send a copy of the veteran's DD214 (certificate of Release or Discharge from Active Duty)- or if the veteran was a World War II or Korean War Veteran, the Report of Separation. If you don't' have a copy of the necessary form you can request it by submitting a Standard Form 180, Request Pertaining to Military Records from the National Archives.

Documents related to any dependent children you're including in your application might be a copy of each child's birth certificate or adoption papers, or school certification of full-time enrollment for children ages 18-23.

If you're a surviving spouse who remarried but is once again single, you should send a copy of the document that legally ended your marriage, including a divorce decree, death certificate or annulment decree.

You should mail the application to:
VHA Office of Community Care
CHAMPVA Eligibility
POX Box 469028
Denver, CO 80246-9028

Exclusions

The following are services not covered under CHAMPVA:

- Acupuncture
- Chiropractic services
- Most dental care
- Non-FDA approved drugs
- Routine hearing exams
- Routine eye exams and glasses
- Laser eye surgery
- Health club memberships
- Experimental and investigational procedures

Key Facts About CHAMPVA

- In most cases, CHAMPVA's allowable amount, which is what the VA pays for specific services and supplies is equal to Medicare/TRICARE rates

- CHAMPVA has an outpatient deductible of $50 per beneficiary per calendar year or a maximum of $100 per family per calendar year and a patient cost share of 25% of the allowable amount up the catastrophic cap which is $3,000 per calendar year.

- If a patient has other health insurance, CHAMPVA pays the lesser of either 75% of the allowable amount after the $50 calendar year deductible is satisfied or the remainder of the charges and the beneficiary will normally have no cost share.

- Although similar, CHAMPVA is a separate program with a different beneficiary population than TRICARE. The programs are administered separately with differences in preauthorization requirements and claim filing procedures.

- Under the CHAMPVA In-House Treatment initiative (CITI_, CHAMPVA beneficiaries may receive cost-free health care services at participating VA facilities.

- Generally, applicants can expect to receive written notification from the VHA Office of Community Care within 45 days after mailing their application. To streamline and speed-up the process, applicants are encouraged to complete the Application for CHAMPVA Benefits in the entirety and attach all required documents.

What to Remember
✓ You may be eligible for health care through CHAMPVA if you don't qualify for TRICARE and at least one of a certain set of criteria is true if you're the surviving spouse of or a child of a veteran with disabilities or a veteran who died.
✓ A veteran who's the qualifying CHAMPVA sponsor for their family may also qualify for the VA health care program based on their own veteran status.
✓ To apply for CHAMPVA benefits you need to complete the Application for CHAMPVA Benefits (VA Form 10-10d)
✓ You also need to complete Other Health Insurance Certification (VA Form 10-7959c)
✓ You'll need to send the page from the VA rating decision showing the veteran is permanently and totally disabled, or the death rating if you're a survivor. You'll also need to send the veteran's DD214 and documents relating to any dependent children you're including in the application.

CHAPTER 15

TRICARE AND TRICARE FOR LIFE

TRICARE is the health care program from the Department of Defense. TRICARE serves 9.6 million active-duty service members, retired service members, National Guard and Reserve members, family members, and survivors. When you're a TRICARE beneficiary, you have access to health care wherever you are.

TRICARE brings together military hospitals and clinics with a network of civilian providers to offer medical, pharmacy and dental options.

TRICARE partners with civilian regional contractors to administer the benefit in two U.S. regions—East and West, as well as one overseas region. Your regional contractor is your go-to resource for information and help.

TRICARE meets the minimum essential coverage requirement under the Affordable Care Act.

Eligibility for TRICARE is determined by the services and shown in the Defense Enrollment Eligibility Reporting System (DEERS). DEERS is a database of service members and dependent worldwide who are eligible for military benefits. To use TRICARE, first make sure your DEERS record is up to date.

TRICARE Covered Services

Generally, you have the same covered services, which includes preventative, maternity, pharmacy and mental health services with any TRICARE program option. Copayments or cost-sharing may apply for certain services depending on your program option and beneficiary status.

Your health care options can change if you move, have a life event like getting married or have a status change like a sponsor retiring from service.

Automatic Enrollment

If you're an Active-Duty Service Member, a family member of a new ADSM, a new family member of a current ADSM or your military sponsor has been called to active duty, you're automatically enrolled in TRICARE Prime if you live in a Prime Service Area (PSA). Otherwise, active-duty family members are automatically enrolled in TRICARE Select.

Active-duty service members must remain enrolled in TRICARE Prime. All others automatically enrolled have up to 90 days to change enrollment if eligible for other TRICARE plans.

TRICARE Prime

TRICARE Prime is a health care option for active-duty service members (ADSMs), retirees, family members and certain others. TRICARE Prime is similar to a managed care or health maintenance organization option, meaning your access to specialty care is managed by your primary care manager (PCM).

Other TRICARE Prime options include:
- TPR which is a TRICARE Prime option for ADSMs living in remote locations outside of a PSA and their family members and USFHP, which is a TRICARE Prime option where care is provided through networks of community-based not-for-profit health care systems in six areas of the United States.

To get TRICARE Prime, you must live in a PSA. You may also get TRICARE Prime if you live within 100 miles of an available PCM and waive your drive-time access standards.

Enrolling in TRICARE Prime
- ADSMs must use TRICARE Prime or TPR
- ADFMs can choose to enroll in TRICARE Prime, TPR, USFHP or Tricare Select
- Retirees and retiree family members may enroll in TRICARE Prime or USFHP. If neither is available, you can enroll in TRICARE Select.
- You can only enroll in or change enrollment to TRICARE Prime (if you live in a PSA) following a Qualifying Life Event (QLE) or during the annual fall TRICARE Open Season.

You have three options to enroll in a TRICARE Prime program:
- Log into milConnect (https://milconnect.dmdc.osd.mil/milconnect/) and click on the "Manage Health Benefits" button
- Call your regional contractor
- Submit a Tricare Prime Enrollment, Disenrollment and Primary Care Manager (PCM) Change Form (DD Form 2876) to your regional contractor.

When you enroll in a TRICARE Prime option, you'll get most of your routine care from an assigned or selected PCM. Your PCM may be a military hospital or clinic, a civilian TRICARE network provider, or a primary care provider under USFHP. Referrals and pre-authorizations may be required for some services.

ADSMs, ADFMs and transitional survivors (surviving spouses during the first three years and surviving dependent children) pay no enrollment fees. Retirees, their families and others pay yearly enrollment fees.

ADSMs have no out-of-pocket costs for covered health care services from a PCM or with the appropriate referral and pre-authorization. ADFMs have no out-of-pocket costs for covered health care services from a network provider in their enrolled TRICARE region, or with the appropriate referral and pre-authorization.

Retirees pay copayments or cost shares for covered health services from network providers in their enrolled TRICARE region. When following the rules of the TRICARE Prime program option, out-of-pocket costs are limited to the catastrophic cap amount for that calendar year.

Point-of-Service Option

The Point-of-Service option (POS) allows non ADSMs to see any TRICARE-authorized provider without a referral. This means you pay more up front to get nonemergency health care from any TRICARE-authorized provider without a referral. The costs you pay under this option don't count toward your yearly catastrophic cap.

TRICARE Prime Remote

TRICARE Prime Remote (TPR) is a managed care option available in remote areas in the United States. By law, you can only use TPR if both your sponsor's home and work addresses are more than 50 miles or one hour's drive time from a military hospital or clinic. Enrollment is required, and there are no enrollment fees.

You may be reimbursed for reasonable travel expenses under the Prime Travel Benefit if you meet certain requirements.

TPR is available in designated remote locations in the U.S. for:

- Active-duty service members
- Activated National Guard and Reserve members called or ordered to active-duty service for more than 30 days in a row
- Pre-activation, pre-mobilization, pre-deployment TDY orders aren't eligible
- Active-duty family members who live with TPR-enrolled sponsor
- Activated National Guard/Reserve family members if they live in a designated remote location when the sponsor is activated and continued to reside at that address

Under this plan, you'll get most care from your primary care manager. You may have a network PCM is available. If not, you select any TRICARE-authorized provider as your PCM.

Your PCM refers you to specialists for care they can't provide and works with your regional contractor for referrals/authorization. Your PCM also helps find a specialist in the network and files claims for you.

You don't pay enrollment fees. You don't pay out-of-pocket costs for any type of care as long as the care is from your PCM or with a referral. Care received without a referral is subject to point of service fees.

TRICARE Prime Overseas

TRICARE Prime Overseas is a managed care option in overseas areas near military hospitals and clinics. Enrollment is required and there are no enrollment fees.

Active-duty service members, command sponsored active-duty family members and activated National Guard/Reserve members can participate. Command-sponsored family members of activated National Guard/Reserve members can also participate. Retirees and their families can't enroll in TRICARE Prime Overseas.

You'll get most of your care from your assigned primary care manager (PCM) at a military hospital or clinic. Your PCM refers you to a specialist for care he or she can't provide and works with International SOS for authorization when needed.

You don't pay enrollment fees. You don't pay copayments for any type of care as long as it's received from your PCM or with a referral. Care you receive without a referral is subject to point-of-service fees.

TRICARE Prime Remote Overseas

TRICARE Prime Remote Overseas is a managed care option in designated remote overseas location including Eurasia-Africa, Latin America and Canada and Pacific.

Active-duty service members can enroll as can command-sponsored active-duty family members. Activated National Guard/Reserve members can enroll, and so can command-sponsored family members of activated National Guard/Reserve members.

Retirees and their families can't enroll in TRICARE Prime Remote Overseas.

Your assigned primary care manager refers you to specialists for care they can't provide and works with International SOS for authorization when needed.

When you enroll in RICARE Prime Remote Overseas, you can call an overseas point of contact for help (POC). They can help you enroll, schedule appointments at overseas network facilities and file your medical and dental claims. They can also answer questions about coverage options and benefits and navigate the electronic self-service options.

TRICARE Select Options

TRICARE Select is for TRICARE-eligible beneficiaries who can't, or choose not to, enroll in a TRICARE Prime option and who aren't entitled to Medicare with the exception of ADFMs. This program lets you manage your own health care and get care from any TRICARE-authorized provider without a referral.

Like TRICARE Prime, enrollment is required and ADSMs can't use TRICARE Select.

Enrolling in a TRICARE Select Option

You have to take action to enroll in TRICARE Select options.

- ADFMs, retirees and retiree family members can choose to enroll in Tricare Select
- You can only enroll in or change enrollment to TRICARE Select following a Qualifying Life Event or during the annual fall TRICARE Open Season
- To enroll, you can log into milConnect and click on the "Manage health benefits" button.
- You can call your regional contractor
- You can submit a TRICARE Select Enrollment, Disenrollment and Change form (DD Form 3043) to your regional contractor

Costs and Claims

There is no yearly enrollment fee or ADFMs. Retirees, their families and others pay enrollment fees. With TRICARE Select, you pay a yearly deductible and per-visit copayments or cost shares. You'll fall into one of two groups based on when you or your sponsor entered the uniformed service.

This group will determine your costs.

When following the rules of your program option, your out-of-pocket expenses will be limited to your catastrophic cap. Nonparticipating non-network providers may charge up to 15% above the TRICARE-allowable amount, which won't apply to your catastrophic cap. You're responsible for this amount, plus your deductible and copayments or cost shares.

Submit claims to the regional contractor for the area where you live. In the U.S. and U.S. territories, claims must be filed within one year of the date of service or date of inpatient discharge. You're responsible for confirming your claims are received. If you need help, you should contact your regional contractor.

If your provider isn't TRICARE-authorized, but wants to see TRICARE patients, your provider can do so without signing a contract with your regional contractor. Most providers with a valid professional license can become TRICARE-authorized. Then, in doing so, TRICARE will pay them for covered services.

Premium-based health care plans that work like TRICARE Select with the same copayments or cost shares and a choice of providers can be purchased by individuals who qualify. These plans include TRICARE Reserve Select (TRS), TRICARE Retired Reserve (RR) and the Continued Health Care Benefit Program (CHCBP).

TRICARE Select Overseas

TRICARE Select Overseas provides comprehensive coverage in all overseas areas. You must show as being eligible for TRICARE in the Defense Enrollment Eligibility Reporting System. Enrollment is required.

Participants can include:
- Active-duty family members
- Retired service members and their families
- Family members of activated Guard/Reserve members
- Non-activated Guard/Reserve members and their families who qualify for care under the Transitional Assistance Management Program
- Retired Guard/Reserve members at age 60 and their families
- Survivors
- Medal of Honor recipients and their families
- Qualified former spouses
- Qualified former spouses

Costs are based on the sponsor's military status. You'll pay an annual outpatient deductible and cost shares for covered services. You'll also pay enrollment fees.

TRICARE Reserve Select (TRS) and Tricare Retired Reserve (TRR)

TRS and TRR are premium-based, global health care plans for some qualified Selected Reserve or Retired Reserve members, their family members, and survivors. TRS and TRR offer comprehensive health care coverage that's similar to TRICARE Select.
- Enrollment is required.
- TRICARE Open Season doesn't apply to TRS and TRR. These plans offer continuous open enrollment throughout the year.
- There is an initial two-month premium payment due when you enroll.
- Monthly premiums, a yearly deductible and copayments or cost shares apply.
- Get care from any TRICARE-authorized provider without a referral.
- Certain services require pre-authorization.
- Note: When your National Guard or Reserve sponsor is activated for more than 30 days for a preplanned mission or in support of a contingency operation, the coverage stops while you get active-duty benefits.

TRICARE For Life

If you're entitled to Medicare Part A, you typically must have Medicare Part B to keep TRICARE, regardless of your age or where you live. This is a requirement, based on federal law. There are exceptions for ADSMs and ADFMs. If you're eligible for TRICARE and have Medicare Part A and Part B, you're automatically covered by TRICARE for Life.

Medicare Part A covers inpatient hospital care, hospice care, inpatient skilled nursing facility care and some home health care. Medicare Part B covers provider services, outpatient care, home health care, durable medical equipment and some preventative services.

TRICARE for Life covers inpatient and outpatient wraparound coverage, and coverage for overseas care.

TRICARE for Life Costs
There are no enrollment fees or forms for TFL, but you must have Medicare Part A and Part B. Medicare B has a monthly premium.
- If you're covered by TRICARE and Medicare, Medicare pays the Medicare-

allowed amount. TRICARE pays the remaining amount, and you pay nothing.

- If you're covered by only Medicare, Medicare pays the Medicare-allowed amount. TRICARE pays nothing. You may the Medicare deductible and cost share.
- If you're only covered by TRICARE, Medicare pays nothing, and TRICARE pays the TRICARE-allowable amount. You pay the TRICARE deductible and cost-share.
- If you're not covered by either, Medicare pays nothing and TRICARE pays nothing. You're billed charges which may be more than the Medicare or TRICARE-allowed amount.

TRICARE for Life and U.S. Department of Veterans Affairs Benefits
If you're eligible for both Tricare for Life and U.S. Department of Veterans Affairs Benefits and choose to use your TFL benefit for health care not related to a service-connected injury or illness, you'll pay more to see a VA provider than you would pay to see a civilian Medicare provider.

TRICARE will only pay up to 20% of the TRICARE-allowable amount for care you get at a VA facility, and you may be responsible for the rest.

With TFL, the least expensive option is to see a Medicare-participating or Medicare non-participating provider. Medicare participating providers agree to accept the Medicare-allowed amount as payment in full.

Medicare non-participating providers don't accept the Medicare-allowed amount as payment in full and may charge up to 15% above the Medicare-allowed amount. The additional cost may be covered by TFL.

Other TRICARE Program Options
TRICARE offers other coverage options for people with eligibility changes, like children aging out of regular TRICARE coverage or sponsors separating from service. If you've lost all TRICARE eligibility you may qualify to buy coverage under CHCBP.

TRICARE Plus
TRICARE Plus is a primary care option at some military hospitals and clinics. You must enroll to participate, and each hospital or clinic leader decides if TRICARE Plus is available. You can enroll in TRICARE Plus if you're TRICARE-eligible and not enrolled in a TRICARE Prime Plan, the US Family Health Plan or a civilian or Medicare Health Maintenance Organization. You can also enroll if you're a dependent parent or parent-in-law.

TRICARE gives you access to get primary care and you pay nothing out of pocket. TRICARE Plus doesn't cover specialty care.

If you're a dependent parent or parent-in-law, TRICARE won't pay for care by civilian providers, even if the military hospital or clinic refers you for their care. You're responsible for the full cost of the care.

TRICARE Young Adult
TRICARE Young Adult is a premium-based health care plan for qualified adult children who age out of TRICARE. TRICARE Young Adult Prime and TRICARE Young Adult Select are offered worldwide. Your location and sponsor's status determine whether you qualify for TYA Prime and/or TYA Select.

TYA includes medical and pharmacy benefits, but not dental or vision coverage. Coverage, provider choice and costs for TYA are the same as for TRICARE Prime and TRICARE Select.

You can generally buy TYA coverage if you're an adult child of a TRICARE-eligible sponsor; unmarried; at least age 21 but not yet age 26 and not otherwise eligible for TRICARE or employer-based coverage. TRICARE Open Season doesn't apply to TYA. TYA Prime and TYA Select offer continuous enrollment throughout the year.

Transitional Coverage Options

TRICARE offers benefits to help service members and their families transition to civilian life.

First is the Transitional Assistance Management Program (TAMP). TAMP offers 180 days of premium-free health care after your sponsor separates from the military. If you're eligible TAMP starts the day after the sponsor separates from service.

Continued Health Care Benefit Program (CHCBP) is a premium-based health care program managed by Humana Military. CHCBP offers continued health coverage for 18 to 36 months after TRICARE coverage ends. Certain former spouses who haven't remarried before age 55 may qualify for an unlimited duration of coverage.

If you qualify you can purchase CHCBP coverage within 60 days of loss of TRICARE or TAMP coverage, whichever is later. TRICARE Open Season doesn't apply to CHCBP. CHCBP offers continuous open enrollment throughout the year.

While in CHCBP you aren't eligible to receive care at a military hospital or clinic, except in a medical emergency.

The TRICARE Pharmacy Program provides prescription drugs through military pharmacies, TRICARE Pharmacy home deliver, TRICARE retail network pharmacies and non-network pharmacies. Your options for filling your prescription depend on the type of drug your provider prescribes. Express Scripts Inc. manages the TRICARE pharmacy benefit for all TRICARE-eligible beneficiaries.

US Family Health Plan

The US Family Health Plan is an additional TRICARE Prime option available through networks of community-based not-for-profit health care systems in six areas of the U.S. The U.S. Family Health Plan is available to the following beneficiaries who live in a designated US Family Health Plan area:
* Active-duty family members
* Retired service members and their families
* Family members of Activated National Guard/Reserve members
* Non-activated National Guard/Reserve members and their families who qualify for care under the Transitional Assistance Management Program
* Retired National Guard/Reserve members at age 60 and their families
* Survivors
* Medal of Honor recipients and their families
* Qualified former spouses

You must live in one of the designated US Family Health Plan Service areas to enroll.

US Family Health Plan Service Area	Designated Provider
Maryland Washington D.C. Parts of Pennsylvania, Virginia, Delaware and West Virginia	John Hopkins Medicine 1-800-808-7347
Maine New Hampshire	Martin's Point Health Care 1-888-241-4556

Vermont Upstate and western New York Northern tier of Pennsylvania	
Massachusetts including Cape Code Rhode Island Northern Connecticut	Brighton Marine Health Center 1-800-818-8589
New York City Long Island Southern Connecticut New Jersey Philadelphia and area suburbs	St. Vincent Catholic Medical Centers 1-800-241-4848
Southeast Texas Southeast Louisiana	CHRISTUS Health 1-800-678-7347
Western Washington State Parts of eastern Washington State Northern Idaho Western Oregon	Pacific Medical Centers (PacMed Clinics) 1-888-958-7347

Dental and Vision Options

Retirees, their eligible family members and active-duty family members enrolled in a TRICARE health plan may qualify to purchase vision coverage through the Federal Employees Dental and Vision Insurance Program, offered by the U.S. Office of Personnel Management. When eligible, you can enroll during the fall Federal Benefits Open Season. There are three dental options that are separate from TRICARE health care options. The TRICARE Active-Duty Dental Program (ADDP) is for active-duty service members, and National Guard and Reserve members called or ordered to active duty for more than 30 days for a preplanned mission or contingency operation.

The TRICARE Dental Program is for active-duty family members, survivors, National Guard and Reserve members and their families, and individual Ready Reserve members and their family members.

The Federal Employees Dental and Vision Insurance Program is offered by the U.S. Office of Personnel Management. The program is for retired service members and eligible family members, retired National Guard and Reserve members and their eligible family members, certain survivors, and Medal of Honor recipients and their immediate family members and survivors.

TRICARE Regions and Assistance

TRICARE is available worldwide and managed regionally. There are two TRICARE regions in the U.S.—TRICARE East and TRICARE West. Your TRICARE benefit is the same regardless of where you are, but there are different customer service contacts for each region.

TRICARE East Region
Humana Military
1-800-444-5445

Tricare West Region
Health Net Federal Services, LLC
1-844-866-WEST

What to Remember
✓ TRICARE is the Department of Defense health care program for uniformed service members, retirees and their family members
✓ As of December 15, 2021, CVS Pharmacy is part of the TRICARE retail pharmacy network. Walmart, Sam's Club and some community pharmacies are no longer part of the TRICARE network.
✓ There are a number of TRICARE plans available.
✓ Eligibility depends on who you and your sponsor are.
✓ Depending on your plan, you'll get care from either a military hospital or clinic, or a civilian provider. There are two TRICARE regions in the U.S.— TRICARE East and TRICARE West.

CHAPTER 16

VA'S FIDUCIARY PROGRAM

VA's Fiduciary Program was established to protect Veterans and other beneficiaries who, due to injury, disease, or due to age, are unable to manage their financial affairs. VA will only determine an individual to be unable to manage his or her financial affairs after receipt of medical documentation or if a court of competent jurisdiction has already made the determination.

Upon determining a beneficiary is unable to manage his or her financial affairs, VA will appoint a fiduciary. The fiduciary, normally chosen by the beneficiary, must undergo an investigation of their suitability to serve. This investigation includes a criminal background check, review of credit report, personal interview, and recommendations of character references. Only after a complete investigation is a fiduciary appointed to manage a beneficiary's VA benefits.

The fiduciary is responsible to the beneficiary and oversees financial management of VA benefit payments. Generally, family members or friends serve as fiduciaries for beneficiaries; however, when friends and family are not able to serve, VA looks for qualified individuals or organizations to serve as a fiduciary.

What is the Fiduciary Program?
The purpose of the Department of Veterans Affairs (VA) Fiduciary Program is to protect Veterans and beneficiaries who are unable to manage their VA benefits through the appointment and oversight of a fiduciary.

If you have been determined unable to manage your VA benefits, the VA will conduct a field examination to appoint a fiduciary to assist you.

VA Field Examination
A VA field examination will be scheduled for the purpose of appointing a fiduciary to assist you in managing your VA benefits. During the field examination, please have the following information available for review by the field examiner:
- Photo identification.
- The source and amount of all monthly bills, recurring expenses (annual, bi-annual, quarterly, etc.) and income.
- A list of all assets; bank accounts, owned property, stocks, bonds, life insurance, burial plans,etc.
- A list of all current medications.
- Name, phone number, and address of your primary care doctor. Name, phone number, and address of your next of kin.

Selection Process

During the selection process, the VA will first seek to qualify the individual you desire to serve as your fiduciary.

The fiduciary selection is based on an assessment of the qualifications of the proposed fiduciary. When seeking a fiduciary, the following individuals may be considered:

- A spouse or family member
- Court-appointed fiduciaries
- Another interested party,or
- A professional fiduciary

An assessment of the qualifications of a proposed fiduciary includes, but is not limited to:

- The willingness to serve and abide by all agreements
- An interview with a VA representative
- Credit report review
- An inquiry into the criminal background, and
- Interviews with character witnesses

What Are My Rights?

The determination that you are unable to manage your VA benefits does not affect your non-VA finances, or your right to vote or contract.

You have the right to appeal VA's decision finding that you are unable to manage your VA benefits. You also have the right to appeal VA's selection of the fiduciary. If you disagree with the VA on either of these matters, you may:

- Appeal to the Board of Veterans' Appeals (Board) by telling them you disagree with their decision and want the Board to review it, or
- Give them evidence they do not already have that may lead us to change their decision.

How to Apply

To become a fiduciary for a family member or friend, submit a request with the beneficiary's name and VA file number, and your name and contact information to the VA regional office nearest you.

To become a professional fiduciary, submit your resume with a cover letter to the following e-mail address: VA_Fiduciary@va.gov. Include your name, the name of your organization (if applicable), mailing address, and e-mail address with your request.

Responsibilities

Any individual appointed as a VA fiduciary is responsible for managing the beneficiary's VA income and ensuring the beneficiary's just debts are paid. Additional, responsibilities of the fiduciary include, but are not limited to the following:

1. Utilizing the funds for the daily needs (e.g., food, clothing, housing, medical expenses, and personal items) of the beneficiary and his/her recognized dependents.
2. Never borrowing, loaning, or gifting funds belonging to the beneficiary.
3. Reporting any of the following changes to the Fiduciary Activity immediately:
   1. Change in address or phone number of the beneficiary or fiduciary
   2. Change in income or dependents
   3. Incarceration of the beneficiary
   4. Hospitalization of the beneficiary in a VA or state facility, and
   5. Death of the beneficiary or the beneficiary's dependents

4. Establishing a properly titled bank account as follows: (Beneficiary's Name) by (Fiduciary's Name, Federal Fiduciary).
5. Never commingling the beneficiary's funds with those of another.
6. Never withdrawing cash from the beneficiary's account by counter check or ATM withdrawal.
7. Timely submitting periodic accountings when required.
8. Keeping accurate, complete records and receipts, regardless of any requirement to submit periodic accountings.
9. Conserving excess funds in a federally or state insured interest-bearing account or United States savings bonds.
10. Registering saving bonds to reflect proper ownership and the existence of the fiduciary relationship, as follows: (Beneficiary's Name), (Social Security No.), under custodianship by the designation of the Department of Veterans Affairs.
11. Reporting any event that affects the beneficiary's payment or entitlement to benefits and promptly returning any payment that the beneficiary is not due.
12. Notifying VA of any changes or circumstances that would affect your performance as a payee or your decision to continue to serve as a payee (e.g., you sell or transfer your business).
13. Returning any funds owed by the beneficiary to VA if you stop serving as the fiduciary.
14. Notifying the VA if the beneficiary's condition improves to a point where you believe he or she no longer needs a fiduciary.
15. Protecting the beneficiary's funds from the claims of creditors since the beneficiary 's funds are protected by law.

What to Remember
✓ VA's fiduciary program was established to protect veterans and other beneficiaries who are unable to manage their financial affairs.
✓ VA will only determine an individual is unable to manage their financial affairs after receiving medical documentation or if a court of competent jurisdiction has already made the determination.
✓ After determining a beneficiary is unable to manage their financial affairs, VA appoints a fiduciary.
✓ The fiduciary has to undergo an investigation to show they're suitable to serve. The investigation includes a criminal background check, review of credit report, personal interview, and recommendations of character references.
✓ Generally, family members or friends serve as beneficiaries, but when they can't, VA will look for qualified organizations or individuals.

CHAPTER 17

CAREERS & EMPLOYMENT

If a veteran has a service-connected disability limiting their ability to work, or one that prevents them from working, the Veteran Readiness and Employment program can help. The program was formerly called Vocational Rehabilitation and Employment, and it's also known as Chapter 31 or VR& E.

The program helps veterans explore employment options, address training and education needs, and in some cases, family members can qualify for benefits.

There are several tracks available.

Reemployment Track

This track is available to veterans with a service-connected disability. It's meant to help a veteran return to their former job and support their employer in meeting their needs. Veterans are protected under the Uniformed Services Employment and Reemployment Rights Act or USERRA. You can't be disadvantaged in your civilian career because of your service.

A Vocational Rehabilitation Counselor or VRC can help provide a veteran with a full range of services and can direct the veteran to the Department of Labor to start the process.

You may be eligible if you're a veteran with a service-connected disability, and you meet all of the following requirements:
- You have an employment barrier or handicap, and
- Your enrolled in Veteran Readiness and Employment (VR&E), and
- You'd like to return to your former job.

Note: Having an employment handicap means your service-connected disability limits your ability to prepare for, obtain and maintain suitable employment (a job that doesn't make your disability worse, is stable and matches your abilities aptitudes and interests).

You may be eligible for benefits related to helping you return to the job you held before you deployed.

You'll need to first apply for VR&E benefits. Then, you'll work with your Vocational Rehabilitation Counselor.

You can get training and hands-on work experience through programs including:
- The VR&E Special Employer Incentive Program or SEI which is for eligible veterans who face challenges getting a job.
- The VR&E Non-Paid Work Experience program or NPWE for eligible veterans and service members who have an established career goal and learn easily in a hands-on environment or are having trouble getting a job due to a lack of work experience.

Rapid Access to Employment Track

If a veteran wants to follow an employment path using their existing skill set, the Rapid Access to Employment Track can help them. Counseling and rehabilitation services are available.

You may be eligible if you're a servicemember or veteran with a service-connected disability and you meet all of the following requirements:

- You have an employment handicap or barrier, and
- You're enrolled in VR&E, and
- You already have experience, education or training in your field of interest.

Available benefits include tools to help with the job search, professional or vocational counseling, help with writing a resume and preparing for interviews, and help determining if you're eligible for Veterans' Preference.

You'll first need to apply for VR&E benefits, and then work with your Vocational Rehabilitation Counselor.

Self-Employment Track

If a servicemember or veteran has a service-connected disability and employment barrier, but the strong desire, skills and drive to run a business, they could be interested in the Self-Employment track.

You maybe be eligible if you're a servicemember or veteran with a service-connected disability, and you meet all of the following requirements:

- You have an employment barrier or handicap, and
- You're enrolled in VR&E, and
- Your service-connected disability makes it hard for you to prepare for, obtain and maintain suitable employment. Suitable employment is a job that doesn't worsen your disability, is stable, and matches your abilities, aptitudes and interests.

Benefits available through the VR&E Self-Employment Tract include:

- Coordination of services and help in the creation of a proposed business plan
- Analysis of your business concept
- Training in small business operations, marketing and finances
- Guidance in getting the right resources to implement a business plan

After you develop a business plan, the VA will review it and evaluate whether self-employment and the proposed business are viable options.

To get these benefits, first you have to apply for VR&E benefits. Then you begin to work with your Vocational Rehabilitation Counselor.

Employment Through Long-Term Services Track

If you have a service-connected disability that makes it hard to succeed in your employment path, you might be interested in the Employment Through Long-Term Services track. The VA can help you get education and training needed to find work in a different filed, better suiting your current interests and abilities.

You may be eligible for both VR&E training and GI bill benefits, so you have to decide which benefit you want to use.

Available benefits through this VR&E tract include:

- A complete skills assessment
- Career guidance

- Job market evaluation
- Education and training for a professional or vocational field that's a good fit for you
- Apprenticeship, on-the-job training and volunteer opportunities
- Employment assistance

To be eligible, the following must be true:

- You have an employment barrier or handicap, and
- You're enrolled in VR&E, and
- Your service-connected disability makes it hard for you to prepare for, obtain and maintain suitable employment.

Independent Living Track

Under the VR&E Independent Living Track, if your service-connected disability limits your ability to perform activities of daily living such as bathing, dressing or interacting with others, and you can't return to work right away, you might qualify for independent living services through the Independent Living track. You might also receive the services as you work on finding a job, if that's your goal that you've created with your Vocational Rehabilitation Counselor (VRC). In both scenarios, your VRC can help restore your activities of daily living.

You may be eligible for independent living services if you are a servicemember or veteran with a service-connected disability who's eligible for VR&E benefits, and you meet all of the following requirements:

- You have a serious employment handicap (SEH), and
- Your disabilities prevent you from looking for or returning to work, and
- You're in need of services to live as independently as possible

Depending on your needs, services may include:

- Evaluation and counseling to identify needs and goal
- Referral to support resources
- Evaluation to see if you're eligible for the VR&E home adaptation grant. The grant is part of a rehabilitation plan to improve accessibility features in the home.
- Guidance to help you understand if you're eligible for the VA's adaptive housing programs. These programs help people make changes to their home, such as widening doorways or adding ramps, so they can live more independently.

The services generally last up to 24 months. In some cases, you could be able to use services for longer.

Support for a Veteran-Owned Small Business

If someone has a Veteran-Owned Small Business, they may qualify for advantages when bidding on government contracts and access to other resources and support through the Vets First Verification Program.

The Vets First Verification Program is run by the Office Small & Disadvantaged Business Utilization (OSDBU). Registering through OSDBU will let the veteran work with the VA.

You may be eligible if you're a veteran and you or another veteran at your company meet the following requirements.
All of the following must be true of you or another veteran at your company:

- One of you owns 51% or more of the company you want to register, and
- One of you has full control over the day-to-day management, decision-making and strategic policy of the business, and
- One of you has the managerial experience needed to manage the business, and

- One of you is the highest-paid person in the company or can provide a written statement explaining why your taking lower pay helps the business, and
- One of you works full time for the business, and
- One of you holds the highest officer position in the company

To be considered a veteran, at least one of these must be true:
- You served on active duty with the Army, Air Force, Navy, Marine Corps or Coast Guard for any length of time and didn't receive a dishonorable discharge, or
- You served as a Reservist or member of the National Guard and were called to federal active duty or disabled from a disease or injury that started or got worse in the line of duty or while in training status

You may be eligible to register your business as a Service-Disabled Veteran-Owned Small Business (SDVOSB) if you meet all of the requirements of a VOSB listed above, and either you or another veteran owner of the company meet at least one requirement listed below:
- A disability rating letter from VA confirming you have a service-connected disability rating between 0 and 100%, or
- A disability determination from the Department of Defense

If you received an other than honorable, bad conduct or dishonorable discharge you may not be eligible for VA benefits. You can try to apply for a discharge upgrade or learn about the VA Character of Discharge Review Process.

The U.S. Small Business Administration (SBA) can also provide you with resources to help you grow or start a business, and many universities or nonprofits also offer free or low-cost entrepreneur and business-focused courses for veterans and military family members.

Boots to Business
The Boots to Business program can help if you're interested in business ownership or self-employment. You can learn how to develop a business plan and explore key business concepts. This program is open to servicemembers, including members of the National Guard and Reserves. It's offered through Syracuse University.

Syracuse offers other programs for veterans as well, including the Entrepreneurship Bootcamp for Veterans, and The Veteran Women Igniting the Spirit of Entrepreneurship Program.

SBA Support for Veteran-Owned Businesses
The SBA offers support for veterans as they enter the world of owning a business.

The Office of Veterans Business Development is devoted exclusively to the promotion of veteran entrepreneurship. The OVBD facilitates the U.S. of all SBA programs by veterans, service-disabled veterans, reservists, active-duty service members, transitioning service members, and their dependents or survivors.

SBA programs provide access to capital and preparation for small business opportunities, and they can connect veteran small business owners with federal procurement and commercial supply chains.

The Veterans Business Outreach Center Program is an initiative that oversees Veteran Business Outreach Centers (VBOC) across the country.

You can use SBA tools including Lender Match to connect with lenders.

SBA also makes special consideration for veterans through programs including the Military Reservist Economic Injury Disaster Loan Program. This program provides loans of up to $2 million to cover operating costs that can't be met due to the loss of an essential employee called to active duty in the Reserves or National Guard

Veteran Contracting Assistance Programs

Every year the federal government awards a portion of contracting dollars specifically to businesses owned by veterans. Small businesses owned by veterans may also be eligible to buy surplus property from the federal government.

Veteran Small Business Certification (VetCert) Program
Certification with SBA allows service-disabled veteran-owned small businesses (SDVOSBs) to compete for federal sole-source and set-aside contracts across the federal government. Certified veteran-owned small businesses (VOSBs) have additional opportunities to pursue sole-source and set-aside contracts at the VA under the VA's Vet First Program.

SBA's new Veteran Small Business Certification program implements changes from the National Defense Authorization Act for Fiscal Year 2021 (NDAA 2021) which transfers the certification function from the VA to SBA as of January 1, 2023.

The transfer provides veterans with central support for their small business certification needs, and the final rule was published in the Federal Register on November 29, 2022.

Benefits include the ability to compete for sole-source and set-aside contracts across the federal government. Self-certified firms seeking these restriction competition opportunities have to apply to SBA for certification by December 31, 2023.

To apply for certification with SBA as a VOSB or SDVOSB, a firm has to meet the following requirements:
- Be considered a small business as defined by the size standard corresponding to any NAICS code listed in the businesses' SAM profile
- Have no less than 51% of the business owned and controlled by one or more veterans
- For certification as a SDVOSB, have no less than 51% of the business owned and controlled by one or more veterans rated as service-disabled by the VA
- For veterans who are permanently and totally disabled and unable to manage the daily business operations of their business, their business may still qualify for their spouse or appointed, permanent caregiver is assisting in the management.

Certification Transfer from the VA and One-Year Extension
- Firms certified by the VA Center for Verification and Evaluation (CVE) as of January 1, 2023, are automatically granted certification by the SBA for the remainder of the firm's eligibility period.
- SBA intends to grant a one-time, one-year extension of certification to current VOSBs and SDVOSBs verified by the VA as of the transfer date on January 1, 2023. The additional year will be added to the existing eligibility period of a current participant.
- New applications certified by SBA after January 1, 2023, will receive the standard three-year certification period.

Grace Period for Self-Certified Firms
- The NDAA 2021 grants a one-year grace period for self-certified SDVOSBs until January 1, 2024. During the grace period, self-certified businesses will have one year to file an application for SDVOSB certification and may continue to rely on their self-certification to compete for non-VA SDVOSB set-aside contracts.
- Self-certified SDVOSBs that apply before January 1, 2024, will keep eligibility through the expiration of the grace period until SBA issues a final eligibility decision.

- VOSBs and SDVOSBs seeking sole-source and set-aside opportunities with the VA must be certified. There is no grace period.

How to Apply for Certification
SBA will start accepting applications for certification on January 9, 2023. You can go through the application portal available from the SBA.

Service-Disabled Veteran-Owned Small Business Program
The federal government aims to award at least 3% of all federal contracting dollars to SDVOSBs every year. Competition is limited for certain federal contract opportunities to businesses participating in the SDVOSB program. SVODBs may self-certify their status through 2023 to compete for set-aside contracts at most federal agencies, but the VA doesn't recognize SDVOSB self-certification. Self-certified SDVOSDBs have to apply for SBA certification by January 1, 2023, to be able to compete for set-aside contracts with the federal government.

Surplus Personal Property for Veteran-Owned Small Business Programs
Veteran-owned small businesses can access federally owned personal property no longer in use through the General Service Administration's Federal Surplus Personal Property Donation Program. GSA oversees the reuse and donation of federal personal property.

State Agency for Surplus Property operations manage surplus property disbursement.

VOSBs may get federal surplus property in the state where the property will be primarily located and used. You have to agree in writing that your VOSB:

- Is located and operated within the state
- Is unconditionally owned and controlled by one or more eligible veterans, service-disabled veterans or surviving spouses
- Has registered and is in verified status in the VA's Vets First Verification Program database
- After January 1, 2023, to be eligible, a VOSB has to show certified status in SBA's VetCert database
- Will use the property in the normal conduct of its business activities—personal or not-business use isn't allowed
- Will not sell, transfer, loan, lease, encumber or otherwise dispose of the property during the period of restriction unless with express written authorization
- Will get permission from donating SASP before permanently removing the property from the state
- Will use the property as intended within one year of receipt
- Will maintain VOSB eligibility with VA and SASP for the duration of the applicable federal period of restriction for donated property
- Will give SBA, GSA and SASP access to inspect the property and all pertinent records

Subsistence Allowance

In some cases, veterans who participate in the VR&E program might receive a subsistence allowance while they pursue their educational or training program to prepare for a future career. The subsistence allowance is paid every month and is based on the rate of attendance in a training program—full time, three-quarter time or half-time. It's also based on the number of dependents and the type of training.

If a veteran qualifies for the Post-9/11 GI Bill, they might be eligible to receive the Basic Allowance for Housing (BAH) rate for subsistence.

The following Subsistence Allowance rates are paid for training in an Institution of Higher Learning

Number of Dependents	Full Time	¾ Time	½ Time	¼ Time
No Dependents	$723.56	$543.68	$363.77	$181.86
One Dependent	$897.51	$674.11	$450.71	$225.39
Two Dependents	$1057.65	$790.75	$529.79	$264.90
Each Additional Dependent	$77.07	$59.30	$39.55	$19.73

Subsistence Allowance is paid for full time training only, in the following training programs: Non-pay or nominal pay on-job training in a federal, state, local, or federally recognized Indian tribe agency; training in the home; vocational course in a rehabilitation facility or sheltered workshop; institutional non-farm cooperative.

Number of Dependents	Full Time
No Dependents	$723.56
One Dependent	$897.51
Two Dependents	$1057.65
Each Additional Dependent	$77.07

Subsistence Allowance is paid for full time training only in the following training programs: Farm cooperative, apprenticeship, or other on-job training

Number of Dependents	Full Time
No Dependents	$632.61
One Dependent	$765.04
Two Dependents	$881.69
Each Additional Dependent	$57.33

Subsistence Allowance is paid at the following rates for combined training programs: Combination of Institutional and On-Job Training (Full Time Rate Only).

Number of Dependents	Institutional Greater than one half
No Dependents	$723.56
One Dependent	$897.51
Two Dependents	$1057.65
Additional Dependents	$77.07

Subsistence Allowance is paid at the following rates for Non-farm Cooperative Training: Non-farm Cooperative Institutional Training and Non-farm Cooperative On-Job Training - Full Time Rate Only.

Number of Dependents	FT Non-farm Coop/Institutional
No Dependents	$632.61
One Dependent	$765.04
Two Dependents	$881.69
Each Additional Dependent	$57.33

Subsistence Allowance is paid at the following rates for Independent Living programs: A subsistence allowance is paid each month during the period of enrollment in a rehabilitation facility when a veteran is pursuing an approved Independent Living Program plan. Subsistence allowance paid during a period of Independent Living Services is based on rate of pursuit and number of dependents.

	Full-Time	¾ Time	½ Time	¼ Time
No dependents	$723.56	$543.68	$363.77	$181.86
One dependent	$897.51	$674.11	$450.71	$225.39
Two dependents	$1057.65	$790.75	$529.79	$264.90
Each additional dependent	$77.07	$59.30	$39.55	$19.73

Programs for Unemployable Veterans

Veterans awarded 100% disability compensation based upon unemployability may still request an evaluation, and, if found eligible, may participate in a vocational rehabilitation program and receive help in getting a job. A veteran who secures employment under the special program will continue to receive 100% disability compensation until the veteran has worked continuously for at least 12 months.

Election of Alternate Subsistence Allowance Under Public Law 111- 377

This option is included for your use in comparing subsistence allowance rates that may be available to certain Veterans participating in the Chapter 31 program. The VA is authorized to allow a Veteran, entitled to both a Chapter 31 subsistence allowance and Post 9/11 GI Bill Chapter 33 educational assistance, to elect to receive a replacement in an alternate amount instead of the regular Chapter 31 subsistence allowance.

The alternate payment will be based on the military basic allowance for housing (BAH) for an E-5 with dependents residing in the zip code of the training facility. Training in foreign institutions and training that is solely on-line or in-home will be based on the national average BAH.

Paralympic Veteran Benefit

Some veterans in training for the U.S. Paralympics may qualify for a monthly subsistence allowance from VA. The allowance is pegged to the subsistence allowance for participants in a fulltime institutional program under chapter 31 of title 38 of the U.S. Code. According to the rule, the VA will pay the allowance to a veteran with a service-connected or non-service-connected disability if the veteran is invited by the U.S. Paralympics to compete for a slot on the U.S. Paralympic team or is residing at the U.S. Paralympic training center for training or competition.

Applications for the allowance must be submitted through the U.S. Paralympics. Through the program, VA will pay a monthly allowance to a veteran with a service-connected or non-service-connected disability if the veteran meets the minimum military standard or higher (e.g., Emerging, Talent Pool, National Team) in his or her respective sport at a recognized competition. Besides making the military standard, an athlete must also be nationally or internationally classified by his or her respective sport federation as eligible for Paralympic competition within six or 12 months of a qualifying performance.

Athletes must also have established training and competition plans and are responsible for turning in monthly and quarterly reports in order to continue receiving a monthly assistance allowance. The monthly allowance rate for an athlete approved for monetary assistance is the same as the 38 U.S.C. Chapter 31 Vocational Rehabilitation & Employment Rate (VR&E), which in FY 2022 starts at $670.77 and increases depending on the number of dependents.

Vocational Training for Children with Spina Bifida

To qualify for entitlement to a vocational training program an applicant must be a child:

- To whom VA has awarded a monthly allowance for spina bifida; and
- For whom VA has determined that achievement of a vocational goal is reasonably feasible.

A vocational training program may not begin before a child's 18th birthday, or the date of completion of secondary schooling, whichever comes first. Depending on the need, a child may be provided up to 24 months of full-time training.

Vocational Training for Children of Female Vietnam Veterans Born with Certain Birth Defects

Section 401 of P.L. 106-416, which became law on November 1, 2000 directed the Secretary of VA to identify birth defects of children of female Vietnam veterans that: (1) are associated with service during the Vietnam era; and (2) result in the permanent physical or mental disability of such children. The law excludes from such defects familial or birth- related defects or injuries. The law further directs the Secretary to provide to such children a monthly monetary allowance, as well as necessary health care to address the defect and any associated disability.

Veterans' Preference

Veterans' Preference gives eligible veterans preference in appointment over many other applicants. Veterans' Preference applies to all new appointments in the competitive service and many in the excepted service. Veterans' preference doesn't guarantee veterans a job, and it doesn't apply to internal agency actions like promotions, transfers, reassignments or reinstatements.

See Chapter 31 for more details.

Applying for Benefits

Interested veterans can apply by filling out "VA Form 28-1900", "Application for Veteran Readiness and Employment for Claimants with Service-Connected Disabilities," and mailing it to the VA regional office serving his or her area.

What to Remember
✓ The Veteran Readiness and Employment (VR&E) program was formerly known as Vocational Rehabilitation and Employment.
✓ Services are available to help with job training, employment, accommodations, resume development and job seeking skills.
✓ Other services may be provided to help veterans start their own business or secure independent living services if they are severely disabled and can't work in traditional employment.
✓ Five tracks are part of the VR&E support program. These tracks are: reemployment; rapid access to employment; self-employment, employment through long-term services; and independent living.
✓ To apply, you don't need to wait for a rating. You can go ahead and fill out VA Vocational Rehabilitation-Getting Ahead After You Get Out (VA Form 28-0588). Severely injured active-duty service members can automatically receive VR&E benefits before VA issues a disability rating because of Sec. 1631(b) of the National Defense Authorization Act (PL 110-181).

CHAPTER 18

VET CENTERS

Vet Centers serve veterans and their families by providing a continuum of quality care that adds value for veterans, families, and communities. Care includes:

- Professional readjustment counseling;
- Professional counseling for posttraumatic stress disorder;
- Marital and family counseling;
- Substance abuse information and referral;
- Community education;
- Outreach to special populations, including disenfranchised and unserved veterans;
- The brokering of services with community agencies;
- Provides a key access link between the veteran and other services in the U.S. Department of Veterans Affairs;
- Promotion of wellness activities with veterans to help them reach quality health and life goals

Eligibility

All veterans are encouraged to seek Vet Center services. Any veterans and current service members, including members of the National Guard and Reserve components, are eligible if they have served on active military duty in any combat theater or area of hostility, or have experienced a military sexual trauma.

Veterans and current servicemembers are also eligible if they provided direct emergent medical care or mortuary services while serving on active duty to the casualties of war or served as a member of an unmanned aerial vehicle crew providing direct support to operations in a combat zone or area of hostility. Vietnam-era veterans who have access care at a Vet Center prior to January 2, 2013, are also eligible.

Vet Center services are also provided to family members of veterans and service members for military-related issues when they aid in the adjustment of those who have served. This includes bereavement counseling for families who experienced an active-duty death.

Sexual Trauma and Harassment Counseling

Vet Center services include individual readjustment counseling, referral for benefits assistance, group readjustment counseling, liaison with community agencies, marital and family counseling, substance abuse information and referral, job counseling and placement, sexual trauma counseling, and community education.

At the VA, Veterans can receive free, confidential treatment for mental and physical health conditions related to Military Sexual Trauma (MST). You may be able to receive this MST-related care even if you are not eligible for other VA services. To receive these services, you do not need a VA service-connected disability rating, to have reported the incident when it happened, or have other documentation that it occurred.

Knowing that MST survivors may have special concerns, every VA healthcare facility has an MST Coordinator who can answer any questions you might have about VA's MST services. VA has a range of services available to meet Veterans where they are at in their recovery:

Every VA healthcare facility has providers knowledgeable about treatment for problems related to MST. Many have specialized outpatient mental health services focusing on sexual trauma. Vet Centers also have specially trained sexual trauma counselors. VA has almost two dozen programs nationwide that offer specialized MST treatment in residential or inpatient setting. These programs are for Veterans who need more intense treatment and support.

Because some Veterans do not feel comfortable in mixed-gender treatment settings, some facilities have separate programs for men and women. All residential and inpatient MST programs have separate sleeping areas for men and women.

Vet Center Locations

ALABAMA
Birmingham Vet Center
400 Emery Drive, Suite 200
Hoover, AL 35244
(205) 989-6415

Huntsville Vet Center
415 Church Street, Bldg H, Suite 101
Huntsville, AL 3580
(256) 539-5775

Mobile Vet Center
3221 Springhill Ave Bldg 2, Suite C
Mobile, AL 36607
(251) 478-2237

Montgomery Vet Center
4405 Atlanta Highway
Montgomery, AL 36109
(334) 273-7796

ALASKA
Anchorage Vet Center
4400 Business Park Blvd.
Suite B-34
Anchorage, AK 99503
(907) 563-6966

Fairbanks Vet Center
540 4th Avenue
Suite 100
Fairbanks, AK 99701
(907) 456-4238

Wasilla Vet Center
851 East West Point Ave,
Suite 102
Wasilla, AK 99654
(907) 376-4318

ARIZONA
Lake Havasu Vet Center
1720 Mesquite
Suite 101
Lake Havasu, AZ 86403
(928) 505-0394

Mesa Vet Center
1303 South Longmore
Suite 5
Mesa, AZ 85202
(480) 610-6727

Phoenix Vet Center
4020 North 20th St.
Suite 110
Phoenix, AZ 85016
(602) 640-2981

Prescott Vet Center
3180 Stillwater Dr.
Suite A
Prescott, AZ 86305
(928) 778-3469

Tucson Vet Center
2525 E. Broadway Blvd.
Suite 100
Tucson, AZ 85716
(520) 882-0333

West Valley Vet Center
14050 N. 83rd Ave.
Suite 170
Peoria, AZ 85381
(623) 398-8854

Yuma County Vet Center
1450 East 16th St.
Suite 103
Yuma, AZ 85365
(928) 726-1343

ARKANSAS
Fayetteville Vet Center
1416 N. Collage Ave.
Fayetteville, AR 72703
(910) 488-6252

Little Rock Vet Center
10800 Financial Centre Parkway
Suite 175
North Little Rock, AR 72703
(501) 918-1800

CALIFORNIA
Antelope Valley Vet Center
38925 Trade Center Dr.
Suites I/J
Palmdale, CA 93551
(661) 267-1026

Bakersfield Vet Center
1110 Golden State Ave.
Bakersfield, CA 93301
(661) 325-8387

Chatsworth Vet Center
20946 Devonshire St.
Suite 101
Chatsworth, CA 91311
(818) 576-0201

Chico Vet Center
280 Cohasset Road
Chico, CA 95926
(530) 899-8549

Chula Vista Vet Center
180 Otay Lakes Rd.
Suite 108
Bonita, CA 91902
(877) 618-6534

Citrus Heights Vet Center
5650 Sunrise Blvd.
Suite 150
Citrus Heights, CA 95610
(916) 535-0420

Concord Vet Center
1333 Willow Pass Rd.
Suite 106
Concord, CA 94520
(925) 680-4526

Corona Vet Center
800 Magnolia Avenue
Suite 110
Corona, CA 92879-8202
(909) 734-0525

East Los Angeles Vet Center
5400 E Olympic Blvd
Suite 140
Commerce, CA 90022
(323) 728-9966

Eureka Vet Center
2830 G Street
Suite A
Eureka, CA 95501
(707) 444-8271

Fresno Vet Center
1320 E. Shaw Ave
Suite 125
Fresno, CA 93710
(559) 487-5660

High Desert Vet Center
15095 Amargosa Rd.
Suite 107
Victorville, CA 92394

(706) 261-5925

North Bay Vet Center
6010 Commerce Blvd.
Suite 145
Rohnert Park, CA 94928
(707) 586-5966

North Orange County Vet Center
12453 Lewis St.
Suite 101
Garden Grove, CA 92840
(714) 776-0161

Los Angeles Vet Center
1045 W. Redondo Beach Blvd.,
Suite 150
Gardena, CA 90247
(310) 767-1221

Modesto Vet Center
1219 N. Carpenter Rd.,
Suite 12
Modesto, CA 95351
(209) 569-0713

Oakland Vet Center
7700 Edgewater Dr.
Suite 125
Oakland, CA 94621
(510) 562-7906

Peninsula Vet Center
345 Middlefield Rd.
Building 1
Menlo Park, CA 94025
(650) 617-4300

Sacramento Vet Center
1111 Howe Avenue
Suite 390
Sacramento, CA 95825
(916) 566-7430

San Bernardino Vet Center
356 East Vanderbilt Way
San Bernardino, CA 92408
(909) 801-5762

San Diego Vet Center
2790 Truxtun Rd.
Suite 130
San Diego, CA 92106
(858) 642-1500

San Francisco Vet Center
505 Polk Street
San Francisco, CA 94102
(415) 441-5051

San Jose Vet Center
5855 Silver Creek Valley Place
3rd Floor 3A
San Jose, CA 95138

San Luis Obispo Vet Center
1070 Southwood Dr.
San Luis Obispo, CA 93401
(805) 782-9101

San Marcos Vet Center
One Civic Center Dr.
Suite 150
San Marcos, CA 92069
(760) 744-6914

Santa Cruz County Vet Center
1350 41st Ave,
Suite 104
Capitola, CA 95010
(831) 464-4575

South Orange County Vet Center
26431 Crown Valley Parkway
Suite 100
Mission Viejo, CA 92691
(949) 348-6700

Temecula Vet Center
40935 County Center Drive
Suite A/B
Temecula, CA 92591
(951) 302-4849

Ventura Vet Center
790 E. Santa Clara St.
Suite 100
Ventura, CA 93001
(805) 585-1860

West Los Angeles Vet Center
5730 Uplander Way
Suite 100
Culver City, CA 90230
(310) 641-0326

COLORADO
Boulder Vet Center
4999 Pearl East Circle
Suite 106
Boulder, CO 80301
(303) 440-7306

Colorado Springs Vet Center
602 South Nevada Ave.
Colorado Springs, CO 80903
(719) 471-9992

Denver Vet Center
7465 East First Ave.
Suite B
Denver, CO 80230
(303) 326-0645

Ft. Collins Vet Center
702 W. Drake
Building C
Fort Collins, CO 80526
(970) 221-5176

Grand Junction Vet Center
561 25 Road
Suite 101
Grand Junction, CO 81505
(970) 245-4156

Pueblo Vet Center
1515 Fortino Blvd.
Suite 130
Pueblo, CO 81008
(719) 583-4058

CONNECTICUT
Danbury Vet Center
457 North Main St., 1st Floor
Danbury, CT 06811
(203) 790-4000

Hartford Vet Center
25 Elm St. Suite A
Rocky Hill, CT 06067
(860) 563-8800

New Haven Vet Center
291 South Lambert Rd.
Orange, CT 06477
(203) 795-0148

Norwich Vet Center
2 Cliff Street
Norwich, CT 06360
(860) 887-1755

DELAWARE
Sussex County Vet Center
20653 Dupont Blvd, Suite 1
Georgetown, DE 19947
(302) 225-9110

Wilmington Vet Center
2710 Centerville Rd. Suite 103
Wilmington, DE 19808
(302) 994-1660

DISTRICT OF COLUMBIA
Washington, D.C. Vet Center
1296 Upshsur Street Northwest
Washington, DC 20011
(202) 726-5212

FLORIDA
Bay County Vet Center
3109 Minnesota Ave, Suite 101
Panama City, FL 32405
(850) 522-6102

Clearwater Vet Center
29259 US Hwy 19 North
Clearwater, CL 33761-2102
(727) 549-3600

Clermont Vet Center
1655 East Highway 50, Suite 102
Clermont, FL 34711
(352) 536-6701

Daytona Beach Vet Center
1620 Mason Ave., Suite C
Daytona Beach, FL 32117
(386) 366-6600

Fort Lauderdale Vet Center
3666 W. Oakland Park Blvd.
Lauderdale Lakes, FL 33311
(954) 714-2381

Fort Myers Vet Center
2891 Center Point Drive, Suite 100
Ft. Myers, FL 33916
(239) 652-1861

Gainesville Vet Center
105 NW 75th Street, Suite #2
Gainesville, FL 32607
(352) 331-1408

Jacksonville Vet Center
3728 Phillips Highway, Suite 31
Jacksonville, FL 32207
(904) 399-8351

Jupiter Vet Center
6650 W. Indiantown Rd., Suite 120
Jupiter, FL 33458
(561) 422-1220

Lakeland Vet Center
1370 Ariana St.
Lakeland, FL 33803
(863) 284-0841

Melbourne Vet Center
2098 Sarno Road
Melbourne, FL 32935
(321) 254-3410

Miami Vet Center
8280 NW 27th St Suite 511
Miami, FL 33122
(305) 718-3712

Naples Vet Center
2705 Horseshoe Dr. South. Suite #204
Naples, FL 34104
(239) 403-2377

Ocala Vet Center
3300 SW 34th Avenue, Suite 140
Ocala, FL 34474
(352) 237-1947

Okaloosa County Vet Center
6 11th Avenue, Suite G-1
Shalimar, FL 32579
(850) 651-1000

Orlando Vet Center
5575 S. Semoran Blvd., Suite #30
Orlando, FL 32822
(407) 857-2800

Palm Beach Vet Center
4996 10th Ave North, Suite 6
Greenacres, FL 33463
(561) 422-1201

Pasco County Vet Center
5139 Deer Park Drive
New Port Richey, FL 34653
(727) 372-1854

Pensacola Vet Center
4504 Twin Oaks Drive
Pensacola, FL 32506
(850) 456-5886

Pompano Beach Vet Center
2300 West Sample Road, Suite 102
Pompano, FL 33073
(954) 984-1669

Sarasota Vet Center
4801 Swift Rd. Suite A
Sarasota, FL 34231
(941) 927-8285

St. Petersburg Vet Center
6798 Crosswinds Drive, Bldg A
St. Petersburg, FL 33710
(727) 549-3633

Tallahassee Vet Center
2002 Old St. Augustine Road
Bldg A/100
Tallahassee, FL 32301
(850) 942-8810

Tampa Vet Center
Fountain Oaks Business Plaza;
3637 W. Waters Ave., Suite 600
Tampa, FL 33614-2783
(813) 228-2621

GEORGIA
Atlanta Vet Center
1800 Phoenix Boulevard, Building 400,
Suite 404 Box 55
College Park, GA 30349
(404) 417-5414

Augusta Vet Center
2050 Walton Way, Suite 100
Augusta, GA 30904
(706) 729-5762

Columbus Vet Center
2601 Cross Country Drive, Condominium
B-2 Suite 900
Columbus, GA 31906
(706) 596-7170

Lawrenceville Vet Center
930 River Centre Place
Lawrenceville, GA 30043
(404) 728-4195

Macon Vet Center
750 Riverside Drive
Macon, GA 31201
(478) 477-3813

Marietta Vet Center
40 Dodd St., Suite 700
Marietta, GA 30060
(404) 327-4954

Savannah Vet Center
321 Commercial Dr
Savannah, GA 31406
(912) 961-5800

GUAM
Guam Vet Center
400 Route 8, Suite 301
Maite, GU 96910
(671) 472-7161

HAWAII
Hilo Vet Center
70 Lanihuli Street, Suite 102
Hilo, HI 96720
(808) 969-3833

Honolulu Vet Center
1680 Kapiolani Blvd. Suite F-3
Honolulu, HI 96814
(808) 973-8387

Kailua-Kona Vet Center
73-4976 Kamanu St
Kailua-Kona, HI 96740
(808) 329-0574

Kauai Vet Center
4485 Pahe'e St., Suite 101
Lihue, HI 96766
(808) 246-1163

Maui Vet Center
157 Ma'a Street
Kahului, HI 96732
(808) 242-8557

Western Oahu Vet Center
885 Kamokila Boulevard, Suite 105
Kapolei, HI 96707
(808) 674-2414

IDAHO
Boise Vet Center
2424 Bank Drive, Suite 100
Boise, ID 83705
(208) 342-3612

Pocatello Vet Center
1800 Garrett Way, Suite 47
Pocatello, ID 83201
(208) 232-0316

ILLINOIS

Aurora Vet Center
750 Shoreline Drive, Suite 150
Aurora, IL 60504
(630) 585-1853

Chicago Heights Vet Center
1010 Dixie Hwy, 2nd Floor
Chicago Heights, IL 60411
(708) 754-8885

Chicago Vet Center
3348W. 87th Street, Suite 2
Chicago, IL 60652-3767
(773) 962-3740

Metro East Vet Center
228 West Pointe Drive
Swansea, IL 62226
(618) 825-0160

Evanston Vet Center
1901 Howard St
Evanston, IL 60202
(847) 332-1019

Oak Park Vet Center
1515 South Harlem
Forest Park, IL 60130
(708) 383-3225

Orland Park Vet Center
8651 W.159th Street, Suite 1
Orland Park, IL 60462
(708) 444-0561

Peoria Vet Center
8305 N. Allen Road, Suite 1
Peoria, IL 61615
(309) 689-9708

Quad Cities Vet Center
465 42nd Ave (Avenue of the Cities) Suite #140
East Moline, IL 61244
(309) 755-3260

Rockford Vet Center
7015 Rote Road, Suite 105
Rockford, IL 61107
(815) 395-1276

Springfield, IL Vet Center
2980 Baker Drive
Springfield, IL 62703
(217) 492-4955

INDIANA

Evansville Vet Center
1100 N. Burkhardt Rd
Evansville, IN 47715
(812) 473-5993

Fort Wayne Vet Center
3810 Coldwater Road
Fort Wayne, IN 46805
(260) 460-1456

Gary Area Vet Center
107 E. 93rd Ave
Crown Point, IN 46307
(219) 736-5633

Indianapolis Vet Center
8330 Naab Road, Suite 103
Indianapolis, IN 46260
(317) 988-1600

South Bend Vet Center
4727 Miami Street
South Bend, N 46614
(574) 231-8480

IOWA

Cedar Rapids Vet Center
4250 River Center Court NE, Suite D
Cedar Rapids, IA 52402
(319) 378-0016

Des Moines Vet Center
1821 22nd Street #115
Des Moines, IA 50266
(515) 284-4929

Sioux City Vet Center
4280 Sergeant Road, Suite 104
Sioux City, IA 51106
(712) 255-3808

KANSAS

Manhattan Vet Center
1133 College Avenue, Building C, Suite 200
Manhattan, KS 66502
(785) 587-8257

Wichita Vet Center
251 N. Water St.
Wichita, KS 67202
(316) 265-0889

KENTUCKY
Lexington Vet Center
1500 Leestown Rd Suite 104
Lexington, KY 40511
(859) 253-0717

Louisville Vet Center
1347 South 3rd Street
Louisville, KY 40208
(502) 287-6710

LOUISIANA
Alexandria Vet Center
5803 Coliseum Blvd., Suite D
Alexandria, LA 71303
(318) 466-4327

Baton Rouge Vet Center
7850 Anselmo Lane, Suite B
Baton Rouge, LA 70810
(225) 761-3140

New Orleans Vet Center
1250 Poydras Street, 4th Floor,
Suite 400
New Orleans, LA 70113
(504) 507-4977

Shreveport Vet Center
1255 Shreveport Barksdale Highway
Shreveport, LA 71105
(318) 861-1776

MAINE
Bangor Vet Center
615 Oldin Road, Suite 3
Bangor, ME 04401
(207) 947-3391

Lewiston Vet Center
35 Westminster St.
Lewiston, ME 04240
(207) 783-0068

Northern Maine Vet Center
456 York Street
Caribou, ME 04736
(207) 496-3900

Portland Vet Center
475 Stevens Ave.
Portland, ME 04103
(503) 688-5361

Sanford Vet Center
628 Main Street
Springvale, ME 04083
(207) 490-1513

MARYLAND
Annapolis Vet Center
100 Annapolis Street
Annapolis, MD 21401
(410) 605-7826

Baltimore Vet Center
1777 Reisterstown Road Suite 199
Baltimore, MD 21208
(410) 764-9400

Dundalk Vet Center
1553 Merritt Blvd
Dundalk, MD 21222
(410) 282-6144

Elkton Vet Center
103 Chesapeake Blvd. Suite A
Elkton, MD 21921
(410) 392-4485

Prince George's County Vet Center
7905 Malcolm Road, Suite 101
Clinton, MD 20735-1734
(301) 856-7173

Silver Spring Vet Center
2900 Linden Lane Suite 100
Silver Spring, MD 20910
(301) 589-1073

MASSACHUSETTS
Boston Vet Center
7 Drydock Ave, Suite 2070
Boston, MA 02210-2303
(617) 424-0665

Brockton Vet Center
1041L Pearl St. Suite L
Brockton, MA 02301
(508) 580-2730

Cape Cod Vet Center
474 West Main Street
Hyannis, MA 02601
(508) 778-0124

Lowell Vet Center
10 George Street
Lowell, MA 01852
(978) 453-1151

New Bedford Vet Center
73 Huttleton Ave. Unit 2
Fairhaven, MA 02719
(508) 999-6920

198

Springfield Vet Center
95 Ashley Avenue, Suite A
West Springfield, MA 01089
(413) 737-5167

Worcester Vet Center
255 Park Ave Suite 900
Worcester, MA 01609
(508) 753-7902

MICHIGAN
Dearborn Vet Center
19855 Outer Drive, Suite 105 W
Dearborn, MI 48124
(313) 277-1428

Detroit Vet Center
11214 East Jefferson Avenue
Detroit, MI 48214
(313) 822-1141

Escanaba Vet Center
3500 Ludington Street, Suite # 110
Escanaba, MI 49829
(906) 233-0244

Grand Rapids Vet Center
2050 Breton Rd SE
Grand Rapids, MI 49546
(616) 285-5795

Macomb County Vet Center
42621 Garfield Rd. Suite 105
Clinton Township, MI 48038-5031
(586) 412-0107

Pontiac Vet Center
44200 Woodward Avenue, Suite 108
Pontiac, MI 48341
(248) 874-1015

Saginaw Vet Center
5360 Hampton Place
Saginaw, MI 48604
(989) 321-4650

Traverse City Vet Center
3766 N US 31 South
Traverse City, MI 49684
(231) 935-0051

MINNESOTA
Brooklyn Park Vet Center
7001 78th Avenue North, Suite 300
Brooklyn Park, MN 55445
(763) 503-2220

Duluth Vet Center
4402 Haines Road
Duluth, MN 55811
(218) 722-8654

St. Paul Vet Center
550 County Road D West, Suite 10
New Brighton, MN 55112
(651) 644-4022

MISSISSIPPI
Biloxi Vet Center
288 Veterans Ave
Biloxi, MS 39531
(228) 388-9938

Jackson Vet Center
1755 Lelia Dr. Suite 104
Jackson, MS 39216
(601) 985-2560

MISSOURI
Columbia Vet Center
4040 Rangeline Street, Suite 105
Columbia, MO 65202
(573) 814-6206

Kansas City Vet Center
4800 Main Street, Suite 107
Kansas City, MO 64112-2501
(816) 753-1866

Springfield, MO Vet Center
3616 S. Campbell
Springfield, MO 65807
(417) 881-4197

St. Louis Vet Center
287 N. Lindbergh Blvd
Creve Coeur, MO 63141
(314) 894-5739

MONTANA
Billings Vet Center
2795 Enterprise Ave., Suite 1
Billings, MT 59102
(406) 657-6071

Great Falls Vet Center
615 2nd Avenue North
Great Falls, MT 59401
(406) 452-9048

Kalispell Vet Center
690 North Meridian Road, Suite 101
Kalispell, MT 59901
(406) 257-7308

Missoula Vet Center
910 Brooks St.
Missoula, MT 59801
(406) 721-4918

NEBRASKA
Lincoln Vet Center
3119 O Street, Suite A
Lincoln, NE 68510
(402) 476-9736

Omaha Vet Center
3047 S 72nd Street
Omaha, NE 68124
(402) 346-6735

NEVADA
Henderson Vet Center
400 North Stephanie, Suite 180
Henderson, NV 89014
(702) 791-9100

Las Vegas Vet Center
7455 W. Washington Ave, Suite 240
Las Vegas, NV 89128
(702) 791-9170

Reno Vet Center
5580 Mill St. Suite 600
Reno, NV 89502
(775) 323-1294

NEW HAMPSHIRE
Berlin Vet Center
515 Main Street
Gorham, NH 03581
(603) 752-2571

Manchester Vet Center
1461 Hooksett Rd, B7
Hooksett, NH 03106-1882
(860) 646-5170

NEW JERSEY
Bloomfield Vet Center
2 Broad St. Suite 703
Bloomfield, NJ 07003
(973) 748-0980

Lakewood Vet Center
1255 Route 70 Suite 22N
Lakewood, NJ 08701
(908) 607-6364

Secaucus Vet Center
110A Meadowlands Parkway, Suite 102
Secaucus, NJ 07094

(201) 223-7787

South Jersey Vet Center
2900 Fire Road Suite 101
Egg Harbor Township, NJ 08234
(609) 487-8387

Trenton Vet Center
934 Parkway Ave. Suite 201
Ewing, NJ 08618
(609) 882-5744

NEW MEXICO
Albuquerque Vet Center
2001 Mountain Road NW
Albuquerque, NM 87104
(505) 346-6562

Farmington Vet Center
4251 E. Main Suite A
Farmington, NM 87402
(505) 327-9684

Las Cruces Vet Center
1120 Commerce Drive, Suite B
Las Cruces, NM 88001
(575) 523-9826

Santa Fe Vet Center
2209 Brothers Road Suite 110
Santa Fe, NM 87505
(505) 988-6562

NEW YORK
Albany Vet Center
17 Computer Drive West
Albany, NY 12205
(518) 626-5130

Babylon Vet Center
100 West Main Street
Babylon, NY 11702
(631) 661-3930

Binghamton Vet Center
53 Chenango Street
Binghamton, NY 13901
(607) 722-2393

Bronx Vet Center
2471 Morris Ave., Suite 1A
Bronx, NY 10468
(718) 367-3500

Brooklyn Vet Center
25 Chapel St. Suite 604
Brooklyn, NY 11201
(718) 630-2830

Buffalo Vet Center
2372 Sweet Home Road Suite 1
Amherst, NY 14228
(716) 862-7350

Harlem Vet Center
2279 3rd Avenue 2nd Floor
New York, NY 10035
(646) 273-8139

Manhattan Vet Center
32 Broadway 2nd Floor - Suite 200
New York, NY 10004-1637
(785) 350-4920

Middletown Vet Center
726 East Main Street, Suite 203
Middletown, NY 10940
(845) 342-9917

Nassau Vet Center
970 South Broadway
Hicksville, NY 11801-5019
(516) 348-0088

Queens Vet Center
75-10B 91 Avenue
Woodhaven, NY 11421-2824
(718) 296-2871

Rochester Vet Center
2000 S. Winton Road, Bldg 5, Suite 201
Rochester, NY 14618
(585) 393-7608

Staten Island Vet Center
60 Bay Street
Staten Island, NY 10301
(718) 816-4499

Syracuse Vet Center
109 Pine Street Suite 101
Syracuse, NY 13210
(315) 478-7127

Watertown Vet Center
210 Court Street Suite 20
Watertown, NY 13601
(315) 782-5479

White Plains Vet Center
300 Hamilton Ave. Suite C 1st Floor
White Plains, NY 10601
(914) 682-6250

NORTH CAROLINA
Charlotte Vet Center
2114 Ben Craig Drive Suite 300
Charlotte, NC 28262
(704) 549-8025

Fayetteville Vet Center
2301 Robeson St., Suite 103
Fayetteville, NC 28305
(910) 488-6252

Greensboro Vet Center
3515 West Market Street Suite 120
Greensboro, NC 27406
(336) 323-2660

Greenville, NC Vet Center
1021 WII Smith Blvd. Suite 100
Greenville NC 27834
(252) 355-7920

Jacksonville Vet Center
110A Branchwood Dr
Jacksonville, NC 28345
(910) 577-1100

Raleigh Vet Center
8851 Ellstree Lane, Suite 122
Raleigh, NC 27617
(919) 361-6419

NORTH DAKOTA
Bismarck Vet Center
619 Riverwood Drive, Suite 105
Bismarck, ND 58504
(701) 224-9751

Fargo Vet Center
3310 Fiechtner Drive. Suite 100
Fargo, ND 58103-8730
(701) 237-0942

Minot Vet Center
3300 South Broadway
Minot, ND 58701
(701) 852-0177

OHIO
Cincinnati Vet Center
4545 Montgomery Road
Norwood, OH 45212
(513) 763-3500

Cleveland Vet Center
5310 1/2 Warrensville Center Rd
Maple Heights, OH 44137
(216) 707-7901

Columbus Vet Center
30 Spruce Street
Columbus, OH 43215
(614) 257-5550

Dayton Vet Center
3085 Woodman Drive Suite 180
Kettering, OH 45420
(937) 296-0489

Parma Vet Center
5700 Pearl Road Suite 102
Parma, OH 44129
(440) 845-5023

Stark County Vet Center
601 Cleveland Ave N, Suite C
Canton, OH 44702
(330) 454-3120

Toledo Vet Center
1565 S. Byrne Road, Suite 104
Toleo, OH 43614
(419) 213-7533

OKLAHOMA
Lawton Vet Center
1016 SW C Avenue, Suite B
Lawton, OK 73501
(580) 585-5880

Oklahoma City Vet Center
6804 N. Robinson Avenue
Oklahoma City, OK 73116
(405) 456-5184

Tulsa Vet Center
14002 E. 21st Street
Tulsa, OK 74134-1422
(918) 628-2760

OREGON
Central Oregon Vet Center
1645 NE Forbes Rd. Suite 105
Bend, OR 97701
(541) 749-2112

Eugene Vet Center
190 East 11th Avenue
Eugne, OR 97401
(541) 465-6918

Grants Pass Vet Center
211 S.E. 10th St.
Grants Pass, OR 97526
(541) 479-6912

Portland Vet Center
1505 NE 122nd Ave.
Portland, OR 97230
(503) 688-5361

Salem Vet Center
2645 Portland Road NE, Suite 250
Salem, OR 97301
(503) 362-9911

PENNSYLVANIA
Bucks County Vet Center
2 Canal's End Road Suite 201B
Bristol, PA 19007
(215) 823-4590

City Center Philadelphia Vet Center
801 Arch Street Suite 502
Philadelphia, PA 19107
(215) 923-2284

DuBois Vet Center
100 Meadow Lane, Suite 8
DuBois, PA 15801
(814) 372-2095

Erie Vet Center
240 West 11th Street Suite 105
Erie, PA 16501
(814) 453-7955

Harrisburg Vet Center
1500 North Second Street Suite 2
Harrisburg, PA 17102
(717) 782-3954

Lancaster Vet Center
1817 Olde Homestead Lane, Suite 207
Lancaster, PA 17601
(717) 283-0735

Norristown Vet Center
320 East Johnson Hwy, Suite 201
Norristown, PA 19401
(215) 823-5245

Northeast Philadelphia Vet Center
101 East Olney Avenue Suite C-7
Philadelphia, PA 19120
(215) 924-4670

Pittsburgh Vet Center
2500 Baldwick Rd, Suite 15
Pittsburgh, PA 15205
(412) 920-1765

Scranton Vet Center
1002 Pittston Ave.
Scranton, PA 18505
(570) 344-2676

White Oak Vet Center
2001 Lincoln Way, Suite 280
White Oak, PA 15131
(412) 678-7704

Williamsport Vet Center
49 East Fourth Street Suite 104
Williamsport, PA 17701
(570) 327-5281

PUERTO RICO
Arecibo Vet Center
50 Gonzalo Marin St
Arecio, PR 00612
+1 (787) 879-4510

Ponce Vet Center
35 Mayor Street, Suite 1
Ponce, PR 00730
+1 (787) 841-3260

San Juan Vet Center
7th Tabonuco Street Suite 4
Guaynabo, PR 00968-4702
(787) 749-4410

RHODE ISLAND
Providence Vet Center
2038 Warwick Ave
Warwick, RI 02889
(401) 739-0167

SOUTH CAROLINA
Charleston Vet Center
3625 West Montague Avenue
North Charleston, SC 29418
(843) 789-7000

Columbia Vet Center
1710 Richland Street, Suite A
Columbia, SC 29201
(803) 765-9944

Greenville, SC Vet Center
3 Caledon Court, Suite B
Greenville, SC 29615
(864) 271-2711

Myrtle Beach Vet Center
2024 Corporate Centre Dr, Suite 103
Myrtle Beach, SC 29577
(843) 232-2441

SOUTH DAKOTA
Rapid City Vet Center
621 6th St, Suite 101
Rapid City, SD 57701
(605) 348-0077

Sioux Falls Vet Center
3200 W 49th Street
Sioux Falls, SD 57106
(605) 330-4552

TENNESSEE
Chattanooga Vet Center
1300 Premier Drive, Suite 180
Chattanooga, TN 37421
(423) 855-6570

Johnson City Vet Center
2203 McKinley Road, Suite 254
Johnson City, TN 37604
(423) 928-8387

Knoxville Vet Center
1645 Downtown West Blvd #28
Knoxville, TN 37919
(865) 633-0000

Memphis Vet Center
1407 Union Ave., Suite 410
Memphis, TN 38104
(901) 522-3950

Nashville Vet Center
1420 Donelson Pike Suite A-5
Nashville, TN 37217
(615) 883-2333

TEXAS
Abilene Vet Center
3564 N. 6th St.
Abilene, TX 79603
(325) 232-7925

Amarillo Vet Center
3414 Olsen Blvd. Suite E
Amarillo, TX 79109
(806) 354-9779

Arlington Vet Center
3337 West Pioneer Parkway
Pantgo, TX 76013
(817) 274-0981

Austin Vet Center
1524 South IH 35 Suite 100
Austin, TX 78704
(512) 416-1314

Beaumont Vet Center
990 IH10 North, Suite 180
Beaumont, TX 77702
(409) 347-0124

Corpus Christi Vet Center
4646 Corona Suite 250
Corpus Christi, TX 78411
(361) 854-9961

Dallas Vet Center
8610 Greenville Ave., Suite 125
Dallas, TX 75243
(214) 361-5896

El Paso Vet Center
1155 Westmoreland Suite 121
El Paso, TX 79925
(915) 772-0013

Fort Worth Vet Center
6620 Hawk Creek Avenue
Westworth Village, TX 76114
(817) 921-9095

Houston Southwest Vet Center
10103 Fondren Road, Suite 470
Houston, TX 77096
(713) 770-1803

Houston West Vet Center
701 N. Post Oak Road Suite 102
Houston, TX 77024
(713) 682-2288

Killeen Heights Vet Center
302 Millers Crossing, Suite #4
Harker Heights, TX 76548
(254) 953-7100

Laredo Vet Center
6999 McPherson Road Suite 102
Laredo, TX 78041
(956) 723-4680

Lubbock Vet Center
3106 50th St. Suite 400
Lubbock, TX 79413
806-792-9782

McAllen Vet Center
2108 S M Street, MedPoint Plaza
Unit 2
McAllen, TX 78503
(956) 631-2147

Mesquite Vet Center
502 West Kearney, Suite 300
Mesquite, TX 75149
(972) 288-8030

Midland Vet Center
4400 N. Midland Drive, Suite 540
Midland, TX 79707
(432) 697-8222

San Antonio Northeast Vet Center
9504 IH 35 N, Suite 214
San Antonio, TX 78233
(210) 650-0422

San Antonio Northwest Vet Center
9910 W Loop 1604 N, Suite 126
San Antonio, TX 78254
(210) 688-0606

Spring Vet Center
14300 Cornerstone Village Dr.,
Suite 110
Houston, TX 77014
(281) 537-7812

UTAH
Major Brent Taylor Vet Center
Outstation
2357 N 400 E
North Ogden, UT 84414

Provo Vet Center
360 S. State Street Bldg C Suite 103
Orem, UT 84770-4494
(801) 377-1117

Saint George Vet Center
1664 South Dixie Drive, Suite C-102
St. George, UT 84770-4494
(435) 673-4494

Salt Lake City Vet Center
22 West Fireclay Avenue
Murray, UT 84107
(801) 266-1499

VERMONT
South Burlington Vet Center
19 Gregory Drive Suite 201
South Burlington, VT 05403
(802) 862-1806

White River Junction Vet Center
118 Prospect Street Suite 100
White River Junction, VT 05001
(802) 295-2908

VIRGINIA
Alexandria Vet Center
6940 South Kings Highway Suite 204
Alexandria, VA 22310
(703) 360-8633

Norfolk Vet Center
3132 Lynnhurst Blvd
Chesapeake, VA 23321
(757) 623-7584

Richmond Vet Center
4902 Fitzhugh Avenue
Richmond, VA 23230
(804) 353-8958

Roanoke Vet Center
1401 Franklin Rd SW Suite 200
Roanoke, VA 24016
(540) 342-9726

Virginia Beach Vet Center
324 Southport Circle, Suite 102
Virginia Beach, VA 23452
(757) 248-3665

WASHINGTON
Bellingham Vet Center
3800 Byron Ave Suite 124
Bellingham, WA 98229
(360) 733-9226

Everett Vet Center
1010 SE Everett Mall Way Suite 207
Everett, WA 98208
(425) 252-9701

Federal Way Vet Center
32020 32nd Ave South Suite 110
Federal Way, WA 98001
(253) 838-3090

Seattle Vet Center
305 S. Lucille St.
Seattle, WA 98108
(206) 764-5130

Spokane Vet Center
13109 E Mirabeau Parkway
Spokane, WA 99216
(509) 444-8387

Tacoma Vet Center
4916 Center St. Suite E
Tacoma, WA 98409
(253) 565-7038

Walla Walla Vet Center
1104 West Poplar
Walla Walla, WA 99362
(509) 526-8387

Yakima Valley Vet Center
2119 W. Lincoln
Yakima, WA 98302
(509) 453-7161

WEST VIRGINIA
Beckley Vet Center
201 Grey Flatts Road
Beckley, WV 25801
(877) 927-8387

Charleston Vet Center
200 Tracy Way
Charleston, WV 25311
(304) 343-3825

Huntington Vet Center
3135 16th Street Road Suite 11
Huntington, WV 25701
(304) 523-8387

Martinsburg Vet Center
300 Foxcroft Ave. Suite 100A
Martinsburg, WV 25401-5341
(304) 263-6776

Morgantown Vet Center
34 Commerce Drive Suite 101
Morgantown, WV 26501
(304) 291-4303

Princeton Vet Center
1511 North Walker Street
Princeton, WV 24740
(304) 425-8098

Wheeling Vet Center
1058 East Bethlehem Blvd.
Wheeling, WV 26003
(304) 232-0587

WISCONSIN
Green Bay Vet Center
1600 S. Ashland Ave
Green Bay, WI 54304
(920) 435-5650

La Crosse Vet Center
20 Copeland Ave.
La Crosse, WI 54603
(608) 782-4403

Madison Vet Center
1291 N. Sherman Ave
Madison, WI 53704
(608) 264-5342

Milwaukee Vet Center
7910 N. 76th Street, Suite 100
Milwaukee, WI 53223
(414) 902-5561

WYOMING
Casper Vet Center
1030 North Poplar, Suite B
Casper, WY 82601
(307) 261-5355

Cheyenne Vet Center
3219 E Pershing Blvd
Cheyenne, WY 82001
(307) 778-7370

CHAPTER 19

EDUCATION BENEFITS OVERVIEW

The Department of Veterans Affairs provides education benefits to eligible servicemembers, veterans and certain dependents and survivors. You may receive financial support for undergraduate and graduate degrees, technical training, licensing and certification tests, apprenticeships, on-the-job training and more.

You may be eligible for one or more of the following VA education benefit programs:
- Post-9/11 GI Bill
- Montgomery GI Bill-Active Duty (MGIB-AD)
- Montgomery GI Bill-Selected Reserve (MGIB-SR)
- Reserve Educational Assistance Program (REAP
- Post-Vietnam Era Educational Assistance Program (VEAP)
- National Call to Service (NCS)
- Survivors' and Dependents' Educational Assistance (DEA)

You may be eligible to receive funds for:
- School tuition and fees for public, private or foreign schools, flight programs, correspondence training and distance learning
- Books and supplies
- License or certification tests
- National exams including SATs, ACTs, GMATs and LSATs
- On-the-job and apprenticeship training
- Distance learning/correspondence school
- Vocational/technical training
- Relocating from highly rural areas
- A monthly housing allowance

Each program offers different amounts of financial assistance and has varying eligibility requirements. If you are eligible for more than one education benefit, choose the one that's right for you. You can't choose more than benefit program at a time to receive payment under.

Post-9/11 GI Bill

The Post-9/11 GI Bill provides up to 36 months of education benefits to eligible servicemembers and veterans. The benefits may include financial support for school tuition and fees, books and supplies, and housing. You might also receive reimbursement for license or certification tests such as CPA, national tests like SAT or GMAT, or assistance for apprenticeships/on-the-job training. A one-time payment to help you relocate from certain rural areas to attend school is also available.

Each type of benefit, such as tuition or books, has a maximum rate. Based on the length of your active service, you are entitled to a percentage of the maximum total benefit.

You may be eligible for the Post-9/11 GI Bill benefits if you:
- Have at least 90 aggregate days of qualifying active service after September 10, 2001, or
- Were honorably discharged from active duty for a service-connected disability after serving at least 30 continuous days after September 10, 2001

If you are in the Armed forces, you may be able to transfer your Post-9/11 GI Bill benefits to your spouse and children. Generally, you must use these benefits within 15 years from your last period of active service of at least 90 days.

Montgomery GI Bill-Active Duty

Montgomery GI Bill-Active Duty (MGIB-AD) or Chapter 30, provides up to 36 months of financial assistance for educational pursuits including college, vocational, or technical training, correspondence courses, apprenticeships and on-the-job training, flight training, high-tech training, licensing and certification tests, entrepreneurship training, and national exams. Generally, your MGIB-AD benefits are paid directly to you each month.

You may be eligible for MGIB-AD benefits while you are on or after you separate from active duty. At a minimum you must have a high school diploma or GED. To receive benefits after separating, you must have received an honorable discharge. You generally have 10 years from your last date of separation from active duty to use your MGIB-AD benefits.

Montgomery GI Bill-Selected Reserve

Montgomery GI Bill-Selected Reserve (MGIB-SR) or Chapter 1606 provides up to 36 months of financial assistance for educational pursuits. MGIB-SR benefits are available to members of the Selected Reserve of the army, Navy, Air Force, Marine Corps or Coast Guard or the Army or Air National Guard. Generally, MGIB-SR benefits are paid directly to you on a monthly basis.

You may be eligible for MGIB-SR benefits if you have a six-year obligation to serve in the Selected Reserve, complete your Initial Active Duty for Training, serve in a drilling unit and remain in good standing, and obtain a high school diploma or equivalency.

The Guard and Reserves decide if you are eligible, while the VA makes the payments for the program. Generally, your eligibility for MGIB-SR benefits end on the day you leave the Selected Reserve.

Reserve Educational Assistance Program (REAP)

REAP or Chapter 1607 was an education program providing up to 36 months of education benefits to members of the Selected Reserves, Individual Ready Reserve (IRR) and National Guard who are called or ordered to active service in response to a war or national emergency, as declared by the President or Congress.

The National Defense Authorization Act of 2016 ended REAP on November 25, 2015. Some individuals remained eligible for benefits until November 25, 2019. The Post-9/11 GI Bill has largely replaced REAP, because it also provides educational assistance to Reserve and National Guard members. In many cases, it provides more benefit than REAP.

Post-Vietnam Era Educational Assistance Program

The Post -Vietnam Era Educational Assistance Program or Chapter 32 is an education benefit available if you made contributions from your military pay before April 1, 1987, to participate in the program. Your contributions are matched on a $2 for $1 basis by the government. You can use the benefits for degree, certificate, correspondence, apprenticeships/on-the-job training, and vocational flight training programs.

Benefit entitle Is 1-36 months, depending on the number of contributions you made to VEAP. You have 10 years from your release from active duty to use your VEAP benefits. If you have not completely used your entitlement after 10 years, your remaining contributions will be automatically refunded.

National Call to Service
The National Call to Service Program is a Department of Defense program administered by VA. It offers certain educational benefits to individuals who serve in a military occupational specialty designated by the Secretary of Defense. To be eligible you must serve on active duty for 15 months after completing initial entry training. After this, and without a break in service, you must serve either an additional period of active duty or a period of 24 months in an active status in the Selected Reserve. Without a break in service, you must serve the remaining period of obligated service:

- On active duty in the Armed Forces
- In the Selected Reserve
- In the Individual Ready Reserve
- In AmeriCorps or another domestic national service program jointly designated by the Secretary of Defense and the head of such a program.

If you meet the service requirements, you can choose one of the following benefits:

- A cash bonus of $5,000
- Repayment of a qualifying student loan not to exceed $18,000
- Entitlement to an allowance equal to the 3-year monthly MGIB-AD rate for 12 months
- Entitle to an allowance equal to 50% of the less-than-three-year monthly MGIB-AD rate for 36 months

VA only administers the education options. VA does not issue the cash bonus or repay student loans.

Survivors' and Dependents' Educational Assistance
Survivors' and Dependents' Educational Assistance (DEA) or Chapter 35, provides assistance for degree and certificate programs, apprenticeships/on-the-job training, correspondence courses, and other programs. You must be the son, daughter or spouse of:

- A veteran who died or is permanently and totally disabled as the result of a service-connected disability. The disability must arise from active service in the Armed Forces.
- A veteran who died from any cause while such permanent and total service-connected disability existed.
- A servicemember who died during active military service.
- A servicemember missing in action or captured in the line of duty by a hostile force.
- A service member forcibly detained or interned in the line of duty by a foreign government or power.
- A servicemember who is hospitalized or receiving outpatient treatment for a service-connected permanent and total disability and is likely to be discharged for that disability.

The program offers up to 45 months of education benefits. If you are the child, you generally must use your benefits between the ages of 18 and 26. If you are the spouse, your benefits end 10 years from the date VA finds you eligible or from the date of death of your spouse.

A surviving spouse who meets the criteria below may be eligible for benefits for 20 years from the date of death:

- VA rated your spouse permanently and totally disabled with an effective date within three years from discharge or
- Your spouse died on active duty

Veteran Rapid Retraining Assistance Program (VRRAP)

The Veteran Rapid Retraining Assistance Program (VRRAP) offers education and training for high-demand jobs to veterans who are unemployed because of the COVID-19 pandemic.

To be eligible, you have to meet all of the following requirements:

- You're at least 22 years old, but not older than 66 and
- You're unemployed because of the COVID-19 pandemic, and
- You're not rated as totally disabled because you can't work, and
- You're not enrolled in a federal or state jobs program

Note: You can't receive VRRAP benefits at the same time as you're receiving unemployment benefits, including CARES Act Benefits. You also can't be eligible for other education benefits or VR&E at the time you apply for VRRAP. You can get VRRAP benefits if you were at one time eligible for the Post-9/11 GI Bill but you've transferred all of your benefits to family members.

VRRAP covers education and training programs approved under the GI Bill and Veteran Employment Through Technology Education Courses that lead to high-demand jobs. The Department of Labor determines what a high-demand job is.

If you're eligible for the program, you can get up to 12 months of tuition and fees, and a monthly housing allowance based on Post-9/11 GI Bill rates.

VRRAP will be available until December 11, 2022. VA will stop making payments on that date or when they reach the $386 million funding limit or the 17,250-participant limit.

If you have questions, you can call (888) 442-4551.

Edith Nourse Rogers STEM Scholarship

The Edith Nourse Rogers Science Technology Engineering Math (STEM) Scholarship allows some eligible veterans and dependents in high-demand fields to extend their Post-9/11 GI Bill or Fry Scholarship benefits.

You may be eligible for this scholarship as a veteran or a Fry Scholar if you meet at least one of the below requirements:

- You're currently enrolled in an undergraduate STEM degree program or qualifying dual-degree program, or
- You've earned a post-secondary degree or graduate degree in an approved STEM degree field and are enrolled in a covered clinical training program for health care professionals, or
- You've earned a post-secondary degree in an approved STEM degree field and are working toward a teaching certification

If you're currently enrolled in an undergraduate STEM degree or qualifying dual-degree program, all of the following must be true:

- You're enrolled in a qualifying undergraduate STEM degree program that requires at least 120 standard semester credit hours (or 180 quarter credit hours) to complete, and
- You've completed at least 60 standard credit hours (or 90 quarter credit hours)

toward your degree, and
- You have six months or less of your Post-9/11 GI Bill (or Fry Scholarship) benefits left

At this time, you can't use the STEM scholarship for graduate degree programs.

If you're enrolled in a covered clinical training program for health care professionals, all of the following must be true:
- You've earned a qualifying degree in a STEM field, and
- You've been accepted or are enrolled in a covered clinical training program for health care professionals, and
- You have six months or less of your Post-9/11 GI Bill (or Fry Scholarship) benefits left.

If you're working toward a teaching certification, all of the following must be true:
- You've earned a qualifying post-secondary degree in a STEM field, and
- You've been accepted or are enrolled in a teaching certification program, and
- You have six months or less of your Post-9/11 GI Bill (or Fry Scholarship) benefits left.

If you meet eligibility requirements, VA can't guarantee you'll receive the Rogers STEM Scholarship. VA gives priority to veterans and Fry Scholars who are eligible for the maximum Post-9/11 GI Bill benefit and require the most credit hours compared to other applicants.

You can apply for the scholarship online through the VA. VA usually decides about each scholarship within 30 days and they award scholarships on a monthly basis. If more information is needed to decide, the VA sends a letter.

If the VA approves your application, you'll get a Certificate of Eligibility in the mail. This is also called a decision letter. You should bring this to the VA certifying official at your school.

If your application isn't approved, you'll get a denial letter by mail.

What to Remember
✓ Education benefits help veterans, service members and qualified family members pay college tuition, get career counseling and find the right school or training program.
✓ Since 1944, the GI Bill has helped qualifying veterans and family members cover the costs of school or training.
✓ The Post-9/11 GI Bill is to help pay for school or job training if you served on active duty after September 10, 2001.
✓ Other programs are available if you aren't eligible for the Post-9/11 GI Bill including The Montgomery GI Bill Active Duty and The Montgomery GI Bill Selected Reserve.
✓ If you performed a period of national service, you may also be eligible for the National Call to Service program.

CHAPTER 20

EDUCATION BENEFITS - MONTGOMERY G.I. BILL (MGIB), CHAPTER 30

The Montgomery GI Bill Active Duty (MGIB-AD) can help pay for education and training programs for veterans who served at least two years on active duty. You may be eligible for education benefits through this program if you were honorably discharged and you meet the requirements of one of the following categories.

Category I
All of these are true:
- You have a high school diploma, GED or 12 hours of college credit, and
- You entered active duty for the first time after June 30, 1985, and
- You had your military pay reduced by $100 a month for the first 12 months of service

And you served continuously for at least one of these time periods:
- 3 years, or
- 2 years if that was your agreement when you enlisted, or
- 4 years if you entered the Selected Reserve within a year of leaving active duty—this is called the 2 by 4 program

Category II
All of these are true:
- You have a high school diploma, GED, or 12 hours of college credit, and
- You entered active duty before January 1, 1977 (or before January 2, 1978, under a delayed enlistment program contracted before January 1, 1977), and
- You served at least 1 day between October 19, 1984, and June 30, 1985, and stayed on active duty through June 30, 1988 (or through June 30, 1987, if you entered the Selected Reserve within 1 year of leaving active duty and served 4 years, and
- You had at least 1 day of entitlement left under the Vietnam Era GI Bill (Chapter 34) as of December 31, 1989

Category III
All of these are true:
- You have a high school diploma, GED, or 12 hours of college credit, and
- You don't qualify for MGIB under categories I or II, and
- You had your military pay reduced by $1,200 before separation

And one of these is true:
- You were on active duty on September 30, 1990, and involuntary separated not by your choice after February 2, 1991, or

- You were involuntarily separated on or after November 30, 1993, or
- You chose to voluntarily separate under either the Voluntary Separation Initiative (VSI) program or the Special Separation Benefit (SSB) program

Category IV
Both of these are true:
- You have a high school diploma, GED or 12 hours of college credit, and
- You had military pay reduced by $100 a month for 12 months or made a $1,200 lump sum contribution

And one of these is true:
- You were on active duty on October 9, 1996, had money left in a VEAP account on that date, and you chose MGIB before October 9, 1997, or
- You entered full-time National Guard duty under title 32, USC, between July 1, 1985, and November 28, 1989, and chose MGIB between October 9, 1996 and July 9, 1997

The amount a veteran will receive through this education program depends on:
- Their length of service
- The type of education or training program they choose
- Their category (I, II, III or IV)
- Whether they qualify for a college fund or kicker
- How much the veteran paid into the $600 Buy-Up program

You usually have 10 years to use your MGIB-AD benefits, but this can change depending on your situation.

The transferability option under the Post-9/11 GI Bill allows Servicemembers to transfer all or some unused benefits to their spouse or dependent children. The request to transfer unused GI Bill benefits to eligible dependents must be completed while serving as an active member of the Armed Forces. The Department of Defense (DoD) determines whether or not you can transfer benefits to your family. Once the DoD approves benefits for transfer, the new beneficiaries apply for them at VA. To find out more, visit DoD's website or apply now.

Type of Assistance

Eligible Servicemembers may transfer all 36 months or the portion of unused Post-9/11 GI Bill benefits (unless DoD or the Department of Homeland Security has limited the number of transferable months). If you're eligible, you may transfer benefits to the following individuals:
- Your spouse
- One or more of your children
- Any combination of spouse and child

Available Benefits and Eligibility

Family members must be enrolled in the Defense Eligibility Enrollment Reporting System (DEERS) and be eligible for benefits at the time of transfer to receive transferred benefits.

The option to transfer is open to any member of the armed forces active duty or Selected Reserve, officer or enlisted who is eligible for the Post-9/11 GI Bill, and meets the following criteria:
- Has at least six years of service in the armed forces (active duty and/or Selected Reserve) on the date of approval and agrees to serve four additional years in the armed forces from the date of election.

- Has at least 10 years of service in the armed forces (active duty and/or Selected Reserve) on the date of approval, is precluded by either standard policy (by Service Branch or DoD) or statute from committing to four additional years and agrees to serve for the maximum amount of time allowed by such policy or statute.

- Transfer requests are submitted and approved while the member is in the armed forces.

- Effective 7/20/19, eligibility to transfer benefits will be limited to sevicemembers with less than 16 years of active duty or selected reserve service.

Transfer Process. While in the armed forces, transferors use the Transfer of Education Benefits (TEB) website to designate, modify, and revoke a Transfer of Entitlement (TOE) request. After leaving the armed forces, transferors may provide a future effective date for use of TOE, modify the number of months transferred, or revoke entitlement transferred by submitting a written request to VA.

Upon approval, family members may apply to use transferred benefits with VA by applying online or by printing, completing, and mailing the VA Form 22-1990e to your VA Regional processing office of jurisdiction. VA Form 22-1990e should only be completed and submitted to VA by the family member after DoD has approved the request for TEB. Do not use VA Form 22-1990e to apply for transfer of education benefits.

Other Factors to Consider
Marriage and Divorce

- A child's subsequent marriage will not affect his or her eligibility to receive the educational benefit; however, after an individual has designated a child as a transferee under this section, the individual retains the right to revoke or modify the transfer at any time.

- A subsequent divorce will not affect the transferees eligibility to receive educational benefits; however, after an individual has designated a spouse as a transferee under this section, the eligible individual retains the right to revoke or modify the transfer at any time.

Reallocation of Benefits
If a Veteran or Servicemember wants to reallocate transferred benefits they can do so using the TEB Portlet in MilConnect.

Reallocation of Benefits if a Family Member Dies
The Harry W. Colmery Veterans Assistance Act of 2017 allows for designation and transfer of Post-9/11 GI Bill benefits to eligible dependents of the Veteran/servicemember upon the death of the Veteran/servicemember or of a dependent who had unused transferred benefits.

Nature of Transfer
Family member use of transferred educational benefits is subject to the following rules:
Spouses:

- May start to use the benefit immediately

- May use the benefit while the member remains in the Armed Forces or after separation from active duty

- Are not eligible for the monthly housing allowance while the member is serving on active duty

- If servicemember's last discharge was before January 1, 2013, can use the benefit for up to 15 years after the last separation from active duty. If the servicemember's last discharge is after January 1, 2013, there is no time limit to use benefits

Children

- May start to use the benefit only after the individual making the transfer has completed at least 10 years of service in the armed forces
- May use the benefit while the eligible individual remains in the armed forces or after separation from active duty
- May not use the benefit until he or she has attained a secondary school diploma (or equivalency certificate), or he or she has reached age 18
- Is entitled to the monthly housing allowance stipend even though the eligible individual is on active duty
- Is not subject to the 15-year delimiting date, but may not use the benefit after reaching 26 years of age

Discharges and Separations

As previously mentioned, if the veteran is separated from active duty, the character of discharge must specifically be listed as "Honorable." "Under Honorable Conditions," or a "General" discharge do not establish eligibility. A discharge for one of the following reasons may result in a reduction of the required length of active duty to qualify for benefits under the MGIB:

- Convenience of the Government; or
- Disability; or
- Hardship; or
- Medical conditions existing before entry into Service; or
- Force reductions; or
- Medical condition which is not a disability due to misconduct, but which prevents satisfactory performance of duty.

The following types of active duty do not establish eligibility for MGIB benefits:

- Time assigned by the military to a civilian institution to take the same course provided to civilians.
- Time served as a cadet or a midshipman at a service academy.
- Time spent on active duty for training in the National Guard or Reserve.

Please note: Time assigned by the military to a civilian institution, and time served at a service academy does not break the continuity of active duty required to establish eligibility for MGIB benefits. Active duty for training does count toward the four years in the Selected Reserve under the 2 by 4 program.

Approved Courses

This program provides veterans up to 36 months of education benefits. The benefits may be used for:

- Undergraduate or graduate degrees from a college or university;
- Cooperative training programs;
- Accredited independent study programs leading to standard college degrees;
- Courses leading to certificates or diplomas from business, technical or vocational schools;
- Vocational flight training (from September 30, 1990 only – Individuals must have a private pilot's license and meet the medical requirements for a desired license before beginning training, and throughout the flight training program);
- Apprenticeship / job training programs offered by a company or union;
- Correspondence courses;

VA may approve programs offered by institutions outside of the United States, when they are pursued at educational institutions of higher learning, and lead to an associate or higher degree, or the equivalent. Individuals must receive VA approval prior to attending or enrolling in any foreign programs.

Restrictions on Training

- Bartending and personality development courses
- Non-accredited independent study courses;
- Any course given by radio;
- Self-improvement courses such as reading, speaking, woodworking, basic seamanship, and English as a second language;
- Any course which is avocational or recreational in character;
- Farm cooperative courses;
- Audited courses;
- Courses not leading to an educational, professional, or vocational objective;
- Courses an individual has previously completed;
- Courses taken by a Federal government employee under the Government Employees' Training Act;
- Courses paid for in whole or in part by the Armed Forces while on active duty;
- Courses taken while in receipt of benefits for the same program from the Office of Workers' Compensation Programs.

The VA must reduce benefits for individuals in Federal, State or local prisons after being convicted of a felony.

An individual may not receive benefits for a program at a proprietary school if he or she is an owner or official of the school. Benefits are generally payable for 10 years following a veteran's release from active duty.

Part-Time Training

Individuals unable to attend school full-time should consider going part-time. Benefit rates and entitlement charges are pro-rated as follows:

- Individuals who are on active duty or training at less than one-half time, will receive the lesser of:
- The monthly rate based on tuition and fees for the course(s); or
- The maximum monthly rate based on training time.
- Individuals training at less than one-half time will receive payment in one sum for the whole enrollment period.

Remedial, Deficiency and Refresher Training

Remedial and deficiency courses are intended to assist a student in overcoming a deficiency in a particular area of study. In order for such courses to be approved, the courses must be deemed necessary for pursuit of a program of education. Refresher training is for technological advances that occurred in a field of employment. The advance must have occurred while the student was on active duty, or after release. There is an entitlement charge for these courses.

Tutorial Assistance

Students may receive a special allowance for individual tutoring, if attending school at one-half time or more. To qualify, the student must have a deficiency in a subject. The school must certify the tutor's qualifications, and the hours of tutoring. Eligible students may receive a maximum monthly payment of $100.00. The maximum total benefit payable is $1,200.00.

There is no entitlement charge for the first $600.00 of tutorial assistance. To apply for tutorial assistance, students must submit VA Form 22-1990t, "Application and Enrollment

Certification for Individualized Tutorial Assistance." The form should be given to the certifying official in the office handling VA paperwork at the school for completion.

Months of Benefits/Entitlement Charged

Individuals who complete their full period of enlistment may receive up to 36 months of MGIB benefits.

Individuals are considered to have completed their full enlistment period if they are discharged for the convenience of the government after completing 20 months of an enlistment of less than three years; or 30 months of an enlistment of three years or more Individuals will earn only one month of entitlement for each month of active duty after June 39, 1985, if they are discharged for other specific reasons (i.e. service- connected disability, reduction in force, hardship, etc.) before completing the enlistment period.

Individuals will earn one month of entitlement for each four months in the Selected Reserve after June 30, 1985. Individuals qualifying for more than one VA education program may receive a maximum of 48 months of benefits. For example, if a student used 30 months of Dependents' Educational Assistance, and is eligible for chapter 1606 benefits, he or she could have a maximum of 18 months of entitlement remaining.

Individuals are charged one full day of entitlement for each day of full-time benefits paid. For correspondence and flight training, individuals use one month of entitlement each time the VA pays the equivalent of one month of full-time benefits. Individuals pursuing a cooperative program use one month for each month of benefits paid.

For apprenticeship and job-training programs, the entitlement charge during the first 6 months is 75% of full-time. For the second six months, the charge is 55% of full-time. For the rest of the program, the charge is 35% of full-time. VA can extend entitlement to the end of a term, quarter, or semester if the ending date of an individual's entitlement falls within such period. If a school does not operate on a term basis, entitlement can be extended for 12 weeks.

Rates of Educational Assistance After Separation from Active Duty

The basic monthly rates increase October 1 every year with the Consumer Price Index (CPI) increase. While in training, students receive a letter with the current rates when the increase goes into effect each year. The rates may increase at other times by an act of Congress.

Basic Monthly Rates for College and Vocational Training

For approved programs in college and vocational or technical schools, basic payments are monthly, and the rates are based on training time. When students train at less than half time, they will be paid tuition and fees. But if tuition and fees amount to more than would be paid at the half-time rate (or the quarter-time rate if training at quarter-time or less), payments will be limited to the half time (or the quarter-time rate). For on-the-job training (OJT) and apprenticeship programs, rates are monthly and based on the length of time in the program. MGIB rates decrease as the student's wages increase according to an approved wage schedule.

Rates for Other Types of Training

For correspondence courses, students receive 55% of the approved charges for the course. For flight training, students receive 60% of the approved charges for the course. For reimbursement of tests for licenses or certifications, students receive 100% of the charges up to a maximum of $2,000 per test.

Rates of Educational Assistance While on Active Duty

If a service member goes to school while on active duty, he or she may have two options for using MGIB benefits. They may be eligible to receive:

- "Regular" MGIB; or
- Tuition Assistance plus MGIB, or
- Tuition Assistance "Top-Up"

Using "Regular" MGIB While on Active Duty

If a service member uses "regular" MGIB while on active duty, VA can pay whichever is less:

- The monthly rate based on tuition and fees for your course(s); or
- The maximum monthly MGIB rate (basic rate plus any increases he or she may qualify for).

The basic monthly rates increase October 1 every year with the Consumer Price Index (CPI) increase. While in training, students receive a letter with the current rates when the increase goes into effect each year. The rates may increase at other times by an act of Congress.

Using Tuition Assistance Top-Up

If a student is on active duty, he or she may be eligible to receive Tuition Assistance (TA) from his or her branch of service. If the student has been on active duty for two years, he or she may also be eligible to use MGIB to supplement, or "top up," the TA. Top-up covers the remaining percentage of costs approved for TA that TA alone doesn't' cover—**up to specified limits**. For example, if a student's service authorizes 75% of costs, top-up can pay the remaining 25% of costs approved for TA.

DETAILED Rates of Educational Assistance— The rates effective October 1, 2022, are detailed in the following charts:

Basic Monthly Rates Effect October 1, 2022 Montgomery G.I. Bill – Active Duty (MGIB), Chapter 30					
Type of Training	Full-Time	¾ Time	½ Time	Less Then ½ Time but More than 1/4	¼ Time
Institutional	$2210.00	$1657.50	$1105.00	$1105.00	$552.50
Cooperative Training	$2210.00				
Correspondence Training	VA pays 55% of the established cost for the number of lessons completed. You have to submit completed lessons to receive payments.				
Apprenticeship On-The-Job Training-	First six months: $1657.50 for each full month Second six months: $1215.50 for each full month Remainder of program: $773.50 for each full month				
Flight Training	VA will pay 60% of approved charges.				

Basic Monthly Rates Effective October 1, 2022 For persons who's initial active-duty obligation was less than three years and who served less than three years (excluding 2x4 participants)					
Type of Training	Full- Time	¾ Time	½ Time	Less Then ½ Time but More Than ¼	¼ Time
Institutional	$1793.00	$1344.75	$896.50	Tuition & Fees, not to exceed $896.50	$448.25
Correspondence Training	VA will pay 55% of the established cost of lessons completed.				
Apprenticeship On-The-Job Training-	First six months: $1344.75 for each full month Second six months: $986.15 for each full month Remainder of program: $627.55 for each full month				
Flight Training	VA will pay for 60% of the approved charges.				
Cooperative Training	$1793.00 (Full-Time Only)				

Special Notes:
Cooperative Training is full time only.
Individuals taking correspondence courses will receive 55% of the approved charges for the course. Individuals taking flight training will receive 60% of the approved charges for the course, including solo hours.

Increased Above Basic Rates
Individuals may qualify for the following increases above their basic monthly rates. These increases don't apply to correspondence courses, the test for a license or certification, or flight training.

College Fund

Certain branches of service may offer the College Fund. The College Fund money, or "kicker," is an additional amount of money that increases the basic MGIB monthly benefit and is included in the VA payment.

Important: Students can't receive College Fund money without receiving MGIB. A common misunderstanding is that the College Fund is a separate benefit from MGIB. The College Fund is an add-on to the MGIB benefit.

VEAP Kicker

VA pays an additional amount, commonly known as a "kicker," if directed by the Department of Defense (DoD). If an individual is eligible under Category 3 or Category 4 and has a VEAP kicker, he or she can receive the amount of the VEAP kicker contributed by the service department divided by the total months of MGIB eligibility.

Accelerated Payments for Education Leading to Employment in High Technology

One of the provisions contained in the Veterans Education and Benefits Expansion Act of 2001 (Public Law 107-103) called for accelerated payments for education leading to employment in high technology, effective October 1, 2002. Following are questions and answers provided by the VA regarding this provision.

What is an accelerated payment?
An accelerated payment is a lump sum payment of 60% of tuition and fees for certain high cost, high tech programs. If a participant does not have sufficient entitlement to cover 60% of tuition and fees, he or she will receive pay based on the actual remaining entitlement.

VA will make accelerated payments for one term, quarter, or semester at a time. However, if the program is not offered on a term, quarter or semester basis, the accelerated payment is paid for the entire program. To qualify, a participant must be enrolled in a high-tech program and must certify that he or she intends to seek employment in a high-tech industry as defined by VA. Accelerated payment is paid instead of Montgomery GI Bill benefits that would otherwise have been received.

Who qualifies for accelerated payments?
Only individuals eligible for the Montgomery GI Bill - Active Duty (Chapter 30) qualify for accelerated payments.

How high do the tuition and fees have to be?
To receive accelerated payment, the tuition and fees must be more than double the Montgomery GI Bill benefits that a participant would otherwise receive for that term. For example, if the full-time rate is $732 and a participant is enrolled in a 4-month semester, the tuition and fees must be over $5,856 (4 months x $732=$2,928; $5,856=2 x $2,928) before he or she could receive an accelerated payment.

If a participant receives $900 in monthly benefits, the tuition and fees must be over $7,200 (4 months x $900=$3,600; $7,200= 2 x $3,600).

If a participant receives $1,050 in monthly benefits, the tuition and fees must be over $8,400 (4 months x $1,050 = $4,200; 2 x $4,200 = $8,400).

What programs qualify for accelerated payment?

Both degree and non-degree programs qualify. A participant must be enrolled in a program in one of the following categories:

Life science or physical science (but not social science);
Engineering (all fields);
Mathematics;
Engineering and science technology;
Computer specialties; and·
Engineering, science, and computer management

What industries qualify for accelerated payments?

A participant must intend to seek employment in one of the following industries:
Biotechnology;
Life Science Technologies;
Opto-electronics;
Computers and telecommunications; Electronics;
Computer-integrated manufacturing; Material Design;
Aerospace;
Weapons;
Nuclear technology

How does a participant apply for accelerated payments?

The individual must ask the school to include his or her request for accelerated payment to VA when it sends the enrollment information to VA for processing. The individual's request must include his or her certification of intent to seek employment in a high technology industry.

How is the education entitlement charged?

VA will divide the accelerated payment by the amount of the individual's full-time monthly rate (including kickers and additional contributions) and will reduce the entitlement by the resulting number of months and days. Example: Jill received an accelerated payment of $3,600. Her full-time rate is $900. VA will charge her entitlement as follows: $3,600/$900 = 4 months.

When can accelerated payments be made?

Accelerated payments may only be made for terms or other enrollment periods that begin on or after October 1, 2002.

Can school-related expenses (such as books, supplies and living expenses) be counted as tuition and fees for accelerated payments?

No. Only the school's tuition and fees can be considered for accelerated payment.

Can an individual receive accelerated payments for short, non-degree courses?

Yes, as long as they are approved for VA benefits. Short, expensive, IT courses offered by businesses typically are not approved for VA benefits.

221

Can an individual receive accelerated payments for non-technical courses when taking them as part of a high technology program?

Yes. However, the degree or certificate must require the completion of these other non-technical courses.

Is it possible to receive an accelerated payment check before a school term begins?

No. VA needs to verify that the individual has enrolled before sending out the large payment. VA will pay the student as soon after the start of the term as possible. Individuals will receive payment faster if they receive direct deposit.

Does a student have to verify enrollment each month if he or she receives an accelerated payment?

No. After the individual completes his or her enrollment, VA will ask the student to verify that he or she have received the accelerated payment. VA will also ask the individual to indicate how he or she used the accelerated payment (such as toward tuition, fees and books and supplies). VA is asking the latter question for statistical purposes only because the law requires them to collect this information. A student's answer will have no bearing on his or her entitlement to the accelerated payment. The student must respond to these questions within 60 days from the end of the enrollment period or VA will create an overpayment equal to the accelerated payment. As with any course, the student must notify VA of any change in his or her enrollment. The student's school must report any changes as well.

Is there any financial risk with accelerated payment?

Yes. If a student receives a grade, which does not count toward graduation requirements, he or she may have to repay all or part of the accelerated payment, depending on the circumstances. This could be a large amount of money.

Do the accelerated payments have to be paid back if the individual fails to find employment in a high technology industry?

No. The fact that he or she intended to find employment in a high technology industry is sufficient.

List of approved high technology programs

1.09 Animal Sciences
01.0901 Animal Sciences, General
01.0902 Agricultural Animal Breeding
01.0903 Animal Health
01.0904 Animal Nutrition
01.0905 Dairy Science
01.0906 Livestock Management
01.0907 Poultry Science

01.10 Food Science and Technology
01.1001 Food Science
01.1002 Food Technology and Processing
01.1099 Food Science and Technology
01.1101 Plant Sciences General
01.1102 Agronomy and Crop Science
01.1103 Horticultural Science

222

01.1104 Agricultural and Horticultural Plant Breeding
01.1105 Plant Protection and Integrated Pest Management
01.1106 Range Science Management

01.12 Soil Sciences
01.1201 Soil Science and Agronomy General
01.1202 Soil Chemistry and Physics
01.1203 Soil Microbiology
01.1299 Soil Sciences

03.01 Natural Resources Conservation and Research
03.0104 Environmental Science

03.03 Fishing and Fisheries Sciences and Management

Forestry
03.0501 Forestry, General
03.0502 Forest Sciences and Biology
03.0506 Forest Management/Forest Resources Management
03.0508 Urban Forestry
03.0509 Wood Science and Wood Products/Pulp and Paper Technology
03.0510 Forest Resources Production and Management
03.0511 Forest Technology/Technician

Wildlife and Wildlands Science and Management
09.07 Radio, Television, and Digital Communication
09.0702 Digital Communication and Media/Multimedia

11.0101 Computer and Information Sciences, General
11.0102 Artificial Intelligence and Robotics
11.0103 Information Technology
11.0199 Computer and Information Sciences

Computer Programming
11.0201 Computer Programming/Programmer General
11.0202 Computer Programming Specific Applications

Data Processing
11.0301 Data Process and Data Processing Technology/Technician

Information Science/Studies Computer Systems Analysis Computer Science

Computer Software and Media Application
11.0801 Web Page, Digital/Multimedia and Information Resources Design 11.0802 Data Modeling/Warehousing and Database Administration
11.0803 Computer Graphics
11.0899 Computer Software and Media Applications, Other.
Computer System Networking and Telecommunications Computer/Information Technology Administration and Management
11.1001 System Administration/Administrator
11.1002 System, Networking and LAN/WAN Management/Manager
11.1003 Computer and Information Systems Security
11.1004 Web/Multimedia Management and Webmaster

Engineering
Instructional program that prepare individuals to apply mathematical and scientific principles to the solution of practical problems.
*15. Engineering Technologies/Technicians.

Instructional programs that prepare individuals to apply basic engineering principles and technical skills in support of engineering and related projects.

Biological and Biomedical Sciences.
Instructional programs that focus on the biological sciences and the non-clinical biomedical sciences, and that prepare individuals for research and professional careers as biologist and biomedical scientist.

Mathematics and Statistics.
Instructional programs that focus on the systematic study of logical symbolic language and its applications.

*Military Technologies.
A program that prepares individuals to undertake advanced and specialized leadership and technical responsibilities for the armed services and related national security organizations. Includes instruction in such areas as weapons systems and technology, communications, intelligence, management, logistics and strategy.

30.01 Biological and Physical Sciences

30.06 System Science and Theory

30.08 Mathematics and Computer Science Biopsychology

Gerontology

Accounting and Computer Science Behavioral Sciences

Natural Sciences Nutrition Sciences Neuroscience Cognitive Science

*40. Physical Sciences.
Instructional programs that focus on the scientific study of inanimate objects, processes of matter and energy, and associated phenomena.

*41. Science Technologies/Technicians.
Instructional programs that prepare individuals to apply scientific principles and technical skills in support of scientific research and development.

42.11 Physiological Psychology/Psychobiology

42.19 Psychometrics and Quantitative Psychology
24 Psychopharmacology Forensic Psychology

*51.14 Medical Clinical Sciences/Graduate Medical Studies
51.1401 Medical Scientist (MS, PhD)

NOTE: "*" means all programs are considered high technology programs within that discipline.

Eligibility Periods
Benefits generally end 10 years from the date of an individual's last discharge or release from active duty. The VA may extend the 10-year period by the amount of time an individual is prevented from training due to a disability, or the individual is being held by a foreign government or power.

The VA may extend the 10-year period if an individual reenters active duty for 90 days or more after becoming eligible. The extension ends 10 years from the date of discharge or release from the later period. Periods of active duty of fewer than 90 days may qualify for

extensions, only if the discharge or release was for:
- A service-connected disability; or
- A medical condition existing before active duty; or
- Hardship; or
- A reduction in force.

If an individual's discharge is upgraded by the military, the 10-year period begins on the date of the upgrade.

Special Note for Individuals Eligible Under the 2 by 4 Program

If an individual is eligible based upon two years of active duty followed by four years in the Selected Reserve, the individual may have 10 years from release from active duty, or 10 years from the completion of the four-year Selected Reserve obligation to use benefits, whichever is later.

Miscellaneous Information

Any change in educational, professional or vocational objectives is considered a "change of program." The law permits one change of program without prior VA approval, provided an individual's attendance, conduct and progress in the last program were satisfactory. Additional "changes of program" require prior VA approval. VA will not charge a change of program if the individual enrolls in a new program after successful completion of the immediately preceding program.

Once an individual starts receiving benefits, he must maintain satisfactory attendance, conduct and progress. The VA may stop benefits if an individual does not meet the standards set by the school. VA may later resume benefits if the individual reenters the same program at the same school, and the school approves the reentry, and certifies it to VA.

If the individual does not reenter the same program at the same school, VA may resume benefits if the cause of unsatisfactory attendance, conduct or progress has been removed; and the program that the student intends to pursue is suitable to his or her abilities, aptitudes and interests.

Application for Benefits

VA Form 22-1990, "Application for Education Benefits" must be completed. The form may be obtained from individual schools, from any VA regional office, or by calling (888) GIBILL-1.

The completed form should be sent to the VA regional office with jurisdiction over the state in which the individual will train.

If an individual is not on active duty, copy 4 of DD Form 214 ("Certificate of Release or Discharge from Active Duty"), must also be sent to the VA. If an individual is on active duty, enrollment must be approved by the Base Education Services Officer, and service must be verified by the Commanding Officer.

If training has already started, the school, employer, or union should complete VA Form 22-1999 ("Enrollment Certification") and submit it along with the application.

Procedures for Receipt of Monthly Payments

After selecting a school and submitting an application to VA, the school official must complete an enrollment certification, and submit it to the appropriate VA regional office. If a student meets the basic eligibility requirements for benefits, and the program or course is approved, VA will process the enrollment based on certified training time.

If a student is enrolled in a degree program at a college or university, or a certificate or diploma program at a business, technical, or vocational school, they will not receive payment until they have verified their attendance. Students will receive a "Student Verification of Enrollment Form 22-8979" each month and must complete and return it to the appropriate VA regional office. After processing, VA will release a check.

If an individual is in an apprenticeship or job-training program, he or she will receive a form to report the hours worked each month. The form must be signed and given to the certifying official for the company or union. The certifying official must complete the form and send it to the appropriate VA regional office. After processing, VA will release a check.

If an individual is taking a correspondence course, he or she will receive a form each quarter, on which the student must show the number of lessons completed that quarter. The completed form should be sent to the school for certification of the number of lessons serviced during the quarter. The school will send the form to the appropriate VA regional office. After processing, VA will release a check. Payments are based on the number of lessons serviced by the school.

VA will send flight schools a supply of blank monthly certification of flight training forms. The school must complete the form by entering the number of hours, the hourly rate, and the total charges for flight training received during the month. The student should review and sign the completed form and send it to the appropriate VA regional office. After processing, VA will release a check.

NOTE: It is against the law for schools to cash VA checks under a Power of Attorney Agreement.

Timely Receipt of Verification Forms and Checks

Students should receive their verification forms for each month by the fifth of the following month. If it is not received by then, the VA should be immediately contacted so that another form can be issued.

One a completed verification form has been submitted, the student should receive a check within 2 weeks. If a check is not received by then, the VA should immediately be contacted so that appropriate action can be taken.

Advance Payments

An advance payment for the initial month, or partial month and the following month may be made, if:

- The school agrees to handle advance payments; and
- Training is one-half time or more; and
- A request is made by the individual in writing; and
- The VA receives the enrollment certification at least 30 days prior to the start of classes.
- Advance payments are made out to the individual and sent to the applicable school for delivery to the individual registration. VA cannot issue a check more than 30 days before classes start. Before requesting an advance payment, students should verify with the school certifying official that the school has agreed to process advance payments.

Requests for advance payments must be on VA Form 22-1999, "Enrollment Certification," or a sheet of paper attached to the enrollment certification.

Once a student receives an advance payment at registration, the school must certify to VA that the student received the check. If a student reduces enrollment, or withdraws from all courses during the period covered by an advance payment, he or she must repay the overpayment to VA.

If an individual believes that the amount of a VA check is incorrect, the VA should be contacted before the check is cashed.

Direct Deposit

Chapter 30 payments can be sent directly to a student's savings or checking account through Direct Deposit (Electronic Funds Transfer). To sign up for direct deposit by phone, students must call (877) 838-2778.

Student Responsibilities

To ensure timely receipt of correct payments, students should be sure to promptly notify the VA of:

- Any change in enrollment;
- Any change in address;
- Any change in selected reserve status;
- Any changes affecting a student's dependents (if a student is receiving an allowance which includes an additional amount for dependents).

In addition, students should use reasonable judgment when accepting and cashing a check. All letters from VA about monthly rates and effective dates should be read carefully. If a student thinks the amount of a VA check is wrong, VA should be contacted **before** cashing the check. Any incorrect checks should be returned to VA.

If a student cashes a check for the wrong amount, he or she will be liable for repayment of any resulting overpayment.

Recovery of Overpayments

VA must take prompt and aggressive action to recover overpayments. Students have the right to request a waiver of the overpayment, or verification that the amount is correct. If an overpayment is not repaid or waived, VA may add interest and collection fees to the debt. VA may also take one or more of the following actions to collect the debt:

- Withhold future benefits to apply to the debt;
- Refer the debt to a private collection agency;
- Recover the debt from any Federal income tax refund;
- Recover the debt from the salary (if student is a Federal employee);
- File a lawsuit in Federal court to collect the debt;
- Withhold approval of a VA home loan guarantee.

Changes in Enrollment

If a student withdraws from one or more courses after the end of the school's drop period, VA will reduce or stop benefits on the date of reduction or withdrawal. Unless the student can show that the change was due to mitigating circumstances, the student may have to repay **all** benefits for the course.

VA defines mitigating circumstances as "unavoidable and unexpected events that directly interfere with the pursuit of a course, and which are beyond the student's control. Examples of reasons VA may accept include:

- Extended illness;
- Severe illness or death in immediate family;
- Unscheduled changes in employment; and
- Lack of childcare.

Examples of reasons VA may not accept include:
- Withdrawal to avoid a failing grade;
- Dislike of the instructor; and
- Too many courses attempted.

(VA may ask the student to furnish evidence to support the reason for change, such as physician or employer written statements.)

The first time a student withdraws from up to 6 credit hours, VA will "excuse" the withdrawal, and pay benefits for the period attended. If a student receives a grade that does not count toward graduation, all benefits for the course may have to be repaid. Affected students should check the school's grading policy with the office handling VA paperwork.

If a student receives a non-punitive grade, the school will notify VA, and VA may reduce or stop benefits. The student may not have to repay the benefits if he or she can show that the grades were due to mitigating circumstances.

Work-Study Programs

Students may be eligible for an additional allowance under a work-study program that allows students to perform work for VA in return for an hourly wage.

Students may perform outreach services under VA supervision, prepare and process VA paperwork, work at a VA medical facility or National Cemetery, or perform other approved activities.

Students must attend school at the three-quarter of full-time rate. VA will select students for the work study program based on different factors. Such factors include:
- Disability of the student;
- Ability of the student to complete the work-study contract before the end of his or her eligibility for education benefits;
- Job availability within normal commuting distance to the student;
- VA will give the highest priority to a veteran who has a service-connected disability or disabilities rated by VA at 30% or more.

The number of applicants selected will depend on the availability of VA-related work at the school or at VA facilities in the area.

Students may work during or between periods of enrollment and can arrange with VA to work any number of hours during his or her enrollment. However, the maximum number of hours a student may work is 25 times the number of weeks in the enrollment period.

Students will earn an hourly wage equal to the Federal or State minimum wage, whichever is greater. If a student works at a college or university, the school **may** pay the difference between the amount VA pays and the amount the school normally pays other work-study students doing the same job.

Students may elect to be paid in advance for 40% of the number of hours in the work- study agreement, or for 50 hours, whichever is less. After completion of the hours covered by the first payment, VA will pay the student after completion of each 50 hours of service.

Students interested in taking part in a work-study program must complete VA Form 22-8691, "Application for Work-Study Allowance." Completed forms should be sent to the nearest VA regional office.

Educational Counseling

VA can provide services to help eligible individuals understand their educational and

vocational strengths and weaknesses and to plan:
- An educational or training goal, and how the goal can be reached; or
- An employment goal for which an individual qualifies based on present training or experience.

VA can also help plan an effective job search. Counseling is available for:
- Service members who are on active duty, and are within 180 days of discharge, and are stationed in the United States: or
- Veterans with discharges that are not dishonorable, who are within one year from date of discharge.

Vocational Rehabilitation

Veterans may qualify for Training and Rehabilitation under Chapter 31 of Title 38, United States Code, if:
- The veteran has a service-connected disability or disabilities rated by VA at 20% or more; and
- The veteran received a discharge from active duty that was not dishonorable; and
- The veteran has an employment handicap.

Veterans may also qualify with a service-connected disability or disabilities rated by VA at 10%, and:
- The veteran has a serious employment handicap; or
- The veteran first applied for vocational rehabilitation benefits before November 1, 1990, reapplied after that date, and has an employment handicap.

Vocational rehabilitation helps disabled veterans become independent in daily living. Veterans may also receive assistance in selecting, preparing for, and securing employment that is compatible with their interests, talents, skills, physical capabilities, and goals.

To apply for vocational rehabilitation, VA for 28-1900, "Disabled Veterans Application for Vocational Rehabilitation," must be completed and sent to the nearest VA regional office For detailed information on vocational rehabilitation refer to Chapter 14 of this book.

Appeal of VA Decision

VA decisions on education benefits may be appealed within one year of the date an individual receives notice of a VA decision.

What to Remember
✓ The Montgomery GI Bill Active Duty (MGIB-AD) can help you pay for education and training programs if you served at least two years on active duty.
✓ If you take part in the $600 Montgomery GI Bill Buy-Up program, you'll get more money each month through your GI Bill monthly payments. You should fill out DD Form 2366 and take it to your payroll or personnel office.
✓ With a $600 contribution, you can get up to $5,400 more in GI Bill benefits
✓ The program can't be used with the Post-9/11 Gi Bill
✓ You can get up to 36 months of education benefits. The amount you receive depends on your length of service, the type of education or training program you choose, your category, whether you qualify for a college fund or kicker, and how much you paid into the $600 Buy-Up program.

CHAPTER 21

EDUCATION BENEFITS: POST-9/11 G.I. BILL, CHAPTER 33

The Post-9/11 GI Bill (Chapter 33) helps veterans pay for school or job training. If a veteran served on active duty after September 10, 2001, they may qualify.

Most Post-911 GI Bill students have to verify their enrollment.

If you are getting a monthly housing allowance (MHA), college fund (kicker) payments, or both, you must verify your enrollment every month if you're at a non-college degree facility or institution of higher learning.

Eligibility
You may be eligible if you meet at least one of these requirements:
- You served at least 90 days on active duty, either all at once or with breaks in service on or after September 11, 2001, or
- You received a Purple Heart on or after September 11, 2001, and were honorably discharged after any amount of service, or
- You served for at least 30 continuous days, all at once without a break in service on or after September 11, 2001, and were honorably discharged with a service-connected disability, or
- You're a dependent child using benefits transferred by a qualifying veteran or service member.

Note: If you're a member of the Reserves who lost education benefits when the Reserve Educational Assistance Program (REAP) ended in November 2015, you might qualify to receive restored benefits under the Post-911 GI Bill.

You can only use 1 education benefit for a period of service, so you have to choose which one you want to use. After you make the choice, you can't change your mind.

Benefits
Benefits available through the Post-9/11 GI Bill include:
- Tuition and fees: If you qualify for the maximum benefit, the VA will cover the full cost of public, in-state tuition and fees. The VA caps the rates for private and foreign schools, and those rates are updated each year.
- Money for housing: If you're in school for more than half-time, VA will base your monthly housing allowance on the cost of living where your school is located.

- Money for books and supplies: You can receive up to the maximum stipend per school year.
- Money to help you move from a rural area to go to school: You might qualify for a one-time payment if you live in a county with six or fewer people per square mile, and you're either moving at least 500 miles to go to school or have no other option but to fly by plane to get to your school.

Your award of benefits is based on how long you served on active duty or certain other forces. You can start using your Post-911 GI Bill benefits while you're on active duty if you served at least 90 days. You're eligible for 100% of the full benefit if you meet at least one of these requirements:

- You served on active duty for a total of at least 1,095 days (at least 36 months)
- You served on active duty and received a Purple Heart on or after September 11, 2001, or
- You served on active duty for at least 30 continuous days without a break and the VA discharged you because of a service-connected disability

Expiration

If your service ended before January 1, 2013, your Post-9/11 GI Bill benefits will expire 15 years after your last separation date from active service. You must use all of your benefits by that time or you'll lose what's left.

If your service ended on or after January 1, 2013, your benefits won't expire because of a law called the Forever GI Bill—Harry W. Colmery Veterans Educational Assistance Act.

How VA Determines Post-9/11 GI Bill Coverage

Post-9/11 GI Bill benefits cover in-state tuition and rates and public schools at the percentage a veteran is eligible for. Even if someone is an out-of-state student, they may be able to get the in-state tuition rate. Post-9/11 GI Bill benefits cover tuition at private schools at the current national maximum amount.

To determine how much of a veteran's tuition and fees will be covered at a public school, the veteran will need the following information:

- How much the school they want to attend charges for in-state tuition and fees
- Whether the school they want to attend offers the in-state tuition rate to veterans who live out-of-state
- What percentage of Post-9/11 GI Bill benefits you're eligible for

If you want to attend a public school in the state where you live, the state will offer you in-state tuition. Your Post-9/11 GI Bill benefits will cover you at the percentage you're eligible for.

Member Serves	Percentage of Maximum Benefit Payable
At least 36 months	100%
At least 30 continuous days on active duty and must be discharged due to service-connected disability	100%
At least 30 months, but less than 36 months	90%
At least 24 months, but less than 30 months	80%
At least 18 months, but less than 24 months	70%
At least 12 months, but less than 18 months	60%

At least 06 months, but less than 12 months	50%
At least 90 days, but less than 06 months	40%

As an example, if the in-state tuition at a public school is $22,000, the amount that the VA will cover depends on the percentage you're entitled to. If you're entitled to 100% of your Post-9/11 GI Bill benefits, the VA will cover the full cost of tuition and fees. If you're entitled to 70%, the VA would cover $15,400 of your tuition and fees, and you'd be responsible for the rest.

If you want to go to an out-of-state public school that offers VA-approved programs, under the Veterans Choice Act, the school must offer you the in-state rate, and your benefits will cover you at your eligible percentage.

If you want to attend an out-of-state public school that doesn't offer in-state rates to veterans, the school will charge the out-of-state tuition rate. Your Post-9/11 Gi Bill benefits will cover you at the percentage you're eligible for only up to the amount of the in-state tuition rate.

If you qualify for the Yellow Ribbon Program, you may be able to get additional payments to help you cover the remaining amount.

Enrollment Verification

Staring in the 2021-22 school year, Post-9/11 GI Bill students who receive Monthly Housing Allowance and/or kicker payments are required to verify their enrollment to keep receiving their payments.

Enrollment verification is a new requirement to verify that the student has stayed enrolled in the same courses or training every month. A student receiving MHA/kicker payments will have their payments withheld if they don't verify or report they're no longer enrolled in courses or training. The VA provides the option to verify by text message.

This requirement is only for Post-9/11 GI Bill and does not impact other benefit programs, such as the Montgomery GI Bill (MGIB), Veteran Employment Through Technology Education Courses (VET TEC), Survivors' and Dependents' Educational Assistance (DEA), Veterans Rapid Retraining Assistance Program (VRRAP), or the Edith Nourse Rogers STEM Scholarship. This requirement also does not currently apply to GI Bill students in on-the-job, apprenticeship, flight, or correspondence training.

If you are an MGIB student, this new requirement does not represent a change for you, as MGIB students are already required to verify enrollment.

It will become effective in phases. Currently, the monthly verification requirement is only being applied to Post-9/11 GI Bill students at Institutions of Higher Learning (IHL) and certain Non-College Degree (NCD) facilities who also receive MHA and/or kicker payments.

All students, including students who do not receive MHA/kicker payments, will be provided with more information before the enrollment verification requirement applies to them.

NOTE: If you are taking multiple classes and any of them start after the effective date of this requirement, you will be required to verify enrollment even if your other classes started earlier.

Enrolling By Text Message:

Opt in: Students with a U.S. mobile phone number can use text messages as a simple, quick option for verifying monthly enrollment. As your enrollment approaches, you will receive a text informing you that you've been registered to receive texts for enrollment verification. Then, about 24 hours later, you'll receive the following opt-in text: "Post-9/11 GI Bill housing and kicker payments now require monthly enrollment verification. Would you like to submit yours via text? Please reply YES or NO." Reply "YES" to opt in. The text message link will expire 14 days after receipt, so please respond within that time frame.

Verify: After opting in, you can verify your enrollment every month simply by responding to the following text message from VA: "Did you remain enrolled in your courses in MONTH YEAR as certified? Please reply YES or NO. If you have dropped all your courses, you must reply NO." Reply "YES" to verify enrollment for the previous month. If you don't reply within 6 days, the conversation will expire and you will need to call the Education Call Center (ECC) at 1-888-GIBILL-1 (1-888-442-4551) domestically or 001-918-781-5678 internationally to verify your enrollment.

NOTE: When responding to opt-in or verification texts, it may take up to a day to receive a confirmation text in response. Texts will be sent from 44354.

Enrollment verification via text message is safe and secure. VA will never ask for your personal information, such as social security number or bank account information, via text. VA strongly recommends text message verifications for verifying your enrollment.

By Email:

If you opt out of text messages, can't verify by text, or do not have a US mobile phone number, you will be automatically enrolled in email verification at the email address on file with VA. After being opted into email verification, you will receive an email from do-not-reply@notifications.va.gov with the subject line, "Confirmation: You've been enrolled into VA's email verification!"

On the last day of each month, you will receive an email with the subject line, "Action Required: Verify Your Monthly Enrollment". Select "Yes, my enrollment is the same" to verify your enrollment. After selecting your response, you will be taken to a confirmation page thanking you for verifying your monthly enrollment.

If you don't select a response within 14 days, the links in your email will expire and you will need to call the ECC to verify your enrollment.

If your enrollment status has changed, select "No, my enrollment has changed." Please contact your SCO to ensure your enrollment record with VA has been adjusted.

By Phone:

If you are unable to verify via text or email, you will need to contact the ECC at 1-888-GIBILL-1 (1-888-442-4551) and ask a representative to verify your enrollment. NOTE: ECC wait times may be high due to the number of students verifying enrollment each month. If you're not sure if VA has your mobile phone number and/or email on file, you can contact the ECC to update your contact information and ensure you can verify via text or email.

If you fail to verify enrollment for two consecutive months, your MHA/kicker payments will be placed on hold. In order to have your payments released, you will need to call the ECC to verify your enrollment. When you call the ECC to release your payments, you may also enroll in text message verification at that time.

Transferring Post-9/11 GI Bill Benefits

You may be able to transfer your Post-9/11 GI Bill benefits if you're on active duty or in the Selected Reserve and you meet all of these requirements:

- You've completed at least 6 years of service on the date your request is approved, and
- You agree to add 4 more years of service, and
- The person getting benefits has enrolled in the Defense Enrollment Eligibility Reporting System (DEERS)

If DoD approves the Transfer of Entitlement, a spouse or dependent children can apply for up to 35 months of benefits, including money for tuition, housing, books and supplies.

Spouses:

- May use the benefit right away
- May use the benefit while the veteran is on active duty or after they've separated from service
- Don't qualify for the monthly housing allowance while you're on active duty
- May use the benefit for up to 15 years after the veteran's separation from active duty

Children:

- May start to use the benefit only after the veteran has finished at least 10 years of service
- May use the benefit while the veteran is on active duty or after they've separated from service
- May not use the benefit until they've gotten a high school diploma or equivalency certificate or have reached the age of 18
- Qualify for the monthly housing allowance even when you're on active duty
- Don't have to use the benefit within 15 years after your separation from active duty, but can't use the benefit after they've turned 26 years old

Your dependents may still qualify even a child marries or you and your spouse divorce. Servicemembers and veterans can cancel or change a TOE at any time. If a veteran wants to revoke benefits for a dependent and they're still in the service, they should turn in another transfer request for the dependent through milConnect. If a dependent's transfer eligibility has been totally revoked, the veteran can't transfer benefits again to that dependent.

While you're still on active duty, you request a transfer, change or revocation of TOE through milConnect and not the VA. If the DoD approves the TOE, family members may apply for benefits.

If your family member is able to apply for benefits, they can apply online or by mail. They should fill out an Application for Family Member to Use Transferred Benefits (VA Form 22-1990E) and mail it to the nearest VA regional office.

Full Rates for School and Training Programs—Effective August 1, 2022, to July 31, 2023

VA sends tuition and fees directly to the school or training program. The amounts listed are full monthly rates. If you're eligible for a percentage of the full benefit, you should multiple the rate by your percentage to find the amount of your monthly payment.

Public Institution of Higher Learning (Like a State University or Community College)
VA will pay the net tuition and mandatory fees. You may be eligible for in-state tuition rates at a public school even if you haven't lived in the state where the school is located.

Private Institution of Higher Learning
VA will pay the net tuition and mandatory fees up to $26,381.37.

Foreign Institution of Higher Learning (Outside of the U.S. Whether Public or Private)
VA will pay the net tuition and mandatory fees up to $26,381.37 in U.S. dollars.

Non-College Degree Programs
VA will pay the net tuition and mandatory fees up to $26,321.37.

Flight Training
VA will pay the net tuition and mandatory fees up to $15,075.05.

Money For Books and Supplies
You may be eligible for up to $1,000 each academic year for books and supplies. This is based on how many courses you're enrolled in for the year and the percentage of benefits you're eligible for. VA pays the benefit at the start of each term.

Monthly Housing Allowance
You may be eligible for money to help pay your housing while you study. If you're eligible for a monthly housing allowance (MHA) you're paid at the end of each month. The VA pays a percentage of the full monthly housing allowance based on two factors—the percentage of Post-9/11 GI Bill benefits you're eligible for and how many credits you're taking.

If you're taking online classes only, VA will pay a housing allowance that's half the national average.

If you're taking flight or correspondence training, you aren't eligible for MHA.

Contributions
Unlike the MGIB and VEAP, the new MGIB does not require the servicemember to elect or decline, nor make monthly contributions. Unfortunately, if someone has already contributed to the GI Bill, he or she won't get their money back, unless they use all of their new GI Bill entitlements.

If they do, the $1,200 contribution to the MGIB (or a proportional amount, if they used any of their MGIB entitlement) will be added to his or her final new GI Bill education payment. If an individual is currently paying into the MGIB, he or she can stop now, and still be eligible for the new program.

Covered Expenses
The plan covers any approved programs offered by a school in the United States that is authorized to grant an associate (or higher) degree.

You can be reimbursed up to $2,000 per test. Your entitlement will be charged one month for every $2172.71 paid to you rounded to the nearest non-zero whole month; this means even low-cost tests are charged one month of entitlement per test.

If a servicemember transferred to the Post-9/11 GI Bill from MGIB-Active Duty, MGIB-SR, or REAP, he or she may also receive Post-9/11 benefits for flight training, apprenticeship, or on-the-job training programs, and correspondence courses.

College Funds
If the member is eligible for a "kicker," such as the Army or Navy College Fund, or a Reserve "Kicker," he or she will still receive the extra monthly benefit under the new GI bill. This monthly amount will be paid to the member, not to the university.

College Loan Repayment
Individuals, who were previously ineligible for the MGIB because they elected the College Loan Repayment Program (CLRP), are eligible for the new GI Bill, but only active-duty

service performed after their initial active-duty service obligation counts toward the new benefits. In other words, if they were initially enlisted for five years and received the CLRP, they would have to reenlist or extend their enlistment in order to take advantage of the new GI Bill.

VEAP & the Post-9/11 Benefit

If someone is eligible for both VEAP and the Post-9/11 GI Bill he or she may un-enroll from VEAP and receive a refund of contributions or leave the remaining contributions in the VEAP account and remain eligible for both benefit programs.

NOTE: Students may not receive benefits under more than one program at the same time.

Yellow Ribbon Program

The Yellow Ribbon Program can help pay for higher out-of-state, private school, foreign school or graduate school tuition and fees not covered by the Post-9/11 GI Bill.

To be eligible you must qualify for the Post-9/11 Bill at a 100% benefit level, and at least one of the following has to be true:

- You served at least 36 months on active duty—this can be all at once, or with breaks in the service, and you were honorably discharged, or
- You received a Purple Heart on or after September 11, 2001, and were honorably discharged after any amount of service, or
- You served at least 30 continuous days, all at once without a break on or after September 11, 2001, and were discharged or released from active duty for a service-connected disability, or
- You're an active-duty service member who has served at least 36 months on active duty either all at once or with breaks in service, or
- You're a spouse using the transferred benefits of an active-duty service member who has served at least 36 months on active duty, or
- You're a dependent child using benefits transferred by a veteran, or
- You're a Fry Scholar

Your school has to meet certain requirements. All of the following has to be true of your school:

- Your school is an institution of higher learning, and
- Your school offers the Yellow Ribbon Program, and
- Your school hasn't offered the Yellow Ribbon benefit to more than the maximum number of students in their agreement with VA, and
- Your school has certified your enrollment with VA and provided Yellow Ribbon Program information

If you qualify for the Yellow Ribbon program, your school contributes a certain amount toward your additional tuition and fees through a grant, scholarship or similar program. The VA will match the contribution.

If you qualify for Post-9/11 GI Bill benefits, you'll get a Certificate of Eligibility or COE. You turn your COE into your school's certifying official, or the financial aid, military liaison or appropriate office. You then ask to apply for the Yellow Ribbon program at the school.

The school will make a decision based on whether it's already enrolled the maximum number of students for the program period, and how much funding you would receive.

Work-Study Program

Veterans and eligible transfer-of-entitlement recipients who train at the three-quarter rate of pursuit or higher may be eligible for a work-study program in which they work for VA and receive hourly wages. Students under the work-study program must be supervised by a VA employee and all duties performed must relate to VA. The types of work allowable include:

- VA paperwork processing at schools or other training facilities
- Assistance with patient care at VA hospitals or domiciliary care facilities.
- Work at national or state veterans' cemeteries.
- Other VA-approved activities.

REAP Note

If you're a member of the Reserves who lost education benefits when the Reserve Educational Assistance Program (REAP) ended in November 2015, you may qualify to receive restored benefits under the Post-9/11 GI Bill.

What to Remember
✓   The Post 9/11 GI Bill (Chapter 33) helps you pay for school or job training.
✓   You may be eligible if you meet certain requirements such as having served at least 90 days on active duty on or after September 11, 2001.
✓   You can get a maximum of 48 months of VA education benefits, but many applicants are eligible only for 36 months.
✓   If you qualify for the maximum benefit, the VA will cover the full cost of public in-state tuition and fees. The VA caps the rates for private and foreign schools, updating he rates every year.
✓   If you're eligible for other VA education benefits, you have to pick the one you want to use and you can't change your mind later.

CHAPTER 22

EDUCATION BENEFITS - MONTGOMERY G.I. BILL – SELECTED RESERVE (MGIB- SR), CHAPTER 1606

The Montgomery GI Bill-Selected Reserve Program is for members of the Selected Reserve, including the Army Reserve, Navy Reserve, Air Force Reserve, Marine Corps Reserve, Coast Guard Reserve, Army National Guard and Air National Guard. While the reserve components decide who is eligible for the program, VA makes the payments for the program. Chapter 1606 is the first educational program that does not require service in the active Armed Forces in order to qualify.

Eligibility Requirements

You could be eligible for benefits under the MGIB-SR program if one of the following is true:

- You have a 6-year service obligation in the Selected Service, meaning you agreed to serve 6 years, or
- You're an officer in the Selected Reserve and you agreed to serve 6 years in addition to your initial service obligation.

Your obligation must have started after June 30, 1985, or for some types of training after September 3, 1990. All of these have to be true too:

- You complete your initial active duty for training (IADT), and
- You get a high school diploma or certificate of equal value like a GED before finishing IADT—you can't use 12 hours toward a college degree to meet this requirement, and
- You remain in good standing while you're serving in an active Selected Reserve unit—you'll still be eligible if you're discharged from Selected Reserve service due to a disability not caused by misconduct

You can receive benefits until your eligibility period ends or you use all of your entitlement—whichever comes first. Entitlement is the number of months of benefits you may receive. The VA might extend the eligibility period if you're called to active duty. If this were to happen, the VA would extend your eligibility for the amount of time you're mobilized plus four months.

Your eligibility for the program typically ends the day you leave the Selected Reserve, with a few exceptions. You can still qualify for MGIB-SR benefits for up to 14 years from the date of your first 6-year obligation if at least one of the following is true for you:

- You separated because of a disability not caused by misconduct, or
- Your unit was deactivated between October 1, 2007, and September 30, 2014, or
- You involuntarily separated for reasons other than misconduct between October 1, 2007, and September 30, 2014

ROTC (Reserve Officers' Training Corps) scholarship under section 2107 of title 10, Code:

An individual can't be eligible for MGIB – SR if he or she is receiving financial assistance through the Senior ROTC program under this section of the law.

Note: However, an individual may still be eligible for MGIB – SR if he or she receives financial assistance under **Section 2107a** of title 10, U.S. Code. This financial assistance program is for specially selected members of the Army Reserve and National Guard only. Individuals should check with their ROTC advisor for more information.

Note: There's no restriction on service academy graduates receiving MGIB – SR. Service academy graduates who received a commission aren't eligible under MGIB – AD.

If an individual enters Active Guard and Reserve (AGR) status, his or her eligibility for MGIB – SR will be suspended. He or she may be eligible for MGIB – AD. The individual may resume MGIB – SR eligibility after AGR status ends.

Approved Courses

Individuals may receive benefits for a wide variety of training, including:
- Undergraduate degrees from a college or university;
- Beginning November 30, 1993, graduate degrees from a college or university;
- Accredited independent study programs leading to standard college degrees;
- Technical courses for a certificate at a college or university.

Individuals with 6-year commitments beginning after September 30, 1990 may take the following types of training:
- Courses leading to a certificate or diploma from business, technical, or vocational schools;
- Cooperative training;
- Apprenticeship or job training programs offered by companies;
- Correspondence training;
- Independent study programs;
- Flight training (Individuals must have a private pilot's license, and must meet the medical requirements for the desired license program before beginning training, and throughout the flight training program.)

VA may approve programs offered by institutions outside of the United States, when they are pursued at educational institutions of higher learning, and lead to a college degree. Individuals must receive VA approval prior to attending or enrolling in any foreign programs.

Eligibility for this program is determined by the Selected Reserve components. Payments for the program are made by the VA. A state agency or VA must approve each program offered by a school or company. If an individual is seeking a college degree, the school must admit the individual to a degree program by the start of the individual's third term.

Restrictions on Training

Benefits are not payable for the following courses:
- Courses paid by the military Tuition Assistance program, if student is enrolled at

less than ½time;
- Courses taken while student is receiving a Reserve Officers' Training Corps scholarship;
- Non-accredited independent study courses;
- Bartending and personality development courses;
- Any course given by radio;
- Self-improvement courses such as reading, speaking, woodworking, basic seamanship, and English as a 2nd language;
- Any course which is avocational or recreational in character;
- Farm-cooperative courses;
- Audited courses;
- Courses not leading to an educational, professional, or vocational objective;
- Courses previously taken and successfully completed;
- Courses taken by a Federal government employee under the Government Employee's Training Act;
- Courses taken while in receipt of benefits for the same program from the Office of Workers' Compensation Programs.

VA must reduce benefits for individuals in Federal, State, or local prisons after being convicted of a felony.

An individual may not receive benefits for a program at a proprietary school if he or she is an owner or official of the school.

Part-Time Training

Individuals unable to attend school full-time should consider going part-time. Benefit rates and entitlement charges are less than the full-time rates. For example, if a student receives full-time benefits for 12 months, the entitlement charge is 12 months. However, if the student receives ½ time benefits for 12 months, the charge is 6 months. VA will pay for less than ½ time training if the student is not receiving Tuition Assistance for those courses.

Remedial, Deficiency and Refresher Training

Remedial and deficiency courses are intended to assist a student in overcoming a deficiency in a particular area of study. Individuals may qualify for benefits for remedial, deficiency, and refresher courses if they have a 6-year commitment that began after September 30, 1990. In order for such courses to be approved, the courses must be deemed necessary for pursuit of a program of education.

Refresher training is for technological advances that occurred in a field of employment. The advance must have occurred while the student was on active duty, or after release. There is an entitlement charge for these courses.

Tutorial Assistance

Students may receive a special allowance for individual tutoring performed after September 30, 1992, if they entered school at one-half time or more. To qualify, the student must have a deficiency in a subject. The school must certify the tutor's qualifications, and the hours of tutoring. Eligible students may receive a maximum monthly payment of $100.00. The maximum total benefit payable is $1,200.00.

There is no entitlement charge for the first $600.00 of tutorial assistance.

To apply for tutorial assistance, students must submit VA Form 22-1990t, "Application and Enrollment Certification for Individualized Tutorial Assistance." The form should be given to the certifying official in the office handling VA paperwork at the school for completion.

Months of Benefits/Entitlements Charged

Eligible members may be entitled to receive up to 36 months of education benefits. Benefit entitlement ends 10 years from the date the member becomes eligible for the program, or on the day the member leaves the Selected Reserve. (If a member's Reserve or National Guard unit was deactivated during the period October 1, 1991, through September 30, 1999, or if the member was involuntarily separated from service during this same period, eligibility for MGIB-SR benefits is retained for the full 10-year eligibility period. Eligibility for MGIB-SR benefits is also retained if a member is discharged due to a disability that was not caused by misconduct. Eligibility periods may be extended if a member is ordered to active duty.)

Individuals qualifying for more than one VA education program may receive a maximum of 48 months of benefits. For example, if a student used 30 months of Dependents' Educational Assistance, and is eligible for chapter 1606 benefits, he or she could have a maximum of 18 months of entitlement remaining.

Individuals are charged one full day of entitlement for each day of full-time benefits paid. For correspondence and flight training, one month of entitlement is charged each time VA pays one month of benefits. For cooperative programs, one month of entitlement is used for each month of benefits paid.

For apprenticeship and job training programs, the entitlement charge changes every 6 months. During the first 6 months, the charge is 75% of full time. For the second 6 months, the charge is 55% of full time. For the remainder of the program, the charge is 35% of full time.

Rates of Educational Assistance Pay

The following basic monthly rates are effective October 1, 2022:

Basic Monthly Rates Montgomery GI Bill Selected Reserve (MGIB-SR), Chapter 1606			
Type of Training	Full-Time	¾ Time	½ Time	Less Than ½ Time
Institutional	$439.00	$329.00	$219.00	$109.75
Cooperative Training	$439.00 (Full-Time Only)			
Correspondence Training	Entitlement charged at the rate of one month for each $439.00 paid (Payment for correspondence courses is made at 55% of the approved charges for the course.)			
Apprenticeship On-The-Job Training -	First six months: $329.25 for each full month Second six months: $241.45 for each full month Remainder of program: $153.65 for each full month			
Flight Training	Entitlement charged at the rate of one month for each $439.00 paid (Payment for flight training is made at 60% of the approved charges for the course, including solo hours.)			

Eligibility Periods

Under previous law, benefits typically ended upon separation from the Reserves. Section

530 of the National Defense Authorization Act of 2008 **increased the time period that a reservist may use SRMGIB benefits to ten years after separation from the Reserves**, as long as the member's discharge characterization is honorable. **This revision is backdated to October 28, 2004.**

For individuals who separated from the Selected Reserve prior to 10/28/2004, generally benefits end the day of separation. For individuals who stayed in the Selected Reserve, generally benefits ended 14 years from the date the individual became eligible for the program.

Exceptions: If an individual stayed in the Selected Reserve, VA could generally extend the 14-year period if:

- The individual couldn't train due to a disability caused by Selected Reserve service; or
- The individual was activated for service in the Persian Gulf Era (which hasn't ended for purposes of VA education benefits); or
- The individual's eligibility expired during a period of his or her enrollment in training.

Miscellaneous Information

Any change in educational, professional or vocational objectives is considered a "change of program." The law permits one change of program without prior VA approval, provided an individual's attendance conduct and progress in the last program were satisfactory. Additional "changes of program" require prior VA approval. VA will not charge a change of program if the individual enrolls in a new program after successful completion of the immediately preceding program.

Once an individual starts receiving benefits, he must maintain satisfactory attendance, conduct and progress. The VA may stop benefits if an individual does not meet the standards set by the school. VA may later resume benefits if the individual reenters the same program at the same school, and the school approves the reentry, and certifies it to VA.

If the individual does not reenter the same program at the same school, VA may resume benefits if the cause of unsatisfactory attendance, conduct or progress has been removed; and the program that the student intends to pursue is suitable to his or her abilities, aptitudes and interests.

Effective November 1, 2000, VA education benefits can be paid (with some exceptions) for school breaks, if the breaks do not exceed 8 weeks; and the terms before and after the breaks are not shorter than the break. Prior to November 1 2000, VA education benefits could be paid only if the breaks did not exceed a calendar month.

Application for Benefits

When an individual becomes eligible for the program, his or her unit will provide the individual with a "Notice of Basic Eligibility, DD Form 2384," or DD Form 2384-1 (for persons who establish eligibility on or after October 1, 1990). The unit will also code the eligibility into the Department of Defense personnel system.

When the individual finds a school, program, company, apprenticeship or job-training program, there are two important steps that must be followed:

- Make sure the program is approved for VA training. Contact the local VA regional office if there are any questions.
- Complete VA Form 22-1990, "Application for Education Benefits." The completed form should be sent to the VA regional office with jurisdiction over the state where training will occur.

Following receipt of an application, VA will review it and advise if anything else is needed. If an individual has started training, the application of "Basic Eligibility" should be taken to the school, employer or union. The certifying official should complete VA Form 22 1999, "Enrollment Certification," and send all the forms to VA.

Procedures for Receipt of Monthly Payments

After selecting a school and submitting an application to VA, the school official must complete an enrollment certification, and submit it to the appropriate VA regional office. If a student meets the basic eligibility requirements for benefits, and the program or course is approved, VA will process the enrollment based on certified training time.

VA will accept the Notice of Basic Eligibility to pay benefits for 120 days after an individual's eligibility date. If the eligibility date is more than 120 days before the training program starts, VA will not approve the claim unless the Department of Defense personnel system shows that the individual is eligible. Only a student's reserve component can update the DoD personnel system. VA cannot change an individual's eligibility record.

When VA approves a claim, it will issue a letter with the details of the benefits payable. The first payment should be received within a few days of receipt of the letter.

If a student is enrolled in a degree program at a college or university, he or she will receive payment after the first of each month for the training during the preceding month. If a student is enrolled in a certificate or diploma program at a business, technical, or vocational school, he or she will not receive payment until they have verified their attendance. Students will receive a "Student Verification of Enrollment Form 22-8979" each month and must complete and return it to the appropriate VA regional office. After processing, VA will release a check.

If an individual is in an apprenticeship or job-training program, he or she will receive a form to report the hours worked each month. The form must be signed and given to the certifying official for the company or union. The certifying official must complete the form and send it to the appropriate VA regional office. After processing, VA will release a check.

If an individual is taking a correspondence course, he or she will receive a form each quarter, on which the student must show the number of lessons completed that quarter. The completed form should be sent to the school for certification of the number of lessons serviced during the quarter. The school will send the form to the appropriate VA regional office. After processing, VA will release a check. Payments are based on the number of lessons serviced by the school.

VA will send flight schools a supply of blank monthly certification of flight training forms. The school must complete the form by entering the number of hours, the hourly rate, and the total charges for flight training received during the month. The student should review and sign the completed form and send it to the appropriate VA regional office. After processing, VA will release a check.

NOTE: It is against the law for schools to cash VA checks under a Power of Attorney Agreement.

Timely Receipt of Verification Forms and Checks

Students taking courses leading to a degree at a college or university should receive their checks for each month by the fifth of the next month. If it is not received by then, the VA should be immediately contacted so that appropriate action can be taken.

Students taking courses leading to a certificate or diploma from a business, technical, or vocational school should receive their verification forms for each month by the fifth of the following month. If it is not received by then, the VA should be immediately contacted so that another form can be issued.

One a completed verification form has been submitted; the student should receive a check within 2 weeks. If a check is not received by then, the VA should immediately be contacted so that appropriate action can be taken.

Advance Payments

An advance payment for the initial month, or partial month and the following month may be made, if:

- The school agrees to handle advance payments;and
- Training is one-half time or more; and
- A request is made by the individual in writing; and
- The VA receives the enrollment certification at least 30 days prior to the start of classes.

Advance payments are made out to the individual and sent to the applicable school for delivery to the individual registration. VA cannot issue a check more than 30 days before classes start. Before requesting an advance payment, students should verify with the school certifying official that the school has agreed to process advance payments.

Requests for advance payments must be on VA Form 22-1999, Enrollment Certification, or a sheet of paper attached to the enrollment certification.

Once a student receives an advance payment at registration, the school must certify to VA that the student received the check. If a student reduces enrollment, or withdraws from all courses during the period covered by an advance payment, he or she must repay the overpayment to VA.

If an individual believes that the amount of a VA check is incorrect, the VA should be contacted before the check is cashed.

Direct Deposit

Payments can be sent directly to a student's savings or checking account through Direct Deposit (Electronic Funds Transfer). To sign up for direct deposit by phone, students must call (877) 838-2778.

Student Responsibilities

To ensure timely receipt of correct payments, students should be sure to promptly notify the VA of:

- Any change in enrollment;
- Any change in address;
- Any change in selected reserve status (If an individual changes units or components, both the old and new units must report the change to VA through the components' eligibility data systems.)

In addition, students should use reasonable judgment when accepting and cashing a check. All letters from VA about monthly rates and effective dates should be read carefully. If a student thinks the amount of a VA check is wrong, VA should be contacted **before** cashing the check. Any incorrect checks should be returned to VA.

If a student cashes a check for the wrong amount, he or she will be liable for repayment of any resulting overpayment.

If an individual does not participate satisfactorily in the Selected Reserve, his or her eligibility ends. His or her component can require that a penalty be paid, based on a portion of payments received.

Recovery of Overpayments

VA must take prompt and aggressive action to recover overpayments. Students have the right to request a waiver of the overpayment, or verification that the amount is correct. If an overpayment is not repaid or waived, VA may add interest and collection fees to the debt. VA may also take one or more of the following actions to collect the debt:

- Withhold future benefits to apply to the debt;
- Refer the debt to a private collection agency;
- Recover the debt from any Federal income tax refund;
- Recover the debt from the salary (if student is a Federal employee);
- File a lawsuit in Federal court to collect the debt;
- Withhold approval of a VA home loan guarantee.
- An individual's reserve component will act to collect penalties caused by unsatisfactory participation in the reserve.

Changes in Enrollment

If a student withdraws from one or more courses after the end of the school's drop period, VA will reduce or stop benefits on the date of reduction or withdrawal. Unless the student can show that the change was due to mitigating circumstances, the student may have to repay **all** benefits for the course.

VA defines mitigating circumstances as "unavoidable and unexpected events that directly interfere with the pursuit of a course, and which are beyond the student's control.

Examples of reasons VA may accept include:

- Extended illness;
- Severe illness or death in immediate family;
- Unscheduled changes in employment; and
- Lack of childcare.

Examples of reasons VA may not accept include:

- Withdrawal to avoid a failing grade;
- Dislike of the instructor; and
- Too many courses attempted.

(VA may ask the student to furnish evidence to support the reason for change, such as physician or employer written statements.)

The first time a student withdraws from up to 6 credit hours, VA will "excuse" the withdrawal, and pay benefits for the period attended.

If a student receives a grade that does not count toward graduation, all benefits for the course may have to be repaid.

If a student receives a non-punitive grade, the school will notify VA, and VA may reduce or stop benefits. The student may not have to repay the benefits if he or she can show that the grades were due to mitigating circumstances.

Work-Study Programs

Students may be eligible for an additional allowance under a work-study program that allows students to perform work for VA in return for an hourly wage. Students may perform outreach services under VA supervision, prepare and process VA paperwork, work at a VA medical facility or National Cemetery, or perform other approved activities.

Students must attend school at the three-quarter of full-time rate. VA will select students for the work-study program based on different factors. Such factors include:

- Disability of the student;
- Ability of the student to complete the work-study contract before the end of his or her eligibility for education benefits;
- Job availability within normal commuting distance to the student

VA will give the highest priority to a veteran who has a service-connected disability or disabilities rated by VA at 30% or more.

The number of applicants selected will depend on the availability of VA-related work at the school or at VA facilities in the area. Students may work during or between periods of enrollment and can arrange with VA to work any number of hours during his or her enrollment. However, the maximum number of hours a student may work is 25 times the number of weeks in the enrollment period.

Students will earn an hourly wage equal to the Federal or State minimum wage, whichever is greater. If a student works at a college or university, the school **may** pay the difference between the amount VA pays and the amount the school normally pays other work-study students doing the same job.

Students interested in taking part in a work-study program must complete VA Form 22-8691, Application for Work-Study Allowance. Completed forms should be sent to the nearest VA regional office.

Educational Counseling

VA can provide services to help eligible individuals understand their educational and vocational strengths and weaknesses and to plan: An educational or training goal, and the means by which the goal can be reached; or an employment goal for which an individual qualifies on the basis of present training or experience.

VA can also help plan an effective job search. Counseling is available for:

- Service members eligible for VA educational assistance; and
- Service members on active duty and within 180 days of discharge; or
- Veterans with discharges that are not dishonorable, who are within one year from date of discharge.

Vocational Rehabilitation

Veterans may qualify for Training and Rehabilitation under Chapter 31 of Title 38, United States Code, if:

- The veteran has a service-connected disability or disabilities rated by VA at 20% or more; and
- The veteran received a discharge from active duty that was not dishonorable; and
- The veteran has an employment handicap.

Veterans may also qualify with a service-connected disability or disabilities rated by VA at 10%, and:

- The veteran has a serious employment handicap; or
- The veteran first applied for vocational rehabilitation benefits before November 1, 1990, reapplied after that date, and has an employment handicap.

Vocational rehabilitation helps disabled veterans become independent in daily living. Veterans may also receive assistance in selecting, preparing for, and securing employment that is compatible with their interests, talents, skills, physical capabilities, and goals. To apply for vocational rehabilitation, VA for 28-1900, Disabled Veterans Application for Vocational Rehabilitation, must be completed and sent to the nearest VA regional office.

Appeal of VA Decision

VA decisions on education benefits may be appealed within one year of the date an individual receives notice of a VA decision. Examples of VA decisions include:

- Training time,
- Change of program,
- School or course approval.

If a service member disagrees with a decision about his or her basic eligibility, he or she must contact the unit, National Guard Education Officer, or Army Reserve Education Services Officer. VA does not have authority under the law to reverse eligibility determinations. If the eligibility status is corrected, VA will pay benefits for periods during which the individual was eligible.

What to Remember
✓ The Montgomery GI Bill Selected Reserve (MGIB-SR) program offers up to 36 months of education training and benefits.
✓ You may be eligible if you're a member of the Army, Navy, Air Force, Marine Corps or Coast Guard Reserve, army National Guard or Air National Guard.
✓ You can get up to $439 per month in compensation payments for up to 36 months.
✓ If you haven't started training, you can submit your Application for VA Education Benefits (VA Form 22-1990) online. You can also apply by mail, in person or with the help of a trained professional.
✓ If you've already started training, fill out VA Form 22-1990. Take your application and Notice of Basic Entitlement to your school or employer. Ask them to complete VA Form 22-1999 and send all three forms.

CHAPTER 23

LIFE INSURANCE

VA insurance programs were developed to provide insurance benefits for veterans and servicemembers who may not be able to get insurance from private companies because of a service-connected disability or because of the extra risks involved in military service. VA has responsibility for veterans' and servicemembers' life insurance programs. Listed below are the eight life insurance programs managed by VA. The first four programs listed are closed to new issues. The last four are still issuing new policies.

VALife Veterans Affairs Life Insurance
Veterans Affairs Life Insurance (VALife) provides guaranteed acceptance whole life coverage of up to $40,000 to veterans with service-connected disabilities. Lesser amounts are available, in increments of $10,000 Under the plan, the elected coverage will take effect two years after enrollment, as long as premiums are paid during the two-year period.

All veterans aged 80 and under who have a disability rating from the VA of 0-100% are eligible to apply for the program, with no time limit to apply. Veterans who are 81 or older and have applied for VA Disability Compensation before age 81 but who didn't receive the rating for a new service-connected condition until after turning 81 are also eligible if they apply within two years of their rating.

The two-year waiting period that's part of this life insurance program replaces the need for medical underwriting. If the insured dies within the two-year period, the beneficiary receives all premiums they've paid plus interest.

The premium rate you pay each month or annually for coverage depends on your age and the amount of coverage you elect. The premiums for VALife are fixed, and they're based on your age when you enroll. There are no premium waivers for this program.

VALife takes effect January 1, 2023, and the application will become available on the VA's site after that time.

As a result of VALife, S-DVI programs are closing to new enrollment after December 31, 2022. No new applications will be accepted. If you have S-DVI you can apply for VALife. If you apply before December 31, 2025, you can keep S-DVI during the initial two-year enrollment period for VALife. You don't have to switch to VALife—you can stay in the S-DVI program.

VA Reduces Rates for Servicemembers' Group Life Insurance
On July 1, 2019, the VA reduced Servicemembers' Group Life Insurance (SGLI) monthly premium rates from 7 cents per $1,000 to 6 cents per $1,000 of insurance, along with Family SGLI Program (FSGLI) premium rates for spousal coverage at all age brackets.

This lowers the premium for a servicemember with the maximum coverage of $400,000 from $28 to $24 a month. (Note: Servicemembers pay an additional $1 per month for traumatic injury protection.) Additionally, the SGLI premium rate for members who participate in one-day musters or funeral honors duty is also decreasing from $.25 per $1,000 of coverage to $.20 per $1,000 of coverage. Additionally, the SGLI premium rate for members who participate in one-day musters or funeral honors duty is also decreasing from $.25 per $1,000 of coverage to $.20 per $1,000 of coverage.

The Family SGLI spousal coverage premium rates are decreasing for all age brackets on July 1, 2019. Spousal coverage premiums are based on the age and amount of coverage elected by the Servicemember for their spouse.

Service-Disabled Veterans Insurance-S-DVI (1951-Present) (Policy Prefix - RH or ARH)

Note: S-DVI programs close to new enrollment after December 31, 2022. No new applications are being accepted. You can apply for VALife, or you can stay in the S-DVI program.

The only new insurance issued between 1957 and 1965 to either servicemembers or veterans was Service-Disabled Veterans Insurance. This insurance was (and still is) available to veterans with a service-connected disability.

S-DVI, also called "RH Insurance" is available in a variety of permanent plans, as well as term insurance. Policies are issued for a maximum face amount of $10,000.

VI is open to veterans separated from the service on or after April 25, 1951, who receive a service-connected disability rating of 0% or greater. New policies are still being issued under this program.

To apply for Supplemental S-DVI, you must file "VA Form 29-4364, Application for Supplemental Service-Disabled Veterans (RH) Life Insurance" or send a letter requesting this insurance over your signature. You must apply for the coverage within one year from notice of the grant of waiver of premiums.

Eligibility for S-DVI Insurance ("RH")

Veterans are eligible to apply for S-DVI if they meet the following four criteria:
- They have received a rating for a service-connected disability (even if only 0%).
- They were released from active duty under any other condition than dishonorable on or after April 25, 1951.
- They are in good health except for any service-related condition.
- They apply for the insurance within two years from the date service-connection is established.

Eligibility for Supplemental S-DVI ("Supplemental RH")

The Veterans' Benefits Act of 2010, provided for $30,000 of supplemental coverage to S- DVI policyholders. Premiums may not be waived on this supplemental coverage. S-DVI policyholders are eligible for this supplemental coverage if:
- They are eligible for a waiver of premiums.
- They apply for the coverage within one year from notice of the grant of waiver.
- Are under age 65.

Gratuitous S-DVI ("ARH")

Congress enacted legislation in 1959 to protect veterans who become incompetent from a service-connected disability while eligible to apply for S-DVI, but who die before an application is filed. "ARH" insurance is:
- Issued posthumously;

- Payable to a preferred class of the veteran's relatives;
- Payable in a lump sum only.

Premiums for S-DVI Insurance

The premiums charged for this coverage are:

- Based on the rates a healthy individual would have been charged when the program began in 1951.
- Insufficient to pay all of the claims because the program insures many veterans with severe disabilities.
- Waived for totally disabled veterans (27% of S-DVI policyholders).
- Supplemented on an annual basis by Congressional appropriations.

Effective November 1, 2000, the VA "capped" premium rates at the age 70 rate. This means that a term policyholder's premium will never increase over the age 70 -premium rate. There are no reserves or surplus funds in this program. Therefore, dividends are not paid.

Disability Provisions

VI policies (except supplemental coverage) provide for the following disability benefits:

- A waiver of premiums at no extra premium based on the insured's total disability lasting six months or longer and starting before age 65;
- A total disability premium waiver in cases where the disability commenced prior to the effective date of the policy, providing the disability is service-connected.
- The optional Total Disability Income Provision is not available under this program.

Filing a Death Claim

To file a death claim, the beneficiary should complete "VA Form 29-4125, Claim for One Sum Payment."

The completed form should be mailed or faxed, along with a death certificate to: Department of Veterans Affairs Regional Office and Insurance Center
PO Box 7208
Philadelphia, PA 19101
Fax: (215) 381-3561

If the beneficiary desires monthly payments instead of one lump sum, additional information is needed. The beneficiary should call the Insurance Center at (800) 669- 8477 for instructions.

Customer Service

Any questions regarding USGLI should be directed to the VA Life Insurance Program at (800) 669-8477.Servicemembers' Group Life Insurance - SGLI (1965-PRESENT)

SGLI Online Enrollment System

Beginning October 1, 2017, Army members can manage their Servicemembers' Group Life Insurance (SGLI) coverage using the SGLI Online Enrollment System (SOES). Army members join the Navy and Air Force in using the new system. Navy began using SOES on April 5, 2017, and Air Force on August 2, 2017. Air Force, Navy, and Army members should look for information from their service about when they should access SOES to confirm and certify their SGLI elections.

SGLI provides automatic life insurance coverage of $400,000 to Servicemembers upon enlistment. Members with SGLI also get automatic coverage for their dependent children and spouses (unless the child or spouse is insured under SGLI as a Servicemember) under the Family SGLI program and traumatic injury protection (TSGLI). Servicemembers with full-time SGLI coverage will no longer have to complete a paper SGLV-8286 to make changes to their coverage or beneficiary elections. Instead, these Servicemembers can use the online system, SOES, to manage the amount of their SGLI and spouse coverage and to designate or update beneficiaries.

To access SOES, go to www.dmdc.osd.mil/milconnect, sign in, and go to Benefits, Life Insurance SOES- SGLI Online Enrollment System. Servicemembers can log in with their CAC or with their DS Logon using Internet Explorer as soon as they receive notice that SOES access is available. Servicemembers can then make sure their SGLI coverage and beneficiaries are up to date.

SGLI is supervised by the Department of Veterans Affairs and is administered by the Office of Servicemembers' Group Life Insurance (OSGLI) under terms of a group insurance contract.

Effective December 1, 2005, a new program of insurance was created under SGLI, called Traumatic Servicemembers' Group Life Insurance (TSGLI). This coverage provides servicemembers protection against loss due to traumatic injuries and is designed to provide financial assistance to members so their loved ones can be with them during their recovery from their injuries. The coverage ranges from $25,000 to $100,000 depending on the nature of the injury.

(See following section for further information regarding Traumatic Injury Coverage.)

Eligibility for SGLI

Full-time coverage is available for:
- Commissioned, warrant and enlisted members of the Army, Navy, Air Force, Marine Corps and Coast Guard;
- Commissioned members of the National Oceanic and Atmospheric Administration and the Public Health Service;
- Cadets or midshipmen of the four United States Service Academies;
- Ready Reservists scheduled to perform at least 12 periods of inactive training per year.
- Members of the Individual Ready Reserves who volunteer for assignment to a mobilization category.
- Part-time coverage is available for eligible members of the Reserves who do not qualify for full-time coverage.

Coverage Amounts

Effective September 1, 2005, coverage in the SGLI Program increased from $250,000 to $400,000. Coverage must be elected in $50,000 increments.

A special death gratuity of $150,000 was approved for survivors of servicemembers who:
- Died from October 7, 2001, to September 1, 2005, and
- Died while on active duty.

SGLI Coverage Is:

- Automatic at the time of entry into a period of active duty or reserve status.
- Available in $50,000 increments up to the maximum of $400,000 of insurance.

Members may decline coverage or may elect reduced coverage by contacting their personnel officer and completing "Form 8286, SGLI Election and Certificate." If such a member later wishes to obtain or increase coverage, proof of good health will be required.

NOTE: Reservists called to active duty, are **automatically** insured for $400,000 regardless of whether or not they had previously declined coverage or elected a lesser amount of coverage while on reserve duty.

Coverage Periods

Full-time SGLI coverage for members on active duty or active duty for training and members of the Ready Reserve will terminate:

The 120th day after separation or release from duty, or separation or release from assignment to a unit or position of the Ready Reserve;
For members who are totally disabled on the date of separation or release, at the end of the last day of one year following separation or release or at the end of the day on which the insured ceases to be totally disabled, whichever is earlier, but in no event earlier than

- 120 days following separation or release from such duty; (NOTE: Refer to important updated information following #4, below.)
- At the end of the 31st day of a continuous period of: Absence without leave;
- Confinement by military authorities under a court-martial sentence involving total forfeiture of pay and allowances; or
- Confinement by civilian authorities under sentence adjudged by a civilian court.

*Note: Any insurance terminated as the result of the absence or confinement, together with any beneficiary designation in effect at the time the insurance was terminated, will be automatically restored as of the date the member returns to duty with pay.

The last day of the month in which the member files with the uniformed service, written notice of an election not to be insured.

SGLI Disability Extension Increased to Two Years

On June 15, 2006, the President signed P.L. 109- 233, the Veterans' Housing Opportunity & Benefits Improvement Act of 2006. The law extends the period of Servicemembers' Group Life Insurance (SGLI) coverage for totally disabled veterans following separation from active duty or active duty for training to the earlier of:

- The date on which the insured ceases to be totally disabled; or
- The date that is two years after separation or release,

The law makes identical changes with respect to certain reserve assignments in which the individual performs active duty or volunteers for assignment to a mobilization category.

The SGLI Disability Extension is available to you if you are totally disabled at time of discharge. To be considered totally disabled, you must have a disability that prevents you from being gainfully employed OR have one of the following conditions, regardless of your employment status:

- Permanent loss of use of both hands Permanent loss of use of both feet
- Permanent loss of use of both eyes
- Permanent loss of use of one hand and one foot Permanent loss of use of one foot and one eye Permanent loss of use of one hand and one eye Total loss of hearing in both ears
- Organic loss of speech (lost ability to express oneself, both by voice and whisper, through normal organs for speech - being able to speak with an artificial appliance is disregarded in determination of total disability)

Part-Time Coverage Terminates as Follows:

Part-time coverage is in effect only on the days of active duty or active duty for training, and the hours of inactive duty training, including period of travel to and from duty. A temporary termination of coverage occurs at the end of each such period of duty, including travel time, and coverage is resumed at the commencement of the next period of covered duty or travel.

When part-time coverage is extended for 120 days as the result of a disability, the extended coverage terminates at the end of the 120th day following the Reservist active or inactive period during which the disability was incurred or aggravated.

Unless extended for 120 days because of disability as referred to above, eligibility for part-time coverage terminates at the end of the last day of the member's obligation to perform such duty.

If a member files with the uniformed service a written notice of an election not to be insured, coverage terminates on the last day of the period of active duty or active duty for training, or at the end of the period of inactive duty training, including travel time while returning from such duty during which the election is filed. If the election is filed with a member's uniformed service other than during a period of active duty, active duty for training, or inactive duty the coverage is terminated immediately.

Beneficiary Selection
Any beneficiary can be named. If none is selected, the insurance is distributed, by law, in the following order:
* Spouse; or
* Children; or
* Parents; or
* Executor of estate; or
* Other next of kin.

An insured should designate a beneficiary by completing "Form SGVL 8286, Servicemembers' Group Life Insurance Election and Certificate." The completed form should be submitted to the individual's uniformed service.

Options for Payments of Policy Proceeds
SGLI proceeds can be paid in a lump sum or over a 36-month period.

Alliance Account
If the proceeds are to be paid in a lump sum, then beneficiaries of SGLI and VGLI will receive the payment of their insurance proceeds via an "Alliance Account". Rather than the traditional single check for the full amount of insurance proceeds, the beneficiary receives a checkbook for an interest-bearing account from which the beneficiary can write a check for any amount from $250 up to the full amount of the proceeds.

The Alliance Account:
Earns interest at a competitive rate; Is guaranteed by Prudential;
Gives the beneficiary time to make important financial decisions while their funds are secure and earning interest ;
Gives them instant access to their money at all times.

Accelerated Benefits
On November 11, 1998, the President signed legislation authorizing the payment of "Accelerated Benefits" in the SGLI and VGLI programs subject to the following:
* Terminally ill insureds will have access of up to 50% of the face amount of their coverage during their lifetime.
* This money will be available in increments of $5,000.
* An insured must have a medical prognosis of life expectancy of 9 months or less.

Insurance Options After Separation from Service
When released from active duty or the Reserve, members with full-time SGLI coverage can convert their coverage to Veterans Group Life Insurance or to an individual commercial

life insurance policy with any one of 99 participating commercial insurance companies.

Filing Death Claims

A beneficiary may file a claim for VGLI proceeds by submitting Form SGLV 8283, Claim for Death Benefits, to:

Office of Servicemembers' Group Life Insurance (OSGLI)
PO Box 70173
Philadelphia, PA 19176-9912
SGLI Traumatic Injury (TSGLI) Claims

Taxation

SGLI proceeds that are payable at the death of the insured are excluded from gross income for tax purposes. (The value of the proceeds, however, may be included in determining the value of an estate, and that estate may ultimately be subject to tax.)

If SGLI proceeds are paid to a beneficiary in 36 equal installments, the interest portion included in these installments is also exempt from taxation. In addition, delayed settlement interest (interest accrued from the date of the insured's death to the date of settlement) is also exempt from taxation.

A beneficiary is not required to report to the Internal Revenue Service any installment interest or delayed settlement interest received in addition to the proceeds.

Customer Service

Any questions regarding SGLI should be directed to the Office of Servicemembers' Group Life Insurance (OSGLI) at (800)419-1473.

Traumatic Injury Protection Under Servicemembers' Group Life Insurance (TSGLI)

TSGLI is a program that provides automatic traumatic injury coverage to all servicemembers covered under the Servicemembers' Group Life Insurance (SGLI) program. Every member who has SGLI also has TSGLI effective December 1, 2005. This coverage applies to active-duty members, reservists, funeral honors duty and one-day muster duty. The premium for TSGLI is a flat rate of $1 per month for most service members. This benefit is also provided retroactively for members who incur severe losses as a result of traumatic injury between October 7, 2001, and December 1, 2005 if the loss was the direct result of injuries incurred in Operations Enduring Freedom or Iraqi Freedom.

For the purposes of TSGLI only, "incurred in Operation Enduring Freedom or Operation Iraqi Freedom" means that the member must have been deployed outside the United States on orders in support of OEF or OIF or serving in a geographic location that qualified the service member for the Combat Zone Tax Exclusion under the Internal Revenue Service Code.

The service member is the beneficiary of TSGLI. The member cannot name someone other than himself or herself as the TSGLI beneficiary. If the member is incompetent, the benefit will be paid to his or her guardian or attorney-in-fact. If the service member is deceased, the TSGLI payment will be made to the beneficiary or beneficiaries of the member's basic SGLI.

TSGLI coverage will pay a benefit of between $25,000 and $100,000 depending on the loss directly resulting from the traumatic injury.

In 2011 the VA announced that servicemembers with severe injuries to the genitourinary organs are now eligible for Servicemembers' Group Life Insurance Traumatic Injury Protection (TSGLI). The first payments for eligible veterans and servicemembers began December 1, 2011. Eligibility is retroactive to injuries received on or after October 7, 2001.

Expansion of Covered Losses

Loss of Sight
Loss of Sight lasting 120 days or more is considered as "permanent", qualifying the service member for the same payment rate as for permanent loss of sight ($100,000 for both eyes, $50,000 for one eye).

Uniplegia
Uniplegia (complete and total paralysis of one limb) has been added to the schedule of losses with payment at $50,000.

Amputation of the Hand
The definition of amputation of the hand has been expanded to include loss of four fingers (on the same hand) or loss of thumb, with payment remaining at $50,000 for one affected hand and $100,000 for both hands.

Amputation of the Foot
The definition of amputation of the foot has been expanded to include loss of all toes, with the payment remaining at $50,000 for one affected foot and $100,000 for both feet.

Loss of Four Toes
A new category has been created for loss of four toes (on the same foot and not including the big toe) with payment at $25,000 for one affected foot and $50,000 for both feet.

Loss of Big Toe
A new category has been created for the loss of the big toe, with payment at $25,000 for one affected foot and $50,000 for both feet.

Limb Salvage
Coverage has been expanded to include limb salvage (multiple surgeries intended to save a limb rather than amputate) with payment equivalent to amputation.

Burns
The burn standard, currently 3rd degree (full thickness) burns to at least 30% of face or body, has been expanded to include 2nd degree (partial thickness) burns to at least 20% of the face or body.

Hospitalization as a Proxy for ADL Loss
Continuous 15-day inpatient hospital care is deemed a proxy for the first ADL eligibility period for OTI (Other Traumatic Injury) and TBI (Traumatic Brain Injury) claims.

Facial Reconstruction
Facial Reconstruction, required as a result of traumatic avulsion of the face or jaw that causes discontinuity defects, has been added to the schedule of losses, with payment levels of $25,000 to $75,000, depending upon the severity of the injury.

Losses to the Genitourinary System
Effective beginning in 2011, the VA amended its regulations to add certain genitourinary system injuries to the Schedule of Covered Losses. These are injuries that occur to the genitals or urinary system. Payments range from $25,000 to $50,000 and are retroactive to October 7, 2001. The new losses added to the TSGLI Schedule of Covered Losses include the following:

- Anatomical loss of penis
- Permanent loss of use of the penis
- Anatomical loss of one or both of the testicles
- Permanent loss of use of both testicles
- Anatomical loss of the vulva, uterus or vaginal canal
- Permanent loss of use of the vulva or vaginal canal
- Anatomical loss of one or both ovaries
- Permanent loss of use of both ovaries
- Total and permanent loss of urinary system function

Servicemembers' Group Life Insurance Family Coverage

The Veterans' Opportunities Act of 2001 extended life insurance coverage to spouses and children of members insured under the SGLI program, effective November 1, 2001.

Amount of Coverage

SGLI coverage is available in $50,000 increments up to a maximum of $400,000. Covered members receive 120 days of free coverage from their date of separation. Coverage can be extended for up to two years if the Servicemember is totally disabled at separation. Part-time coverage is also provided to Reserve members who do not qualify for full-time coverage (members covered part-time do not receive 120 days of free coverage).

If you are totally disabled at the time of separation (unable to work), you can apply for the SGLI Disability Extension, which provides free coverage for up to two years from the date of separation. At the end of the extension period, you automatically become eligible for VGLI, subject to premium payments.

Eligibility

Family coverage is available for the spouses and children of:
- Active-duty servicemembers; and
- Members of the Ready Reserve of a uniformed service.

Note: Family coverage is available only for members insured under the SGLI program. Family coverage is not available for those insured under the VGLI program.

Premiums

SGLI coverage for children is free. For monthly premiums for spouses, individuals should call (800) 419-1473.

Dependent Child Coverage Extended to Include Stillborn Children

Public law 110-389 expanded the Servicemembers' Group Life Insurance (SGLI) program to include a member's "stillborn child" as an insurable dependent. SGLI dependent coverage provides for a $10,000 payment to the insured service member upon the death of the member's dependent child. The law was enacted October 10, 2008, and on November 18, 2009, regulations implementing section 402 of the Veterans' Benefits Improvement Act of 2008, were published in the Federal Register, and immediately went into effect. This change applies only for stillbirths on or after October 10, 2008, the effective date of PL110-389.

Declining Coverage

If an individual does not want insurance coverage for a spouse or wants a reduced amount of coverage, he or she must complete "form SGLV-82861, Family Coverage Election", and submit it to his or her personnel officer.

Termination of Coverage

Coverage for a spouse ends 120 days after any of the following events:
- The date elected in writing to terminate a spouse's coverage;
- The date elected in writing to terminate the service member's own coverage;
- The date of the service member's death;
- The date the service member's coverage terminates;
- The date of divorce.

A spouse can covert his or her coverage to a policy with a commercial company within 120 days following one of the events listed above. He or she should contact the Office of Servicemembers' Group Life Insurance at (800) 419-1473.

Coverage for a service member's children ends 120 days after any of the following events:
- The date elected in writing to terminate the service member's own coverage;
- The date the service member separates from service;
- The date of the service member's death;
- The date the child is no longer the service member's dependent.

No conversion options are available to children.

Payment of Proceeds

The service member is paid the proceeds due to the death of a spouse or child. If the service member were to die before payment could be made, the proceeds of a spouse or child claim would be paid to the member's beneficiary, as designated by the member.

Veterans' Group Life Insurance – VGLI (1974- Present)

VMLI is issued to those severely disabled veterans and servicemembers who have received grants for specially adapted housing from VA. These grants are issued to veterans and servicemembers whose movement or vision is substantially impaired because of their disabilities.

Policyholders have three options for their VMLI coverage. They may decline the increase and retain their pre-October level of VMLI coverage and premium, accept the maximum amount of VMLI coverage for which they are eligible, or select a different amount of VMLI coverage. Coverage may not exceed the maximum allowed by law, or their mortgage balance, whichever is less.

VGLI, like SGLI, is supervised by the Department of Veterans Affairs, but is administered by the Office of Servicemembers' Group Life Insurance (OSGLI). VGLI provides for the conversion of Servicemembers' Group Life Insurance to a five-year renewable term policy of insurance protection after a service member's separation from service.

Effective April 11, 2011, VGLI insureds who are under age 60 and have less than $400,000 in coverage can purchase up to $25,000 of additional coverage on each five-year anniversary of their coverage, up to the maximum $400,000. No medical underwriting is required for the additional coverage.

Eligibility for VGLI

Full-time coverage is available for the following members:
- Full-time SGLI insureds that are released from active duty or the Reserves, or;
- Ready Reservists who have part-time SGLI coverage, and who, while performing active duty or inactive duty for training for a period of less than 31 days, incur a disability or aggravate a preexisting disability that makes them uninsurable at standard premium rates, or;
- Members of the Individual Ready Reserve (IRR) and Inactive National Guard (ING), or;
- A member of the Public Health Service (PHS) or Inactive Reserve Corps (IRC), or;
- A member who had part-time SGLI and who, while performing duty (or traveling directly to or from duty), suffered an injury or disability which renders him/her uninsurable at standard premium rates.

Coverage Amounts

VGLI is issued in multiples of $10,000 up to a maximum of $400,000, but not for more than the amount of SGLI coverage the member had in force at the time of separation from active duty or the reserves.

VGLI Renewal

Members may renew their VGLI coverage under the following conditions:

- Members who have separated from service may renew their VGLI coverage for life in 5-year term periods.
- Members of the IRR or ING may renew their VGLI for additional 5-year term periods, as long as they remain in the IRR or ING.

Rather than renew, a member also has the right at any time to **convert** VGLI to an individual commercial life insurance policy with any one of 99 participating commercial insurance companies.

Increasing VGLI Coverage

You can increase your VGLI coverage by up to $25,000 on the five-year anniversary date of your coverage policy if you meet all of the following criteria:

- You have an active VGLI policy.
- You currently have less than $400,000 of VGLI.
- You will be under age 60 on your next policy anniversary date.
- You must request the increase during the 120-day period prior to the 5-year anniversary of your policy

For more information on increasing your VGLI coverage, contact the Office of Servicemembers' Group Life Insurance online at osgli.osgli@prudential.com or by calling 1-800-419-1473.

How to Apply for VGLI

VGLI applications are mailed to eligible members, generally within 60 days after separation, and again shortly before the end of the 16-month application period. Applications are mailed to the address shown on the member's DD-Form 214 or equivalent separation orders. It is the member's responsibility, however, to apply within the time limits, even if they do not receive an application in the mail.
Applications should be mailed to:

Office of Servicemembers Group Life Insurance (OSGLI)
PO Box 70173
Philadelphia, PA 19176-9912

Time Limits to Apply for VGLI

To be eligible, a member must apply for VGLI within the following time limits:

Ordinarily, a member must submit an application to the OSGLI with the required premium within 120 days following separation from service.

If a member is totally disabled at the time of separation from active duty and is granted extended free SGLI coverage, he or she may apply for VGLI anytime during the one-year period of extension.

Individuals who are assigned to the IRR and ING have 120 days after assignment to apply, without evidence of good health, and one year after that with evidence of good health. If an application or the initial premium has not been submitted within the time limits above, VGLI may still be granted if an application, the initial premium and **evidence of insurability** (good health) are submitted to OSGLI within 1 year and 120 days following termination of SGLI. **Applications will not be accepted after one year and 120 days.**

An application for an incompetent member may be made by a guardian, committee, conservator or curator. In the absence of a court-appointed representative, the application may be submitted by a family member or anyone acting on the member's behalf.

VGLI Premium Rates

VGLI premium rates are determined by age group and amount of insurance. To lessen the high cost of term insurance at the older ages, a Decreasing Term Option is available, starting at age group 60 to 64. Under this option, an insured pays a level premium for life, while the insurance amount declines by 25% for three subsequent five-year renewals. At that point, coverage remains level at 25% of the original insurance amount.

Payment of Premiums

Once the VGLI application is approved, the OSGLI will send the insured a certificate and a supply of monthly premium payment coupons.
Premiums may be paid:

- Monthly;
- Quarterly;
- Semiannually;
- Annually;
- By monthly allotment from military retirement pay;
- By monthly deduction from VA compensation payments.

If the insured does not pay the premium when it is due or within a grace period of 60 days, the VGLI coverage will lapse. If VGLI lapsed due to failure to pay the premiums on time, the insured will receive a notification of the lapse and a reinstatement form. The insured may apply to reinstate coverage at any time within 5 years of the date of the unpaid premium. If the insured applies for reinstatement within 6 months from the date of lapse, the individual only needs to provide evidence that he or she is in the same state of health on the date of reinstatement as on the date of lapse. Otherwise, the individual may need to provide proof of good health.

Beneficiary Selection

Any beneficiary can be named. If none is selected, the insurance is distributed, by law, in the following order:

- Spouse, or
- Children, or
- Parents, or
- Executor of estate, or
- Other next of kin.

To name a beneficiary, the insured must submit "Form SGLV 8712, Beneficiary Designation-Veterans' Group Life Insurance." The completed form should be sent to:

Office of Servicemembers Group Life Insurance (OSGLI)
PO Box 70173
Philadelphia, PA 19176-9912

When an insured converts SGLI to VGLI following separation from service, a new beneficiary designation form must be completed. If a new form is not filed, the SGLI beneficiary designation will be considered the VGLI designation for up to 60 days after the effective date of the VGLI. If a new beneficiary is not designated after this 60-day period, the proceeds would be paid "By Law" under the order of precedence in the law.

Alliance Account

VGLI proceeds can be paid in a lump sum or over a 36-month period.
If the proceeds are to be paid in a lump sum then beneficiaries of SGLI and VGLI will receive the payment of their insurance proceeds via an "Alliance Account".

Rather than the traditional single check for the full amount of insurance proceeds, the beneficiary receives a checkbook for an interest-bearing account from which the beneficiary can write a check for any amount from $250 up to the full amount of the proceeds.

The Alliance Account:

- Earns interest at a competitive rate;
- Is guaranteed by Prudential;
- Gives the beneficiary time to make important financial decisions while their funds are secure and earning interest;
- Gives them instant access to their money at all times.

Accelerated Benefits

On November 11, 1998, the President signed legislation authorizing the payment of "Accelerated Benefits" in the SGLI and VGLI programs subject to the following:

- Terminally ill insureds will have access of up to 50 percent of the face amount of their coverage during theirlifetime.
- This money will be available in increments of $5,000.
- An insured must have a medical prognosis of life expectancy of 9months or less.

Taxation

VGLI proceeds are exempt from taxation. Any installment interest or delayed settlement interest that a beneficiary receives in addition to the proceed is also exempt from taxation and does not need to be reported to the IRS.

Customer Service

Any questions regarding VGLI should be directed to the office of Servicemembers' Group Life Insurance (OSGLI) at (800) 419-1473.

Veterans' Mortgage Life Insurance – VMLI (1971- Present)

The Veterans' Mortgage Life Insurance (VMLI) program began in 1971 and is designed to provide financial protection to cover eligible veterans' home mortgages in the event of death. VMLI is issued to those severely disabled veterans, under age 70, who have received grants for Specially Adapted Housing from VA. (Refer to Chapter 6 of this book for information regarding Specially Adapted Housing.) **Veterans must apply for VMLI before their 70th birthday.**

The Veterans Benefits Act of 2010 increased the maximum amount of VMLI coverage from $90,000 to $150,000, which became effective October 1, 2011. On January 1, 2012, the amount increased from $150,000 to 200,000, under the same law. The insurance is payable if the veteran dies before the mortgage is paid off. VA will pay the amount of money still owed on the mortgage up to $200,000.

The insurance is payable only to the mortgage lender. The day-to-day operations of the program are handled by the Philadelphia VAROIC.

Coverage Amounts

VMLI coverage decreases as the insured's mortgage falls below $90,000. This reduced coveragecannot be reinstated. However, ifthe home is sold and a newhome is purchased the veteranbecomes eligible once again for the maximum amount of coverage.

Payment of VMLI Proceeds

Certain conditions apply to the payment of VMLI benefits:

- The insurance is payable at the death of the veteran only to the mortgage holder.

- If the title of the property is shared with anyone other than the veteran's spouse, the insurance coverage is only for the percentage of the title that is in the veteran's name.
- No insurance is payable if the mortgage is paid off before the death of the insured or if it was paid off by other mortgage insurance before the VMLI payment is made.

The insurance will be canceled for any of the following conditions:
- The mortgage is paid in full.
- Termination of the veteran's ownership of the property securing the loan.
- The request of the veteran.
- Failure of the veteran to submit timely statements or other required information

Premiums

Premiums are determined by the insurance age of the veteran, the outstanding balance of the mortgage at the time of application, and the remaining length of time the mortgage has to run. Veterans who desire insurance will be advised of the correct premium when it is determined.

Premiums **must** be paid by deduction from the veteran's monthly compensation or pension payments if the veteran is receiving such payments. If such payments are not being received the veteran may make direct payments, on a monthly, quarterly, semiannual, or annual basis, to the VA Insurance Center in Philadelphia, Pennsylvania.

Dividend Options

If a policyholder is eligible for a dividend, he or she may choose from several dividend options that are available:

Cash paid to the policyholder by US Treasury check.

Credit, held in account for the insured with interest. Can be used to prevent policy lapse. Will be refunded upon the insured's request or will be included in the award to the beneficiary(ies) at the time of the insured's death.

Paid-Up Additions (PUA'S), used as a net single premium to purchase additional paid-up insurance. Available only on "V," "RS," "W," "J," "JR," and "JS" policies. PUA's will be whole life insurance if the basic insurance is an endowment policy.

Deposit, held in account for insured with interest. Available only on permanent plan policies. Considered part of the policy's cash value for the purpose of purchasing reduced paid-up insurance, or if the policy lapses, extended insurance (except for "K" or "JS" policies). Will be refunded upon the insured's request. Will be included in the award to the beneficiary(ies) at the time of the insured's death.

Premium, applied to pay premiums in advance.

Indebtedness applied toward a loan or lien on a policy.

Net Cash used to pay an annual premium with any remainder paid to the policyholder under the cash option.

Net PUA, used to pay an annual premium with any remainder used to purchase paid-up additional insurance.

Net Loan-Lien used to pay an annual premium with any remainder used to reduce an outstanding loan or lien.

To change the method in which dividends are paid, individuals should speak with an Insurance Specialist at (800) 669-8477.

Miscellaneous Information About Government Life Insurance Policies

Power of Attorney is not acceptable for executing a change of beneficiary for government life insurance, even if certain state statutes allow it. Only a court-appointed guardian that is recognized by state statutes can execute a beneficiary designation. If the state statute does not give the guardian broad powers to authorize a beneficiary change, a specific court order is needed to effectuate a change.

Assignment of government life insurance is not allowed, for any reason, nor can ownership of a policy be transferred. Only the insured can exercise the rights and privileges inherent in the ownership of the policy.

Policy Loans are available on permanent plans of insurance. The policyholder can take up to 94% of the reserve value of the policy, less any indebtedness. The policy cannot be lapsed, and premiums must be paid or waived at least one year before a policy has a loan value. Changes in interest rates are made on October 1 of each year, if warranted. Rate changes are tied to the "ten-year constant maturities", U.S. Treasury securities index. A policyholder can apply for a loan by filing VA Form 29-1546, Application for Policy Loan.

The completed form can be faxed to (215) 381-3580, or mailed to:

Department of Veterans Affairs Regional Office and Insurance Center
PO Box 7327
Philadelphia, PA 19101

An Annual Insurance Policy Statement is mailed to the insured on the policy anniversary date of each policy. The statement provides the insured with information about his or her VA insurance. The statement should be reviewed for accuracy each year, and the VA should be contacted immediately if there are any discrepancies.

Totally Disabled or Terminally Ill Policyholders

If you have VA life insurance and become terminally ill or totally disabled, you may be eligible for certain benefit options. Your spouse might also be eligible for certain options if diagnosed with a terminal illness.

SGI Extension for Policyholders Who Become Disabled

You may be able to keep your SGLI coverage for up to two years after the date you leave the military if you're disabled when you leave. Both of the following have to be true:

- You're within 2 years of your separation date, and
- You're totally disabled or having certain conditions

Waiver of Premiums for Policyholders Who Become Totally Disabled

A waiver of premiums mean you don't have to pay your life insurance premiums. In most situations premiums can be waived up to one year before ethe receipt of a claim. To get a waiver for your life insurance premiums, you'd have to meet all of these requirements:

- You have a mental or physical disability preventing you from being able to hold a job, and
- You're covered under Service-Disabled Veterans Life Insurance (S-DVI), and
- Your total disability happens before you're 65, but after the effective date of your life insurance policy, and
- Your total disability continues for at least 6 months in a row

There are exceptions to the conditions above, but if you think you could be entitled to a waiver of your premiums, apply as soon as you can, and the VA will determine if you qualify.

To apply, you should fill out and sign a Claim for Disability Insurance (VA Form 29-357), and mail it to the address listed on the form.

Accelerated Benefits for Terminally Ill Policyholders

Veterans and servicemembers who are eligible, as well as covered spouses of servicemembers can get accelerated benefits, meaning you can get up to 50% of the face value of your coverage in increments of $5,000 paid to you before death.

Both must be true:

- You have SGLI, Family SGLI or VGLI life insurance, and
- You or your covered spouse has a written statement from a doctor saying you have or your spouse has nine months or less to live.

Only the insured servicemember or veteran may apply for accelerated benefits. No one can apply on their behalf, and in the case of a terminally ill spouse, only the insured may apply.

If you're an SGLI or VGLI policyholder, you should fill out a Claim for Accelerated Benefits (SGLV 8284). If you're still on active duty or you're a Reservist, you need to turn the form into your service branch, because they have to fill out part of the form.

If you're a veteran, you should have your doctor fill out their portion of the form, and then send the completed form to:

The Prudential Insurance Company of America
PO Box 70173
Philadelphia, PA 19176-0173

If you're covered under Family SGLI, you should fill out a Claim for Accelerated Benefits (SGLV 8284A). You'll fill out one part, and your doctor will fill out another part of the application.

The VA pays the remaining amount of the face value of an insurance policy to your designated beneficiary or beneficiaries upon your death. In the situation with a terminally ill spouse, VA pays the rest of the insurance policy to you upon the death of your spouse.

What to Remember
✓ Different programs cover veterans (VGLI), service members (SGLI), and family members (FSGLI). You may also be able to get short-term financial coverage through TSGLI to help you recover from a severe, traumatic injury.
✓ Veterans Affairs Life Insurance (VALife) provides low-cost coverage to veterans with service-connected disabilities. Your full life insurance coverage starts 2 years after you apply.
✓ If you're ending your military tour of duty soon, you'll need coverage quickly. In some cases, you have to take action within 120 days of leaving the military to ensure no break in coverage.
✓ Veterans' Group Life Insurance (VGLI) may allow you to keep your life insurance coverage after you leave the military, for as long as you continue to pay the premiums.
✓ With VGLI, you can get between $10,000 and $400,000 in term life insurance benefits. The amount you get will be based on how much SGLI coverage you had when you left the military.

CHAPTER 24

(★★★★★★★★)

HOME LOAN GUARANTIES

The VA home loan program helps servicemembers, veterans and eligible surviving spouses become homeowners. The home loan guarantee benefit is available, and there are other housing-related programs to help you buy, build, repair, keep or adapt a home. VA home loans are provided by private lenders, like banks and mortgage companies. VA guarantees a portion of the loan, which enables the lender to provide you with more favorable terms.

Major features of the VA home loan benefit include:

- No required down payment—some lenders may require down payments for some borrowers using the VA home loan guarantee, but VA doesn't require a down payment.
- Competitively low interest rates
- Limited closing costs
- No need for Private Mortgage Insurance (PMI)
- The VA home loan is a lifetime benefit that can be used multiple times

Benefits

Purchase Loans help veterans purchase a home at a competitive interest rate, often without requiring a down payment or private mortgage insurance. Cash-Out Refinance loans let eligible veterans take cash out of their home equity to pay off things such as debt, to, make home improvements or to fund education. To be eligible veterans must have satisfactory credit, sufficient income to meet the expected monthly obligations, and a valid Certificate of Eligibility (COE).

Interest Rate Reduction Refinance Loans (IRRRL) are also called the Streamline Refinance Loan, and they can help veterans obtain a lower interest rate by refinancing an existing VA loan. This can be used only for veterans with an existing VA guaranteed loan on a property.

Native American Direct Loan (NADL) Program helps eligible Native American veterans finance the purchase, construction or improvement of homes on Federal Trust Land or reduce the interest rate on a VA loan. The veterans' tribal organization must participate in VA direct loan program, and the veteran must have a valid COE.

Cash-OUT Refinance Loan is a VA-backed cash-out refinance loan that lets you replace your current loan with a new one under different terms. If you want to take cash out of your home equity or refinance a non-VA loan into a VA-backed loan, a VA-backed cash-out refinance loan might be right for you. You must qualify for a VA-backed home loan Certificate of Eligibility and meet VA's and your lender's standards for credit, income and other requirements and live in the home you're refinancing.

Adapted Housing Grants help veterans with a permanent and total service-connected disability purchase or build an adapted home or to modify an existing home to account for their disability.

Other resources may be available to veterans depending on their state of residency, including property tax reductions.

General Information

The purpose of the VA loan guaranty program is to help veterans and active-duty personnel finance the purchase of homes with competitive loan terms and interest rates.

The VA does not actually lend the money to veterans. VA guaranteed loans are made by private lenders, such as banks, savings & loans, or mortgage companies. The VA guaranty means the lender is protected against loss if the veteran fails to repay the loan.

The VA Loan Guaranty Service is the organization within the VA that has the responsibility of administering the home loan program.

In 2011 the VA announced that those veterans who qualify and submit to a short sale or deed-in-lieu of foreclosure may be eligible to receive $1,500 in relocation assistance. The VA has instructed mortgage lenders to provide the funds in order to help borrowers cover the cost of moving or other expenses incurred during the process.

In 2012 Under Public Law 112-154, veterans in specially adapted housing or those in receipt of VA home loans, which have property hindered or destroyed by a natural disaster, became eligible to receive VA assistance. The law also extended the VA's authority to extend the guarantee of timely payment of principal and interest of mortgage loans. The VA's authority was also extended with regard to the collection of loan fees and adjustment of maximum home loan guarantee amounts.

PL 116-23 Changes to the VA Home Loan Program

The Blue Water Navy Vietnam Veterans Act of 2019 (PL 116-23) extended the presumption of herbicide exposure, such as Agent Orange, to veterans who served in the offshore waters of the Republic of Vietnam between January 9, 1962, and May 7, 2975. Beginning January 1, 2020, veterans who served as far as 12 nautical miles from the shore of Vietnam or who had served in the Korean Demilitarized Zone are presumed to have been exposed to herbicides. Additionally, PL 116-23 made changes to the VA Home Loan Program, including:

* VA-guaranteed home loans will no longer be limited to the Federal Housing Finance Agency Conforming Loan Limits. Veterans are now able to obtain a no-down-payment home loan in all areas, regardless of loan amount.
* The law exempts Purple Heart recipients currently serving on active duty from the VA Home Loan funding fee
* VA removed the loan limit for Native American Veterans seeking to build or purchase a home on Federal Trust Land
* At this time, there is a temporary change to the VA funding Fee, which is a congressionally mandated fee associated with the VA Home Loan. Veterans and service members will see a slight increase of 0.15 to 0,30% in their funding fee, while National Guard and Reserve members will see a slight decrease in their fee to align with the fee paid by 'Regular Military' borrowers. Veterans with service-connected disabilities, some surviving spouses and other potential borrowers are exempt from the VA loan funding fee and won't be impacted by the change.

Features of VA Guaranteed Home Loans

* Equal opportunity for all qualified veterans to obtain a VA guaranteed loan.
* No down payment (unless required by the lender or the purchase price is more than the reasonable value of the property).
* Buyer's interest rate is negotiable.
* Buyer has the ability to finance the VA funding fee (plus reduced funding fees with a down payment of at least 5% and exemption for veterans receiving VA compensation.
* Closing costs are comparable with other financing types (and may be lower).
* No mortgage insurance premiums are necessary.

- An assumable mortgage may be available.
- Buyer has the right to prepay without penalty.

VA Guaranteed Loans Do Not Do the Following:

- Guarantee that a home is free of defects
- VA only guarantees the loan. It is the veteran's responsibility to assure that he or she is satisfied with the property being purchased. Veterans should seek expert advice as necessary, before legally committing to a purchase agreement.
- If a veteran has a home built, VA cannot compel the builder to correct construction defects, although VA does have the authority to suspend a builder from further participation in the VA home loan program.
- VA cannot guarantee the veteran is making a good investment.
- VA cannot provide a veteran with legal service.

Uses for VA Loan Guarantees

VA loan guarantees can be used for the following:

- To purchase, construct, or improve a home.
- To purchase and improve a home concurrently.
- To purchase a residential condominium or townhouse unit in a VA approved project. (If one veteran is purchasing the property, the total number of separate units cannot be more than 4.)
- To purchase a manufactured home or a manufactured home and manufactured home lot.
- To purchase and improve a manufactured home lot on which to place a manufactured home which the veteran already owns and occupies.
- To refinance an existing home loan.
- To refinance an existing VA loan to reduce the interest rate and make energy-efficient improvements.
- To refinance an existing manufactured home loan in order to acquire a lot.
- To improve a home by installing a solar heating and/or cooling system or other energy-efficient improvements.

Veterans must certify that they plan to live in the home they are buying or building in order to qualify for a VA loan guaranty.

VA loan guarantees are available only for property located in the United States, its territories, or possessions (Puerto Rico, Guam, Virgin Islands, American Samoa, and Northern Mariana Islands).

VA loan guarantees are not available for farm loans unless there is a home on the property, which will be personally occupied by the veteran. Non-realty loans for the purchase of equipment, livestock, machinery, etc. are not made. Other loan programs for farm financing may be available through the Farmers Home Administration, which gives preference to veteran applicants. (Interested veterans should refer to the local telephone directory for the phone number of a local office.)

Although business loans are not available through VA, the Small Business Administration (SBA) has a number of programs designed to help foster and encourage small business enterprises, including financial and management assistance. Each SBA office has a veteran's affairs officer available to speak with. (Interested veterans should refer to the local telephone directory for the phone number of a local SBA office, or call (800) 827- 5722.

Eligibility Requirements

Individuals may qualify for VA home loan guaranties if their service falls within any of the following categories:

Vietnam Eligibility Requirements

- Active duty on or after August 5, 1964, and prior to May 8, 1975. (For those serving in the Republic of Vietnam, the beginning date is February 28, 1961.); and
- Discharge or separation under other than dishonorable conditions; and
- At least 90 days of total service, unless discharged earlier for a service- connected disability.
- Unremarried widows of above-described eligible individuals who died as a result of service.
- Widows of above-described eligible individuals who died as a result of service who remarried after age 57.

Post-Vietnam Eligibility Requirements for Veterans with Enlisted Service Between May 8, 1975, AND September 7, 1980 (if enlisted) Or October 16, 1981 (if officer):

- At least 181 days of continuous service, all of which occurred on or after May 8, 1975, unless discharged earlier for a service-connected disability; and
- Discharge or separation under other than dishonorable conditions.
- Unremarried widows of above-described eligible individuals who died as a result of service.
- Widows of above-described eligible individuals who died as a result of service who remarried after age 57.

Post-Vietnam Eligibility Requirements for Veterans Separated from Enlisted Service between September 7, 1980 (October 17, 1981, for officers) and August 1, 1990:

- At least 24 months of continuous active duty, or the full period (at least 181 days) for which individual was called or ordered to active duty, and discharged or separated under other than dishonorable conditions; or
- At least 181 days of continuous active duty, and discharged due to:
  - o a hardship; or
  - o a service-connected, compensable disability; or
  - o a medical condition which preexisted service, and has not been determined to be service connected; or
  - o the convenience of the government as a result of a reduction in force; or
  - o a physical or mental condition not characterized as a disability, and not the result of misconduct, but which did interfere with the performance of duty.
- Early discharge for a service-connected disability.
- Unremarried widows of above-described eligible persons who died as the result of service.
- Widows of above-described eligible individuals who died as a result of service who remarried after age 57.

Persian Gulf War Eligibility Requirements

- At least 24 months of continuous active duty on or after August 2, 1990, or the full period for which the individual was called or ordered to active duty, and discharged or separated under other than dishonorable conditions; or
- At least 90 days of continuous active duty, and discharged due to:
  - o A hardship; or
  - o A service-connected, compensable disability; or
  - o A medical condition which preexisted service, and has not been determined to be service connected; or

- o   The convenience of the government as a result of a reduction in force; or
- o   A physical or mental condition not characterized as a disability, and not the result of misconduct, but which did interfere with the performance of duty.
- Early discharge for a service-connected disability.
- Unremarried widows of above-described eligible individuals who died as the result of service.
- Widows of above-described eligible individuals who died as a result of service who remarried after age 57.

When law or Presidential Proclamation ends the Persian Gulf War, a minimum of 181 days of continuous active duty will be required for those who did not serve during wartime.

Members of the Reserve and National Guard are eligible if activated after August 1, 1990, served at least 90 days, and discharged or separated under other than dishonorable conditions.)

Active-Duty Personnel Requirements
Individuals who are now on regular duty (not active duty for training) are eligible after having served 181 days (90 days during the Gulf War), unless discharged or separated from a previous qualifying period of active-duty service.

Eligibility Requirements for Members of the Selected Reserve
- At least 6 years in the Reserves or National Guard, or discharged earlier due to a service-connected disability; and
- Discharged or separated under other than dishonorable conditions; or
- Placed on the retired list; or
- Transferred to an element of the Ready Reserve other than the Selected Reserve; or
- Continue to serve in the Selected Reserve.
- Unremarried widows of above-described eligible persons who died as the result of service.
- Widows of above-described eligible individuals who died as a result of service who remarried after age 57.

Eligibility Requirements for Other Types of Service
- Certain U.S. citizens who served in the armed forces of a U.S. ally in World War II. Members of organizations with recognized contributions to the U.S. during World War II (Questions about this type of service eligibility can be answered at any VA regional office.)
- Spouses of American servicemen who are listed as missing-in-action, or prisoners-of-war for a total of 90 days or more.

Certificate of Eligibility
The Certificate of Eligibility is the medium by which VA certifies eligibility for A VA loan guaranty.

Individuals may request a Certificate of Eligibility by completing "VA Form 26-1880 Request for a Certificate of Eligibility for VA Home Loan Benefits." The completed form should be submitted to a VA Eligibility Center along with acceptable proof of service.

Veterans separated after January 1, 1950, should submit DD Form 214, Certificate of Release or Discharge from active Duty.

Veterans separated after October 1, 1979, should submit copy 4 of DD Form 214.

Since there is no uniform document like the DD Form 214 for proof of service in the Selected Reserve a number of different forms may be accepted as documentation of service in the Selected Reserve:

- For those who served in the Army or Air National Guard and were discharged after at least 6 years of such service, NGB Form 22 may be sufficient.
- Those who served in the Army, Navy, Air Force, Marine Corps or Coast Guard Reserves may need to rely on a variety of forms that document at least 6 years of participation in paid training periods or have paid active duty for training.
- Often it will be necessary to submit a combination of documents, such as an Honorable Discharge certificate together with a Retirement Points Statement. It is the reservist's responsibility to obtain and submit documentation of 6 years of honorable service.

In addition, if an individual is now on active duty, and has not been previously discharged from active-duty service, he or she must submit a statement of service that includes the name of the issuing authority (base or command) and is signed by or at the direction of an appropriate official. The statement must identify the individual, include the social security number, provide the date of entry on active duty and the duration of any lost time.

The Certificate of Eligibility should be presented to the lender when completing the loan application. (However, if an individual does not have a Certificate, the lender may have the forms necessary to apply for the Certificate of Eligibility.)

Procedures for Obtaining Loans

- Find a real estate professional to work with.
- Locate a lending institution that participates in the VA program. You may want to get "pre-qualified" at this point - that is, find out how big a loan you can afford. Lenders set their own interest rates, discount points, and closing points.
- Obtain your Certificate of Eligibility. The lender can probably get you a certificate online. Or, you can apply online yourself. To get your Certificate of Eligibility (COE) online, please go to the eBenefits portal at this link. If you need any assistance, please call the eBenefits Help Desk at 1-800-983-0937. Their hours are Monday-Friday, 8am to 8pm EST.
- You find a home you want to buy.
- When you negotiate, make sure the purchase and sales agreement contain a "VA option clause."
- You may also want the agreement to allow you to "escape" from the contract without penalty if you can't get a VA loan.
- You formally apply to the lender for a VA-backed loan. The lender will complete a loan application and gather the needed documents such as pay stubs and bank statements.
- The lender orders a VA appraisal and begins to "process" all the credit and income information.
- The lending institution reviews the appraisal and all the documentation of credit, income, and assets. The lender then decides whether the loan should be granted.
- Finally, the closing takes place and the property is transferred. The lender chooses a title company, an attorney, or one of their own representatives to conduct the closing. This person will coordinate the date and time.
- If a lender cannot be located, the local VA regional office can provide a list of lenders active in the VA program.

A VA loan guaranty does not guarantee approval of a loan. The veteran must still meet the financial institution's income and credit requirements. If a loan is approved, the VA guarantees the loan when it's closed.

VA Eligibility Centers

Regional Loan Center	Jurisdiction	Mailing Address	Telephone Number
Atlanta	Georgia North Carolina South Carolina Tennessee	Department of Veterans Affairs VA Regional Loan Center 1700 Clairmont Rd. Decatur, GA 30033-4032 (Mail: P.O. Box 100023, Decatur, GA 30031-7023)	(888) 768-2132
Cleveland	Connecticut Delaware Indiana Maine Massachusetts Michigan New Hampshire New Jersey New York Ohio Pennsylvania Rhode Island Vermont	Department of Veterans Affairs VA Regional Loan Center 1240 East Ninth Street Cleveland, OH 44199	(800) 729-5772
Denver	Alaska Colorado Idaho Montana Oregon Utah Washington Wyoming	Department of Veterans Affairs VA Regional Loan Center 155 Van Gordon Street Lakewood, CO 80228 (Mail: Box 25126, Denver, CO 80225)	(888) 349-7541
Honolulu	Hawaii Guam American Samoa Commonwealth of the Northern Marianas	Department of Veterans Affairs VA Regional Loan Center Loan Guaranty Division (26) 459 Patteson Rd. Honolulu, HI 96819 *Although not an RLC, this office is a fully functioning Loan Guaranty operation for Hawaii.	(888) 433-0481
Houston	Arkansas Louisiana Oklahoma Texas	Department of Veterans Affairs VA Regional Loan Center 6900 Almeda Road Houston, TX 77030-4200	(888) 232-2571
Phoenix	Arizona California New Mexico Nevada	Department of Veterans Affairs VA Regional Loan Center 3333 N. Central Avenue Phoenix, AZ 85012-2402	(888) 869-0194
Roanoke	District of Columbia Kentucky Maryland Virginia West Virginia	Department of Veterans Affairs VA Regional Loan Center 210 Franklin Road, SW Roanoke, VA 24011	(800) 933-5499

St. Paul	Illinois Iowa Kansas Minnesota Missouri Nebraska North Dakota South Dakota Wisconsin	Department of Veterans Affairs VA Regional Loan Center 1 Federal Drive, Ft. Snelling St. Paul, MN 55111- 4050	(800) 827-0611
St. Petersburg	Alabama Florida Mississippi Puerto Rico U.S. Virgin Islands	Department of Veterans Affairs VA Regional Loan Center 9500 Bay Pines Blvd. St. Petersburg, FL 33708 (Mail: P.O. Box 1437, St. Petersburg, FL 33731)	(888) 611-5916

Guaranty or Entitlement Amount

Eligible veterans, service members and survivors with full entitlement no longer have limits on loans over $144,000. This means you won't have to pay a down payment and the VA will guarantee to your lender if you default on a loan that's over $144,000, the VA will pay them up to 25% of the loan amount.

You have full entitlement if at least one of the following is true:
- You've never used your home loan benefit or
- You've paid a previous VA loan in full and sold the property, or
- You've used your home loan benefit, but had a foreclosure or compromise claim and repaid VA in full

If you apply and are eligible for a VA-backed home loan, you'll receive a COE. This is the document that tells private lenders you have VA home loan eligibility and entitlement. Your lender will still need to approve you for a loan. The lender will determine the size of the loan you are eligible for based on credit history, income, and assets.

With remaining entitlement, your VA home loan limit is based on the county loan limit where you live. You may need to make a down payment if you're using remaining entitlement and your loan amount is over $144,000. This is because most lenders require that your entitlement, down payment or a combination of both covers at least 24% of your total loan amount. If you're willing and able to make a down payment, as a result, you may be able to borrow more than the county loan limit with a VA-backed loan.

Hybrid Adjustable-Rate Mortgages

An Adjustable-Rate Mortgage means the interest rate changes with changes in the market. The first-year rate, which is also referred to as a teaser rate, is generally a couple of percentage points below the market rate. The "cap" is the upper limit of the interest rate. If a teaser rate is 4%, and there is a five-point cap, then the highest that an interest rate could go would be 9%. The amount that the interest rate can rise each year is usually limited to one or two percentage points per year, but the frequency at which the rate adjusts can vary. If interest rates go up, an ARM will adjust accordingly.

Down Payments

The VA does not require a down payment be made, provided that:
- The loan is not for a manufactured home or lot (a 5% down payment is required for manufactured home or lot loans); and
- The purchase price or cost does not exceed the reasonable value of the property, as determined by VA; and
- The loan does not have graduated payment features. (Because with a graduated-

payment mortgage, the loan balance will be increasing during the first years of the loan, a down payment is required to keep the loan balance from going over the reasonable value or the purchase price.)

Even though the VA may not require a down payment, the lender may require one.

Closing Costs and Fees

The VA regulates the closing costs that a veteran may be charged in connection with closing a VA loan. The closing costs and origination fees must be paid in cash, and cannot be included in the loan itself, except in the case of refinancing loans. Although some additional costs are unique to certain localities, the closing costs generally include:

VA appraisal Credit report Survey
Title evidence Recording fees
A 1% loan origination fee Discount points

A veteran is charged the customary fees for title search, credit report, appraisal, and transfer fees, etc., the same as any other borrower, but he is not required to pay commission or brokerage fees for obtaining the loan. In home loans, the lender may also charge a reasonable flat charge, called a funding fee, to cover the costs of originating the loan.

Funding Fees

A VA funding fee is payable at the time of loan closing. This fee may be included in the loan and paid from the loan proceeds. The funding fee does not have to be paid by veterans receiving VA compensation for service-connected disabilities, or who but for the receipt of retirement pay, would be entitled to receive compensation for service-connected disabilities, or surviving spouses of veterans who died in service or from a service-connected disability.

The funding fee rates are as follows:

Loan Purpose	Percent of loan for veterans, active-duty service members, and National Guard and Reserve Members
Purchase or construction loan with down payment of less than 5%; or Refinancing loan. or home improvement loan	2.3%
Purchase or construction loan with down payment between 5% and 10%	1.65%
Second or subsequent use without a down payment	3.6%
Assumption of VA guaranteed loan	0.5%
Interest rate reduction loan	0.5%
Manufactured home loan	1.0%
Purchase or construction loan with a down payment of 10% or more	1.4%

Flood Insurance

If the dwelling is in an area identified by the Department of Housing and Urban Development as having special flood hazards, and the sale of flood insurance under the national program is available, such insurance is required on loans made since March 1, 1974. The amount of insurance must be equal to the outstanding loan balance, or the maximum limit of coverage available, whichever is less.

Interest Rates

The interest rate on VA loans varies due to changes in the prevailing rates in the mortgage market. One a loan is made, the interest rate set in the note remains the same for the life of the loan. However, if interest rates decrease, a veteran may apply for a new VA loan to refinance the previous loan at a lower interest rate.

Repayment Period

The maximum repayment period for VA home loans is 30 years and 32 days. However, the exact amortization period depends upon the contract between the lender and the borrower.

The VA will guarantee loans with the following repayment terms:

- Traditional Fixed Payment Mortgage
- Equal monthly payments for the life of the loan
- Graduated Payment Mortgage – GPM
- Smaller than normal monthly payments for the first few years – usually 5 years, which gradually increase each year, and then level off after the end of the "graduation period" to larger than normal payments for the remaining term of the loan. The reduction in the monthly payment in the early years of the loan is accomplished by delaying a portion of the interest due on the loan each month, and by adding that interest to the principal balance.

Buydown

The builder of a new home or seller of an existing home may "buy down" the veteran's mortgage payments by making a large lump sum payment up front at closing that will be used to supplement the monthly payments for a certain period, usually 1 to 3 years.

Growing Equity Mortgage (GEM)

Provides for a gradual annual increase in monthly payments, with all of the increase applied to the principal balance, resulting in early payoff of the loan.

Prepayment of Loan

A veteran or serviceman may pay off his entire loan at any time without penalty or fee or make advance payments equal to one monthly installment or $100, whichever is the lesser amount. Individuals should check with the mortgage holder for the proper procedure.

Loan Defaults

If a veteran fails to make payments as agreed, the lender may foreclose on the property. If the lender takes a loss, the VA must pay the guaranty to the lender, and the individual must repay this amount to the VA. If the loan closed on or after January 1, 1990, the veteran will owe the VA in the event of default, only if there was fraud, misrepresentation, or bad faith on the veteran's part.

The US Department of Veterans Affairs urges all Veterans who are encountering problems making their mortgage payments to speak with their loan servicers as soon as possible to explore options to avoid foreclosure. Depending on a Veteran's specific situation, servicers may offer any of the following options to avoid foreclosure:

- Repayment Plan: The borrower makes regular installment each month plus part of the missed installments.
- Special Forbearance: The servicer agrees not to initiate foreclosure to allow time for borrowers to repay the missed installments. An example would be when a borrower is waiting for a tax refund.
- Loan Modification: Provides the borrower a fresh start by adding the delinquency to the loan balance and establishing a new payment schedule.
- Additional time to arrange a private sale: The servicer agrees to delay foreclosure to allow a sale to close if the loan will be paid off.
- Short Sale: When the servicer agrees to allow a borrower to sell his/her home for a lesser amount than what is currently required to pay off the loan.
- Deed-in-Lieu of Foreclosure: The borrower voluntarily agrees to deed the

property to the servicer instead of going through a lengthy foreclosure process

Release of Liability

Any veteran who sells or has sold a home purchased with a VA loan guaranty may request release from liability to the VA. (If the VA loan closed prior to March 1, 1988, the application forms for a release of liability must be requested from the VA office that guaranteed the loan. If the VA loan closed on or after March 1, 1988, then the application forms must be requested from the lender to whom the payments are made.) The loan must be current, the purchaser must assume full liability for the loan, and the purchaser must sign an Assumption of Liability Agreement. The VA must approve the purchaser from a credit standpoint.

For loans closed on or after March 1, 1988, the release of liability is not automatic. To approve the assumer and grant the veteran release from liability, the lender or VA must be notified, and release of liability must be requested.

If the loan was closed prior to March 1, 1988, the purchaser may assume the loan without approval from VA or the lender. However, the veteran is encouraged to request a release of liability from VA, regardless of the loan's closing date. If a veteran does not obtain a release of liability, and VA suffers a loss on account of a default by the assumer, or some future assumer, a debt may be established against the veteran. Also, strenuous collection efforts will be made against the veteran if a debt is established.

The release of a veteran from liability to the VA does not change the fact that the VA continues to be liable on the guaranty.

Restoration of Entitlement

Veterans who have used all or part of their entitlement may restore their entitlement amount to purchase another home, provided:

- The property has been sold, and the loan has been paid in full; or
- A qualified veteran buyer has agreed to assume the outstanding balance on the loan and agreed to substitute his entitlement for the same amount of entitlement the original veteran owner used to get the loan. (The veteran buyer must also meet the occupancy, income, and credit requirements of the VA and the lender.)
- If the veteran has repaid the VA loan in full but has not disposed of the property securing that loan, the entitlement may be restored ONE TIME ONLY.

Restoration of entitlement does not occur automatically. The veteran must apply for restoration by completing "Form 26, 1880." Completed forms may be returned to any VA regional office or center (A copy of the HUD-1, Closing Statement, or other appropriate evidence of payment in full should also be submitted with the completed Form 26, 1880.) Application forms for substitution of entitlement can be requested from the VA office that guaranteed the loan.

If the requirements for restoration of entitlement cannot be met, veterans who had a VA loan before may still have remaining entitlement to use for another VA loan. The current amount of entitlement available to eligible veterans has been increased over time by changes in the law.

For example, in 1974 the maximum guaranty entitlement was $12,500. Today the maximum guaranty entitlement is $36,000 (for most loans under $144,000). So, if a veteran used the $12,500 guaranty in 1974, even if that loan is not paid off, the veteran could use the $23,500 difference between the $12,500 entitlement originally used and the current maximum of $36,000 to buy another home with a VA loan guaranty.

Direct Home Loans

VA direct home loans are only available to:

- Native American veterans who plan to buy, build, or improve a home on Native American trust land; or
- Certain eligible veterans who have a permanent and total service-connected disability, for specially adapted homes.

Native American Veterans Living on Trust Lands

A VA direct loan can be used to purchase, construct, or improve a home on Native American trust land. These loans may also be used to simultaneously purchase and improve a home, or to refinance another VA direct loan made under this program in order to lower the interest rate. VA direct loans are generally limited to the cost of the home or $80,000, whichever is less.

To qualify for a VA direct loan, the tribal organization or other appropriate Native American group must be participating in the VA direct loan program. The tribal organization must have signed a Memorandum of Understanding with the Secretary of Veterans Affairs that includes the conditions governing its participation in the program.

Veterans should contact their regional VA office for specific information regarding direct home loans.

Resale of Repossessed Homes

The VA sells homes that it acquires after foreclosure of a VA guaranteed loan. These homes are available to veterans and non-veterans.

The properties are available for sale to the general public through the services of private sector real estate brokers. The VA cannot deal directly with purchasers. Real estate brokers receive the keys to the properties and assist prospective purchasers in finding, viewing, and offering to purchase them.

Participating brokers receive instructional material regarding the sales program and are familiar with VA sales procedures. VA pays the sales commission. Offers to purchase VA acquired properties must be submitted on VA forms. Offers cannot be submitted on offer forms generally used in the real estate industry.

VA financing is available for most, but not all, property sales. The down payment requirements are usually very reasonable, the interest rate is established by VA based on market conditions. Any prospective purchaser who requests VA financing to purchase a VA-owned property must have sufficient income to meet the loan payments, maintain the property, and pay all other obligation. The purchaser must have acceptable credit and must also have enough funds remaining for family support.

Anyone interested should consult a local real estate agent to find out about VA-acquired properties listed for sale in the area.

HUD / FHA Loans

Veterans are not eligible for VA financing based on service in World War I, Active Duty for Training in the Reserves, or Active Duty for Training in the National Guard (unless "activated" under the authority of Title 10, U.S. Code). However, these veterans may qualify for a HUD / FHA veteran's loan.

The VA's only role in the HUD / FHA program is to determine the eligibility of the veteran, and issue a Certificate of Veteran Status, if qualified. Under this program, financing is available for veterans at terms slightly more favorable than those available to non-veterans.

A veteran may request a "Certificate of Veteran Status" by completing "VA form 26-8261a." The completed form and required attachments should be submitted to the veteran's regional VA office for a determination of eligibility.

Servicemembers Civil Relief Act (SCRA)

The Servicemembers Civil Relief Act expanded and improved the former Soldiers' and Sailors' Civil Relief Act (SSCRA). It is designed to provide a wide range of protections for people entering the military, being called to active duty, and deployed servicemembers. The goal of the SCRA is to postpone or suspend particular civil obligations, in order for the servicemember to devote full attention to service duties. It is also designed to relieve stress on family members of deployed servicemembers. Among these obligations, a servicemember may be exempt from are mortgage payments.

The new law expands the current law in place to protect servicemembers and their families from eviction from housing while the servicemember is on active duty.

Based on provisions of a recent SCRA Foreclosure Settlement completed by the Federal government, five major servicers (Bank of America, JPMorgan Chase, Ally (GMAC), CitiMortgage, and Wells Fargo) must provide the following relief to service members and veterans:

- Conduct a review of every servicemember foreclosure since 2006 and provide any who were wrongly foreclosed with compensation equal to a minimum of lost equity, plus interest and $116,785;
- Refund money lost because they were wrongfully denied the opportunity to reduce their interest rates to the 6% cap;
- Provide relief for service members who are forced to sell their homes for less than the amount they owe on their mortgage due to a Permanent Change in Station.
- Extend certain foreclosure protections afforded under the Servicemember Civil Relief Act to service members serving in harm's way.

What to Remember
✓ VA helps servicemembers, veterans and eligible surviving spouses with home ownership. As part of this, VA provides a home loan guaranty benefit.
✓ VA home loans are provided by private lenders like banks and mortgage companies.
✓ VA guarantees a portion of the loan so the lender can provide you with more favorable terms.
✓ No down payment is required, the interest rates are competitive and there are limited closing costs. You don't need Private Mortgage Insurance (PMI).
✓ The VA home loan is a lifetime benefit so you can use the guaranty multiple times.

CHAPTER 25

OVERSEAS BENEFITS

If you're a veteran who lives overseas, you remain entitled to benefits and services earned through military service. Most VA benefits are payable regardless of your place of residence or nationality.

Things to know include:

- Once benefit payments are sent from the U.S., your international direct deposit arrival depends on your foreign financial institution's processing time.
- International Treasury Services won't charge you a currency conversion fee for payments sent overseas, but your foreign financial institution may charge fees as applicable.
- Education benefits are offered to eligible veterans to attend an approved program at a foreign school.
- If you're a disabled veteran, you might be eligible for a Specially Adaptive Housing grant to accommodate your disability. You're required to have a substantial ownership interest in the home to be adapted or built. Any improvements or modifications have to be approved by VA first.

Medical Benefits

The Foreign Medical Program (FMP) is a healthcare benefits program for U.S. veterans with VA-rated service-connected conditions who are residing or traveling abroad (except Canada and the Philippines). Services provided in Canada and the Philippines are under separate jurisdictions, as indicated later in this chapter.

Under the FMP, VA assumes payment responsibility for certain necessary medical services associated with the treatment of service-connected conditions.

The Foreign Medical Program, Office in Denver, Colorado, has jurisdiction over all foreign provided services, with the exception of medical services received in Canada and the Philippines. It is responsible for all aspects of the program, including application processing, verification of eligibility, authorization of benefits, and payments of claims.

In September 2011, the VA announced the creation of a new program to provide comprehensive compensation and pension (C&P) examinations to U.S. veterans living overseas. The program includes providing a medical assessment to evaluate veterans' current disabilities that may be related to their military service.

If VA has previously determined a veteran has a service-connected medical condition, the examination helps determine the current severity of the condition, which could affect the amount of VA disability compensation payable or entitlement to additional benefits. During this time, the VA also conducted its first C&P examinations using telehealth technology, which is designed to ease the burden of travel for U.S. veterans living overseas. The VA plans to provide these services in various locations in Europe and Asia.

Individuals who are traveling to or reside in one of the following countries should use the following number to contact the FMP Office in Denver, Colorado: (877) 345-8179: Germany; Panama; Australia; Italy; UK; Japan; Spain.

For individuals who are in Mexico or Costa Rica, first, dial the U.S.A code and then (877) 345-8179. The number also works from the United States.

Generally, as long as the service is medically necessary for the treatment of a VA rated service-connected condition, it will be covered by the FMP. Additionally, the services must be accepted by VA and / or the U.S. Medical community (such as the American Medical Association and the U.S. Food and Drug Administration.)

Exclusions to Medical Benefits
The following services are not covered by the Foreign Medical Program:
- Procedures, treatments, drugs, or devices that are experimental or investigational;
- Family planning services and sterilization;
- Infertility services;
- Plastic surgery primarily for cosmetic purposes;
- Procedures, services, and supplies related to sex transformations;
- Non-acute institutional care such as long-term inpatient psychiatric care and nursing home care;
- Day care and day hospitalization;
- Non-medical home care (aid and attendance);
- Abortions
- Travel, meals, and lodging (including transportation costs to return to the United States)

Prosthesis
If an individual residing in a foreign country requires a prosthesis for a VA rated service-connected condition, and the cost of the prosthetic appliance is less than $300 (U.S. currency), the individual may purchase the prosthetic appliance from a local healthcare provider, and send the invoice to the FMP Office for reimbursement, or the healthcare provider may bill the VA. If the cost of the prosthetic appliance exceeds $300 (U.S. currency), the individual must obtain preauthorization for the VA Foreign Medical Program Office (see address below).

Application Process—Registration
Although pre-registration for eligible veterans is not necessary, veterans who are permanently relocating to a country under the FMP's jurisdiction are encouraged to notify the FMP once a permanent foreign address is established. At that time, FMP will provide detailed program material, such as benefit coverage, benefit limitations, selecting healthcare providers, and claim filing instructions.

Veterans who are simply traveling and are not planning a permanent relocation do not need to notify the FMP of their travel plans. Program materials are available, however, upon request.

The FMP can be contacted at:

VA Health Administration Center Foreign Medical Program (FMP)
P.O. Box 65021
Denver, CO 80206-9021
Phone: (303) 331-7590
Fax: (303) 331-7803

Medical Services in Canada
The VA Medical and Regional Office Center in White River Junction, Vermont, is responsible for determining eligibility of U.S. veterans for reimbursement of medical

treatment while traveling or residing in Canada. The local offices of Veterans' Affairs-Canada assists veterans in obtaining authorizations for treatment, arranging for treatment (if necessary), and providing information about the medical treatment program.

The same exclusions listed earlier in this chapter also apply to medical services in Canada, and VA assumes payment responsibility only for certain necessary medical services associated with the treatment of service-connected conditions.

To receive reimbursed medical treatment, an authorization must be obtained from the White River Junction office prior to treatment (unless an emergency situation exists).

When required by the VA to support a claim for disability benefits, Veterans Affairs-Canada will make arrangements for disability examinations for veterans residing in Canada. In some instances, arrangements will be made locally in Canada. In other instances, arrangements will be made at bordering VA medical facilities.
Information on how to obtain medical services in Canada, including procedures for filing claims, can be obtained by contacting the following office:

VAM & RO Center (136FC)
North Hartland Road
White River Junction, VT 05009-0001
Fax: (802) 296-5174

Or

Veterans Affairs-Canada Foreign Countries Operations
Room 1055
264 Wellington Street
Ottawa, Ontario, Canada K1A 0P4
(613) 943-7461

Medical Services in the Philippines
The Republic of the Philippines is the only foreign country in which the VA operates a regional office and outpatient clinic.

The same exclusions listed earlier in this chapter also apply to medical services in the Philippines, and VA assumes payment responsibility only for certain necessary medical services associated with the treatment of service-connected conditions.

To receive reimbursed medical treatment, authorization must be obtained prior to treatment (unless an emergency situation exists).

Information on how to obtain medical services in the Philippines, including procedures for filing claims, can be obtained by contacting the following office:

VA Department of Veterans Affairs Manila Regional Office & Outpatient Clinic
1501 Roxas Boulevard 1302
Pasay City, Philippines
TEL: 011-632-318-8387

Virtually all VA monetary benefits, including compensation, pension, educational assistance, and burial allowances, are payable regardless of an individual's place of residence. However, there are some program limitations in foreign jurisdictions, including:
- Home-loan guarantees are available only in the United States and selected territories and possessions.
- Educational benefits are limited to approved degree-granting programs in institutions of higher learning

- Information and assistance are available to U.S. veterans worldwide at American embassies and consulates. In Canada, the local offices of Veterans Affairs-Canada provide information and assistance. Individuals may call toll- free within Canada, (888) 996-2242.

In the Philippines, service is available at the VA Regional Office and Outpatient Clinic in Manila:

VA Regional Office
1131 Roxas Boulevard
Manila, Philippines
TEL: 011-632-521-7521

Direct Deposit

The conventional method of direct deposit is not available outside of the U.S. However, there are foreign banks with branches in the United States, through which direct deposit can be established. Once the funds are received in the U.S. branch through electronic funds transfer, the U.S. branch transfers the money to the foreign branch.

While this process may take a few days longer than direct deposit within the United States, it is still quicker than having checks mailed overseas through the Department of State or International Priority Airmail.

What to Remember
✓ If you're a veteran living overseas, you're entitled to the benefits and services you earned through your military service.
✓ Most benefits are payable from VA regardless of your place of residence or nationality.
✓ If you're a veteran living abroad or traveling, you can receive medical care for VA service-connected disabilities through the VA Foreign Medical Program.
✓ Under the VA Foreign Medical Program, VA assumes payment responsibility for necessary treatment of service-connected disabilities.
✓ You can manage and apply for benefits online using eBenefits. Instead of using your APO address to register on eBenefits, use your last recorded United States address.

CHAPTER 26

TOXIC EXPOSURE

During military service, some servicemembers may have come in contact with hazardous materials and chemical hazards. These types of exposures include:

- **Agent Orange:** If you served in the Republic of Vietnam or in or near the Korean Demilitarized Zone (DMZ) during the Vietnam Era or in certain related jobs, you may have had contact with Agent Orange. Agent Orange is an herbicide, used for clearing trees and plants during war.
- **Asbestos:** If you worked in certain military jobs, you may have had contact with asbestos, which are toxic fibers that were once used in buildings and products.
- **Birth defects like spina bifida:** If you served in the Republic of Vietnam, in Thailand, or in or near the DMZ during the Vietnam Era and your child has spina bifida or certain other birth defects, your child could be eligible for disability benefits.
- **Burn pits and other specific environmental hazards:** If you served in Iraq, Afghanistan or other certain areas, you may have had contact with toxic chemicals in the air, water or soil.
- **Contact with mustard gas or lewisite:** If you served at the German bombing of Bari, Italy in World War II or worked in certain other jobs, you may have had contact with mustard gas.
- **Contaminated drinking water at Camp Lejeune:** If you served at Camp Lejeune or MCAS New River between August 1953 and December 1987, you may be at risk of certain illnesses believed to be caused by contaminants found in the drinking water during that time.
- **Gulf War Illnesses in Southwest Asia:** If you served in the Southwest Asia theater of operations, you may be at risk of certain illnesses or other conditions linked to the region.
- **Gulf War Illnesses in Afghanistan:** If you served in Afghanistan, you may be at risk of certain illnesses linked to this region.
- **Project 112/SHAD:** IF were part of warfare testing for Project 112 or Project Shipboard Hazard and Defense (SHAD) from 1962 to 1974, you may be at risk for illnesses believed to be caused by chemical testing.
- **Radiation exposure:** If you served in the post-WWII occupation of Hiroshima or Nagasaki, were imprisoned in Japan, worked with or near nuclear weapons testing, or served at a gaseous diffusion plant or in certain other jobs, you might be at risk of illnesses believed to be caused by radiation.

The PACT Act

The PACT Act is considered the largest expansion of health care and benefits in the history of VA. The full name of the legislation is The Sergeant First Class (SFC) Heath Robinson Honor Our Promise to Address Comprehensive Toxics (PACT) Act.

The PACT Act was signed into law on August 10, 2022. The PACT Act does the following:

- Helps veterans receive high-quality health care screenings and services related to potential toxic exposures. It expands access to VA health care for veterans exposed during their military service. For post-9/11 combat veterans, the PACT

Act extends the period of time they have to enroll in VA health care from five to ten years after discharge. If a veteran doesn't fall within that window, the bill creates a one-year open enrollment period. This means more veterans can enroll in VA health care without having to demonstrate a service-connected disability.

- The PACT Act codifies the new process at the VA for evaluating and determining presumption of exposure and service connection for various chronic conditions when the evidence of exposure and associated health risks are strong overall, but hard to prove individually. The PACT Act requires VA to get independent evaluations of the process, and external input on the conditions it will review with this framework. The process is meant to be transparent and evidence-based, allowing VA to make faster policy decisions on exposure issues.
- The legislation takes away the need for certain veterans and their survivors to prove service connection if they are diagnosed with one of 23 specific conditions. This significantly reduces the paperwork and need for exams that veterans diagnosed with one of the conditions have to complete before getting access to health care and disability compensation. The list includes 11 respiratory conditions and several forms of cancer. Survivors of veterans who died because of one of these conditions may be eligible for benefits now too.
- The PACT Act requires that veterans enrolled in VA health care be screened regularly for toxic exposure-related concerns. The VA has to create an outreach program for veterans about toxic exposure related to support and benefits.

Airborne hazard is a term referring to any contaminant or potentially toxic substance that someone is exposed to through the air they breathe. While on active duty, military service members may have been exposed to many airborne hazards including:

- The smoke and fumes from open burn pits
- Sand, dust and particulate matter
- General air pollution common in some countries
- Fuel, aircraft exhaust and other mechanical fumes
- Smoke from oil well fires

In Iraq, Afghanistan and other areas of the Southwest theater of military operations, open-air combustion of trash and other waste in burn pits was common. Most burn pits are now closed, and the Department of Defense plans to close the rest.

The following are presumptive conditions related to airborne hazards and burn pit exposure:

- Brain cancer
- Gastrointestinal cancer of any kind
- Glioblastoma
- Head cancer of any type
- Kidney cancer
- Lymphatic cancer of any kind
- Lymphoma of any kind
- Melanoma
- Neck cancer
- Pancreatic cancer
- Reproductive cancer of any type
- Squamous cell carcinoma of the larynx
- Squamous cell carcinoma of the trachea
- Adenosquamous carcinoma of the lung
- Large cell carcinoma of the lung
- Salivary gland-type tumors of the lung
- Sarcomatous carcinomas of the lung
- Typical and atypical carcinoid of the lung

- Respiratory cancer of any type
- Asthma diagnosed after service
- Chronic bronchitis
- Chronic obstructive pulmonary disease (COPD)
- Chronic rhinitis
- Chronic sinusitis
- Constrictive bronchiolitis
- Obliterative bronchiolitis
- Emphysema
- Granulomatous disease
- Interstitial lung disease
- Pleuritis
- Pulmonary fibrosis
- Sarcoidosis

Gulf War Era and Post-9/11 Veteran Eligibility

There are more than 20 burn pit and toxic exposure presumptive conditions that are added based on the PACT Act, which are listed above. The change expands benefits for Gulf War era and post-9/11 veterans.

If you served in any of the following locations and time periods, the VA has determined you were exposed to burn pits or other toxins.

On or after September 11, 2001, in any of these locations:
- Afghanistan
- Djibouti
- Egypt
- Jordan
- Lebanon
- Syria
- Uzbekistan
- Yemen
- The airspace above any of these locations

On or after August 2, 1990, in any of these locations:
- Bahrain
- Iraq
- Kuwait
- Oman
- Qatar
- Saudi Arabia
- Somalia
- The United Arab Emirates
- The airspace above any of these locations

Under the PACT Act, the VA has extended and expanded health care eligibility. The VA encourages veterans to apply, no matter their separation date. If you meet the following requirements, you can get free VA health care for any condition related to your service for up to 10 years from the date of your most recent discharge or separation. You can also enroll any time during the period and get the care you need but you may have to make a copay.

At least one of these has to be true of active-duty service:
- You served in a theater of combat operations during a period of war after the

Persian Gulf War, or
- You served in combat against a hostile force during a period of hostilities after November 11, 1998.

You must have also been discharged or released on or after October 1, 2013.

If you meet the requirements listed below, you can receive care and enroll during a special enrollment period between October 1, 2022, and October 1, 2023.

At least one has to be true of your active-duty service:
- You served in a theater of combat operations during a period of war after the Persian Gulf War, or
- You served in combat against a hostile force during a period of hostilities after November 11, 1998

Both of these must be true:
- You were discharged or released between September 11, 2001, and October 1, 2013, and
- You haven't enrolled in VA health care before

If you haven't filed a claim yet for a new presumptive condition, you can do this now, online, in person or with the help of a trained professional. If the VA previously denied a disability claim in the past, and the condition is now presumptive, you can submit a Supplemental Claim and the VA will review it again.

Toxic Exposure Screenings

Toxic exposure screenings are available at VA health care facilities across the country. Every veteran who's enrolled in VA health care will get an initial screening and a follow-up screening at least every five years. A veteran who's not enrolled but meets eligibility requirements will be able to enroll and receive the screening

Information for Survivors

If you're a surviving family member of a veteran, you may be eligible for:
- A monthly VA Dependency and Indemnity Compensation payment (DIC)
- A one-time accrued benefits payment, if you're the surviving spouse, dependent child, or dependent veteran of a parent who the VA owed unpaid benefits at the time of their death
- A survivors pension

Agent Orange Exposure and VA Disability Compensation
Agent Orange was a tactical herbicide the U.S. military used to clear leaves and vegetation for military operations, primarily during the Vietnam War. Veterans who were exposed to Agent Orange may have certain related cancers or other illnesses. If you have a health condition caused by exposure to Agent Orange during military service, you may be eligible for disability compensation, and below is detailed how you can apply.

You may be eligible for VA disability benefits if you meet both requirements:
- You have a health condition caused by exposure to Agent Orange, and;
- You served in a location that exposed you to Agent Orange.

Eligibility is determined based on the facts of each veteran's claim, but the VA assumes that certain conditions and other illnesses are caused by Agent Orange. These are presumptive conditions. The VA assumes veterans who served in certain locations were exposed to Agent Orange, which is presumptive exposure.

Under the PACT Act, two new Agent Orange related conditions have been added. These are:
- Hypertension (high blood pressure)
- Monoclonal gammopathy of undetermined significance (MGUS)

Under the PACT Act, five new Agent Orange presumptive locations were also added, which are detailed below.

Cancers caused by Agent Orange Exposure include:
- Bladder cancer
- Chronic B-cell leukemia
- Hodgkin's disease
- Multiple myeloma
- Non-Hodgkin's lymphoma
- Prostate cancer
- Respiratory cancers (including lung cancer)
- Some soft tissue sarcomas

Not included are osteosarcoma, chondrosarcoma, Kaposi's sarcoma or mesothelioma.
Other illnesses caused by Agent Orange Exposure include:
- AL amyloidosis
- Chloracne or other types of acneiform disease like it (under the VA's rating regulations, the condition must be at least 10% disabling within a year of herbicide exposure)
- Diabetes mellitus type 2
- Hypertension (high blood pressure)
- Hypothyroidism
- Ischemic heart disease
- Monoclonal gammopathy of undetermined significance (MGUS)
- Parkinsonism
- Parkinson's disease
- Peripheral neuropathy, early onset (under the VA rating regulations, the condition must be at least 10% disabling with a year of herbicide exposure)
- Porphyria cutanea tarda (under the VA rating regulation, the condition must be at least 10% disabling within a year of herbicide exposure)

If you have a cancer or illness not listed on the presumptive conditions, but you believe it was caused by exposure to Agent Orange, you can still file a claim for disability benefits, but you'll have to submit more evidence.

Service Requirements

The VA bases eligibility for disability compensation in part on whether you served in a location exposing you to Agent Orange—this is a presumption of exposure. You have a presumption of exposure if you meet the service requirements below.

Between January 9, 1962, and May 7, 1975, you must have served for any length of time in at least one of these locations:
- In the republic of Vietnam, or
- Aboard a U.S. military vessel that operated in the inland waterways of Vietnam, or
- On a vessel operating not more than 12 nautical miles seaward from the demarcation line of the waters of Vietnam and Cambodia

Or you must have served in at least one of these locations that have been added through the passage of the PACT Act:
- Any U.S. or Royal Thai military base in Thailand from January 9, 1962, through June 30, 1976, or
- Laos from December 1, 1965, through September 30, 1969, or
- Cambodia at Mimot or Krek, Kampong Cham Province, from April 16, 1969, through April 30, 1969, or
- Guam or American Samoa or in the territorial waters off Guam or American Samoa from January 9, 1962, through July 31, 1980, or
- Johnston Atoll or on a ship that called at Johnston Atoll from January 1, 1972

through September 30, 1977.

At least one of the following must be true:
- You served in or near the Korean DMZ for any length of time between September 1, 1967, and August 31, 1971, or
- You served on active duty in a regular Air Force unit location where a C-123 aircraft with traces of Agent Orange was assigned and had repeated contact with this aircraft due to your flight, ground or medical duties, or
- Were involved in transporting, testing, storing or other uses of Agent Orange during your military service, or
- You were assigned as a Reservist to certain flight, ground or medical crew duties at one of the locations listed

Eligible Reserve locations, time periods and units include:
- Lockbourne/Rickenbacker Air Force Base in Ohio, 1969 to 1986 (906th and 907th Tactical Air Groups or 355th and 356th Tactical Airlift Squadrons)
- Westover Air Force Base in Massachusetts, 1972 to 1982 (731st Tactical Air Squadron and 74th Aeromedical Evacuation Squadron or 901st Organization Maintenance Squadron)
- Pittsburgh International Airport in Pennsylvania, 1972, to 1982 (758th Airlift Squadron)

The Bluewater (BWN) Vietnam Veterans Act 2019 (Public Law 116-23)

Public Law 116-23, The Blue Water Vietnam Veterans Act of 2019, went into effect on January 1, 2020. According to this new law, veterans who served as far as 12 nautical miles from the shore of Vietnam or who had served in the Korean Demilitarized zone, are presumed to have been exposed to herbicides such as Agent Orange and may be entitled to service compensation for any of the 14 conditions related to herbicide exposure. These conditions are:
- Chronic B-cell leukemia
- Hodgkin's disease
- Multiple myelomas
- Non-Hodgkin's lymphoma
- Prostate cancer
- Respiratory cancers including lung cancer
- Soft tissue sarcomas other than osteosarcoma, chondrosarcoma, Kaposi's sarcoma or mesothelioma
- AL amyloidosis
- Chloracne
- Diabetes mellitus type 2
- Ischemic heart disease
- Parkinson's disease
- Peripheral neuropathy
- Porphyria cutanea tarda

VA is now able to extend benefits to children with spina bifida whose BWN veteran parent may have been exposed while serving. Additionally, PL 116-23 made changes to the VA Home Loan Program. See Chapter 25 for more information about these changes. To be entitled to VA benefits, veterans must have served between January 9, 1962, and May 7, 1975, and have one or more of these conditions listed in section 3.309€ of title 38, Code of Federal Regulations.

You can apply for initial compensation claims with VA Form 21-526EZ. For initial DIC claims, submit a VA Form 21P-534EZ. For previously denied claims, submit a VA Form 20-

0995.

When filing a claim, you should state on your application that you're filing for one of the conditions related to presumed herbicide exposure such as Agent Orange. You should include any evidence you have of service in the offshore waters of the Republic of Vietnam during the required timeframe. Include the name of the vessels and the dates you served within 12 nautical miles of the Republic of Vietnam if you have that information. Provide medical evidence showing a diagnosis of a current condition related to exposure to herbicide such as Agent Orange or tell the VA where you're being treated.

Agent Orange Exposure from C-123 Aircraft

If you flew on—or worked with—C-123 aircraft in Vietnam or other locations, you may have had contact with Agent Orange. The U.S. military used this herbicide to clear trees and plants during the Vietnam War. C-123 aircraft sprayed Agent Orange during the war, and the planes still had traces of the chemical in them afterward while they were being used, up until 1986.

For active-duty service members, you may be able to get disability benefits if the following are true for you. First, you must have an illness the VA believes is caused by contact with Agent Orange (a presumptive disease). You also must have served in a regular Air Force unit location where a C-123 aircraft with traces of Agent Orange was assigned, and your flight, ground or medical duties must have put you in regular and repeated contact with C-123 aircraft that had traces of Agent Orange.

If you have an illness the VA determines to be caused by Agent Orange, you don't have to show your problem started during or got worse because of your military service, because it's a presumptive disease.

For reservists, you may be able to get disability benefits if you have a presumptive disease and you were assigned to certain flight, ground, or medical crew duties at one of the following locations:

You must have been assigned to one of these locations:
Lockbourne/Rickenbacker Air Force Base in Ohio, 1969-1986 (906th and 907th Tactical Air Groups or 355th and 356th Tactical Airlift Squadrons), or Westover Air Force Base in Massachusetts, 1972-1982 (731st Tactical Air Squadron and 74th Aeromedical Evacuation Squadron, or 901st Organizational Maintenance Squadron), or Pittsburgh International Airport in Pennsylvania, 1972-1982 (758th Airlift Squadron)

You'll need to file a claim for disability compensation for these benefits. Supporting documents you can include with your online application are:

- Discharge or separation papers (DD214 or other separation documents)
- USAF Form 2096 (unit where you were assigned at the time of the training action)
- USAF Form 5 (aircraft flight duties)
- USAF Form 781 (aircraft maintenance duties)
- Dependency records (marriage certificate and children's birth certificates
- Medical evidence (like a doctor's report or medical test results)

You can get help filing your claim by calling the C-123 hotline at 800-749-8387 or emailing the St. Paul regional benefit office at VSCC123.VAVBASPL@va.gov.

How to Get Disability Benefits for Agent Orange-Related Claims

If you haven't filed a claim yet for a presumptive condition, you can file a new claim online. You can also file a claim by mail, in person or with the help of a trained professional.

If your claim for disability was denied in the past and now the VA considers your condition presumptive, you can file a supplemental claim, and the VA will review your case again. You'll need to submit these records:

- A medical record that shows you have an Agent Orange-related health condition and
- Military records to show how you were exposed to Agent Orange during your service

If your condition isn't on the list of presumptive conditions, you'll also need to provide at least one of the below types of evidence:

- Evidence showing the problem started during or got worse because of your military service, or
- Scientific or medical evidence stating your condition is caused by Agent Orange. This scientific evidence could include a published research study or article from a medical journal.

You'll need to submit your discharge or separation papers that show your time and location of service, which may include your DD214 or other separation documents. You may also need more supporting documents for certain claims.

For claims related to C-123 aircraft you'll submit one or more of these forms:

- USAF Form 2096 (unit where you were assigned at the time of the training action)
- USAF Form 5 (flight aircraft duties)
- USAF Form 781 (aircraft maintenance duties)

Specific Environmental Hazards

VA disability compensation provides tax-free monthly payments. If you have a health condition caused by exposure to burn pits or other specific hazards in the air, soil or water during your service, you may be eligible.

You'll need to meet the following three requirements:

- You have a diagnosed illness or other health condition caused by exposure to a specific toxic hazard in the air, soil or water, and
- You served on active duty in a location that exposed you to the hazard, and
- You didn't receive a dishonorable discharge

Some of the ways a veteran could have had exposure to specific environmental hazards include:

- Burn pits and other toxic exposures in Afghanistan, Iraq and certain other areas
- A large sulfur fire at Mishraq State Sulfur Mine near Mosul, Iraq
- Hexavalent chromium at the Qarmat Ali water treatment plant in Basra, Iraq
- Pollutants from a waste incinerator near the Naval Air Facility at Atsugi, Japan

More than 20 burn pit and other toxic exposure presumptive conditions were added under the PACT Act passed in 2022. The changes from the PACT Act affects Gulf War era and post-9/11 veterans.

The following cancers are now presumptive:

- Brain cancer
- Gastrointestinal cancer of any type
- Glioblastoma
- Head cancer of any type
- Kidney cancer
- Lymphatic cancer of any type

- Lymphoma of any type
- Melanoma
- Neck cancer of any type
- Pancreatic cancer
- Reproductive cancer of any type
- Respiratory cancer of any type

The following illnesses are now presumptive:
- Asthma diagnosed after service
- Chronic bronchitis
- Chronic obstructive pulmonary disease (COPD)
- Chronic rhinitis
- Chronic sinusitis
- Constrictive bronchiolitis or obliterative bronchiolitis
- Emphysema
- Granulomatous disease
- Interstitial lung disease (ILD)
- Pleuritis
- Pulmonary fibrosis
- Sarcoidosis

If you served in any of the locations and time periods below, the VA has determined you had exposure to burn pits or other toxins, which is a presumption of exposure.

On or after September 11, 2001, in any of the locations below:
- Afghanistan
- Djibouti
- Egypt
- Jordan
- Lebanon
- Syria
- Uzbekistan
- Yemen
- The airspace above any of these locations

On or after August 2, 1990, in any of these locations:
- Bahrain
- Iraq
- Kuwait
- Oman
- Qatar
- Saudi Arabia
- Somalia
- The United Arab Emirates (UAE)
- The airspace above any of these locations

Veterans Asbestos Exposure

Asbestos is a material once used in many buildings and products. If a veteran served in Iraq or other countries in the Middle East and Southeast Asia, they may have had contact with asbestos when old buildings were damaged. This damage could have led to the release of toxic chemicals into the air. A veteran could have also been exposed to asbestos if they worked in certain jobs or settings, such as construction, vehicle repair or shipyards.

You may be eligible for healthcare and disability compensation if you had contact with asbestos while serving in the military, and you didn't receive a dishonorable discharge.

You'll need to submit evidence that includes:
- Medical records stating your illness or disability, and
- Service records that list your job or specialty, and
- A doctor's statement that says there's a connection between your asbestos contact during your military service and your illness or disability

If you worked in certain jobs or with certain products, you should speak to your health care provider about testing for illnesses affecting your lungs.

Get tested if you worked in:
- Mining
- Milling
- Shipyards
- Construction
- Carpentry
- Demolition

Get tested if you worked with products like:
- Flooring
- Roofing
- Cement sheet
- Pipes
- Insulation
- Friction products like clutch facings and brake linings

Mustard Gas or Lewisite Exposure

If you had contact with mustard gas or lewisite, you may have certain related long-term illnesses. Mustard gas is also known as sulfur mustard, yperite, or nitrogen mustard. Lewisite is a natural compound that contains arsenic, a poison.

You may be eligible for disability benefits if you have a disability believed to be caused by contact with mustard gas or lewisite and your military record shows you had contact.

If you were in the Army and served in these places:
- Bari, Italy
- Bushnell, FL
- Camp Lejeune, NC
- Camp Sibert, AL
- Dugway Proving Ground, UT
- Edgewood Arsenal, MD
- Naval Research Lab, Washington, DC
- Ondal, India
- Rocky Mountain Arsenal, CO
- San Jose Island, Panama Canal Zone

If you were in the Navy and served in these places:
- Bari, Italy
- Camp Lejeune, NC
- Charleston, SC
- Great Lakes Naval Training Center, IL

- Hart's Island, NY
- Naval Training Center, Bainbridge, MD
- Naval Research Laboratory, VA
- Naval Research Laboratory, Washington D.C.
- USS Eagle Boat 58

Some service members who took place in testing in Finschhafen, New Guinean or Porton Down, England may be eligible, as can some select merchant seamen exposed at Bari, Italy.

You'll need to file a claim for disability compensation. You have to claim an actual disease or disability—it isn't sufficient to say you were exposed to mustard gas or lewisite during service. You have to apply based on the illnesses believed to be caused by your contact with one of these chemicals. When you send in a claim you should share any military records that show your contact with blistering agents.

Volunteering for Research Involving Chemical and Biological Testing
It's estimated that as many as 60,000 veterans volunteered for medical research for the U.S. Biological and Chemical Program. If you were involved in the research, you can get medical care through the U.S. Army if you meet both of the following requirements:
- You volunteered for research involving chemical and biological testing between 1942 and 1975, and
- You have an injury or disease caused directly by your participation in the testing.

Camp Lejeune Water Contamination
If a veteran served at Marine Corps Base Camp Lejeune or Marine Corps Air Station (MCAS) New River in North Carolina, they may have had contact with contaminants in the drinking water there. Both scientific and medical research show an association between exposure to these contaminants during military exposure, and the development of certain diseases later on.

If you have qualifying service at Camp Lejeune and a current diagnosis of a condition detailed below, you may be eligible for disability benefits.

You may be eligible for disability benefits if you meet all of the requirements below.

Both must be true:
- You served at Camp Lejeune or MCAS New River for at least 30 cumulative days from August 1953 through December 1987, and
- You didn't receive a dishonorable discharge when you separated from the military.

You must have a diagnosis of one or more of the following presumptive conditions:
- Adult leukemia
- Aplastic anemia and other myelodysplastic syndromes
- Bladder cancer
- Kidney cancer
- Liver cancer
- Multiple myeloma
- Non-Hodgkin's lymphoma
- Parkinson's disease

Veterans, reservists and guardsmen are covered. Health care and compensation benefits are available.

A veteran has to file a disability compensation claim, and provide the following

evidence:

- Your military records showing you served at Camp Lejeune or MCAS New River for at least 30 days from August 1953 through December 1987 while on active duty or in the National Guard or Reserves, and
- Medical records stating that you have one or more of the eight illnesses on the presumptive conditions list, which is above.

Family Member Coverage
Veterans who served at Camp Lejeune or MCAS New River for at least 30 cumulative days from August 1953 through December 1987 as well as their family members can get health care benefits. The VA may reimburse you for your out-of-pocket health care costs related to any of the following conditions:

- Bladder cancer
- Breast cancer
- Esophageal cancer
- Female infertility
- Hepatic steatosis
- Kidney cancer
- Leukemia
- Lung cancer
- Miscarriage
- Multiple myeloma
- Myelodysplastic syndromes
- Neurobehavioral effects
- Non-Hodgkin's lymphoma
- Renal toxicity
- Scleroderma

For a family member to get benefits, they need to file a claim for disability compensation, and provide all of the following evidence:

- A document to prove the relationship with the veteran who served on active duty for at least 30 days at Camp Lejeune, like a birth certificate or marriage license, and
- A document to prove you lived at Camp Lejeune or MCAS New River for at least 30 days from August 1953 through December 1987. This evidence could include utility bills, military orders, base housing records or tax forms, as examples, and
- Medical records that show you have one of the 15 conditions listed above, and the date you were diagnosed and that you're being treated or have been treated in the past for the illness.

You'll also need to provide evidence that you paid health care expenses for your claimed condition during one of the time periods listed below:

- Between January 1, 1957, and December 31, 1987 (if you lived on Camp Lejeune during this time, the VA will reimburse you for care received on or after August 6, 2012, and up to two years before the date of your application), or
- Between August 1, 1953, and December 31, 1956 (if you lived on Camp Lejeune during this time period, the VA will reimburse you for care received on or after December 16, 2014, and up to two years before the date you apply for benefits).

Family members might also want to provide a Camp Lejeune Family Member Program Treating Physician Report (VA-Form 10-100068b). A family member has to get a doctor to fill out and sign the form before submission. It's not required by the VA but can provide information that's important to determine eligibility.

The Camp Lejeune Family Member program staff can be reached at (866) 372-1144.

Ionizing Radiation Exposure

There are some illnesses and cancers believed to be caused by contact with radiation during military service.

You may be eligible for benefits if you did not receive a dishonorable discharge, and meet both of the following requirements:
- You have an illness that's on the list of illnesses believed to be caused by radiation or that doctors say may be caused by radiation, and
- Your illness started within a certain period of time

Radiogenic disease is a term used to mean any disease that could be induced by ionizing radiation. This list includes:
- All forms of leukemia except chronic lymphocytic (lymphatic) leukemia
- Thyroid cancer
- Breast cancer
- Lung cancer
- Bone cancer
- Liver cancer
- Skin cancer
- Esophageal cancer
- Stomach cancer
- Colon cancer
- Pancreatic cancer
- Kidney cancer
- Urinary bladder cancer
- Salivary gland cancer
- Multiple myeloma
- Posterior subcapsular cataracts
- Non-malignant thyroid nodular disease
- Ovarian cancer
- Parathyroid adenoma
- Tumors of the brain and central nervous system
- Cancer of the rectum
- Lymphomas other than Hodgkin's disease
- Prostate cancer
- Any other cancer

You must have had contact with ionizing radiation in one of the following ways while serving in the military:
- You were part of atmospheric nuclear weapons testing, or
- You served in the postwar occupation of Hiroshima or Nagasaki, or
- You were a prisoner of war in Japan or,
- You worked as an X-ray technician, in a reactor plant, or in nuclear medicine or radiography (while on active duty or during active or inactive duty for training in the Reserves), or
- You did tasks like those of a Department of Energy employee that make them a member of the Special Exposure Cohort

You may also qualify for disability benefits if you served in at least one of these locations and capacities:
- You were part of underground nuclear weapons testing at Amchitka Island, Alaska, or
- You were assigned to a gaseous diffusion plant at Paducah, Kentucky, or
- You were assigned to a gaseous diffusion plant at Portsmouth, Ohio, or

- You were assigned to a gaseous diffusion plant at Area K-25 at Oak Ridge, Tennessee

Under the PACT Act, three new response efforts were added to the list of presumptive locations. These include:
- Cleanup of Enewetak Atoll, from January 1, 1977, through December 31, 1980.
- Cleanup of the Air Force B-52 bomber carrying nuclear weapons off the coast of Palomares, Spain, from January 17, 1966, through March 31, 1967.
- Response to the fire onboard an Air Force B-52 bomber carrying nuclear weapons near Thule Air Force Based in Greenland from January 21, 1968, to September 25, 1968.

If someone took part in any of the above efforts, the VA presumes they had exposure to radiation.

You'll need to file a claim for disability compensation and provide the following evidence:
- Medical records that show you've been diagnosed with one of the illnesses on the list believed to be caused by radiation, or a condition your doctor states may have been caused by radiation exposure, and
- Service records to show you were part of one of the radiation risk activities described above.

When a veteran files a claim, the VA will ask the military branch they served with, or the Defense Threat Reduction Agency to give a range of how much radiation they think the veteran may have come in contact with. The VA will then use the highest level of the range reported to decide on benefits.

Project 112 / Project Shad Veterans

Project SHAD, an acronym for Shipboard Hazard and Defense, was part of a larger effort called Project 112, which was a comprehensive program initiated in 1962 by the Department of Defense (DoD).

Project SHAD encompassed a series of tests by DoD to determine the vulnerability of U.S. warships to attacks with chemical and biological warfare agents, and the potential risk to American forces posed by these agents.

Project 112 tests involved similar tests conducted on land rather than aboard ships. Project SHAD involved service members from the Navy and Army and may have involved a small number of personnel from the Marine Corps and Air Force. Service members were not test subjects, but rather were involved in conducting the tests. Animals were used in some, but not most, tests.

DoD continues to release declassified reports about sea- and land- based tests of chemical and biological materials known collectively as "Project 112." The Department of Veterans Affairs (VA) is working with DoD to obtain information as to the nature and availability of the tests, who participated, duration and agents used.

To date, there is no clear evidence of specific, long-term health problems associated with a veteran's participation in Project SHAD, but the VA is conducting a follow-up study to a 2007 Institute of Medicine Study. Veterans who believe their health may have been affected by these tests should contact the SHAD helpline at (800) 749-8387.

Gulf War Illnesses Linked to Southwest Asia Service

If a veteran served in the Southwest Asia theater of military operations, they may suffer from illnesses or conditions the VA assumes are related to service in the region.

A veteran may be eligible for disability benefits if they served in the Southwest Asia theater of military operations during the Gulf War period and didn't receive a dishonorable discharge.

The illness or condition also has to meet certain requirements in terms of time period.

If your illness or condition was diagnosed while you were on active duty or before December 31, 2021, you can get benefits if both the descriptions are true for you and you have one of the presumptive conditions.

Both must be true:
- Your condition caused you to be ill for at least six months, and
- It resulted in a disability rating of 10% or more

You must have one of these presumptive diseases:
- Functional gastrointestinal disorders
- Chronic fatigue syndrome
- Fibromyalgia
- Other undiagnosed illnesses, including but not limited to muscle and joint pain, headaches and cardiovascular disease

If your illness or condition was diagnosed within one year of your date of separation you can get disability benefits if you have a disability rating of 10% or more, and you have one of the following presumptive diseases:
- Brucelliosis
- Camplobacter jejuni
- Coxiella burnetii (Q fever)
- Nontyphoid salmonella
- Shigella
- West Nile Virus
- Malaria

If your illness or condition was diagnosed at any time after your date of separation, you can get disability benefits if you have a disability rating of 10% or more, and you have one these presumptive conditions:
- Mycobacterium tuberculosis
- Visceral leishmaniasis

Included in the Southwest Asia theater of military operations are:
- Iraq, Kuwait, Saudi Arabia
- The neutral zone between Iraq and Saudi Arabia
- Bahrain, Qatar and the United Arab Emirates
- Oman
- The Gulf of Aden and the Gulf of Oman
- The waters of the Persian Gulf, the Arabian Sea, and the Red Sea
- The airspace above these locations

Birth Defects Linked to Agent Orange
Spina bifida is a spinal cord birth defect. A baby develops spina bifida while still in the womb. In some cases, a parent's past contact with specific chemicals causes this birth defect. If you served in Vietnam or Thailand, or in or near the Korean Demilitarized Zone (DMZ)—and your child has spina bifida or certain other birth defects—your child may be able to get disability benefits.VA recognizes that certain birth defects among veterans' children are associated with veterans' qualifying service in Vietnam or Korean.

Spina bifida except spina bifida occulta is a defect in the developing fetus, resulting in incomplete closing of the spine. It's associated with exposure to Agent Orange. Birth defects in children of women veterans associated with their military service in Vietnam, but not related to herbicide exposure may also be included. The affected child must have been conceived after the veteran entered Vietnam or the Korean demilitarized zone during the qualifying service period.

2023 Birth Defect Compensation Rates

Vietnam and Korea Veterans' Children with Spina Bifida

Disability Level	Monthly Payment
Level I (least disabling)	$394
Level II	$1,339
Level III (most disabling)	$2,279

Women Vietnam Veterans Children with Certain Other Birth Defects

Disability Level	Monthly Payment
Level I (least disabling)	$184
Level II	$394
Level III	$1,339
Level IV (most disabling)	$2,279

What to Remember
✓ The PACT Act expands benefit access for veterans exposed to burn pits and other toxic substances. The law provides veterans and survivors with health care and benefits.
✓ Agent Orange was a tactical herbicide. Veterans exposed to Agent Orange may have certain health conditions as a result and may be eligible for disability compensation.
✓ If someone served at Marine Corps Base Camp Lejeune or Marine Corps Air Station New River, they may have had contact with contaminants in drinking water and may be eligible for disability benefits.
✓ If you served in the Southwest Asia theater of military operations, you may have an illness or condition the VA assumes is related to service in the region, called presumptive diseases.
✓ You'll need to submit your discharge or separation papers that show your time and location of service. These may include your DD214 or other separation documents.

CHAPTER 27

MENTAL HEALTH

A veteran may experience difficult live experiences or challenges after leaving the military.

VA offers mental health treatment and care in different settings including in-person, through telehealth and online.

If a veteran is already using VA medical services, they can ask their primary care provider to help them make an appointment with a VA mental health provider. If they aren't using VA medical services, they can contact their nearest Vet Center or VA medical center to find out how to enroll. VA also offers mental health support such as mental health apps, and Veteran Training. Veteran Training is a self-help online portal that provides tools for overcoming everyday challenges. The portal has tools to help veterans with problem-solving skills, anger management, parenting skills and more, all in an anonymous environment.

Mental health resources for veterans include:

National Call Center for Homeless Veterans
If you are or know a veteran who's homeless or at risk of becoming homeless, you can contact the National VA Call Center for Homeless Veterans. The resources are also available to veteran family and friends. You can dial 1-877-424-3838 and Press 1. The hotline is available 24/7.

QuitVet
QuitVet is a tobacco quitline. Any veteran receiving healthcare through VA is eligible to use the QuitVet quitline. You can call 1-855-QUIT-VET to talk to a tobacco cessation counselor.

Veterans Crisis Line
The Veterans Crisis Line connects veterans and service members in crisis and their families with qualified, caring VA responders through a confidential toll-free hotline or online chat. You can dial 911 and Press 1. This is available 24/7. You can also text 838255.

War Vet Call Center
You can call 1-877-927-8387 to reach the Vet Call Center. You can talk about concerns you have or your military experience. The team is made up of veterans from several eras and family members of veterans.

Women Veterans Call Center
The Women Veterans Call Center provides VA services and resources to women veterans, their families and caregivers. You can call 1-855-829-6636.

Real Warriors
Real Warriors is a program operated through the Defense Centers of Excellent for Psychological Health and Traumatic Brain injury. It provides resources about psychological health, traumatic brain injury and PTSD. You can call 1-866-966-1020.

Coaching Into Care
The Coaching Into Care program from the VA offers guidance to veterans' family members and friends to encourage a veteran they care about to reach out for support with mental health challenges. The program is available by calling 1-888-823-7458.

Vet Centers

If you're a combat veteran, you can visit a VA Vet Center for free, individual and group counseling for you and your family. You have access to these services even if you aren't enrolled in VA health care and aren't receiving disability compensation. Vet Centers offer services like:
Military sexual trauma (MST) counseling
Readjustment counseling
Bereavement (grief) counseling
Employment counseling
Substance abuse assessment and referral

Mental Illnesses and VA Disability Benefits

The VA considers several categories of mental illness eligible for disability. The VA rates all service-connected mental illnesses by the severity of limitations and symptoms. These are rated according to the VA's Schedule of Rating Disabilities. After the VA establishes a mental health condition is related to military service, it rates the condition based on severity. The VA looks at medical records to determine the severity of clinical symptoms.

All mental health conditions are rated under the same criteria from the Diagnostic and Statistical Manual of Mental Disorders published by the American Psychiatric association.

In addition, the VA considers scores from a diagnostic tool—the Global Assessment of Functioning Scale (GAF). This is to determine the severity of a disability. GAF scores are designed to measure the ability of someone to function at work, socially and emotionally and they range from 0-100. The higher someone's score, the better their ability to function. A lower score would mean a higher rating from the VA. A GAF score is assigned as part of a Compensation and Pension Exam.

Ratings for Mental Illness
VA regulations provide for ratings of 0%, 10%, 30%, 50%, 70% or 100% for psychiatric conditions. All mental illnesses are rated as chronic adjustment orders. A 0% rating doesn't provide compensation or payments, but it does make someone eligible for health care and other benefits.
- 100% rating: Someone is completely unable to function socially or at work with symptoms like severely inappropriate behavior, hallucinations, delusions or a consistent threat of self-harm, or harming others. They might not be able to care for themselves, they're unable to remember basic information, and they experience severe disorientation and confusion.
- 70% rating: An inability to function in most social and work areas. Symptoms can include illogical speech, depression and panic interfering with function, suicidal thinking, a lack of impulse control and neglecting self-care.
- 50% rating: Some impairment in social and work function with a lack of productivity and reliability. Problems with memory, judgement, mood, understanding and relationships, and/or experience one or more panic attacks every week.
- 30% rating: Some problems with functioning at work or socially. Generally able to speak normally and care for oneself. Symptoms might include chronic sleep disturbances, mild memory loss, depression, anxiety, infrequent panic attacks or suspiciousness.
- 10% rating: Mild symptoms that impair someone at work or socially when under significant stress, or symptoms managed successfully with medication.
- 0% rating: Diagnosis of mental illness but symptoms are mild enough they don't

require continuous medication or don't interfere with functioning socially or at work.

The VA doesn't have set guidelines for mental health ratings. The VA instead uses discretion and examines all medical evidence on someone's symptoms and functional limitations to decide on a rating.

If you aren't entitled to a 100% rating under VA criteria because your symptoms aren't severe enough, you might be able to get payment at the 100% rate if you can't work a job that pays you enough to live above the poverty level. This is called Total Disability based on Employability or TDUI. You have to show you can't work as a result of your service-connected mental illness.

What to Remember
✓   VA has a variety of mental health resources.
✓   VA offers supportive therapy options including talk therapy, residential programs and supported work settings.
✓   Mental health care from the VA is available in-person, through telehealth and online.
✓   The Veterans Crisis Line connects veterans and service members and their families and friends with VA responders through a confidential hotline, online chat or text. You can call 1-800-273-8255, chat online through the VA site, or you can text 838255.
✓   The Women Veterans Call Center provides services and resources to women and can be reached at 1-855-829-6636.ra

CHAPTER 28

BENEFITS FOR FORMER PRISONERS OF WAR (POWS)

Former prisoners of war (POW) are veterans who, while on active duty in the military, air, or naval service, were forcibly detained or interned in the line of duty by an enemy government or its agents or a hostile force. Periods of war include World War II, the Korean War, the Vietnam War, and the Persian Gulf War (service after August 2, 1990).

Veterans who, while on active duty during peacetime, were forcibly detained or interned by a foreign government or its agents or a hostile force are also considered former POWs if the circumstances of the internment were comparable to wartime internment (for example Iran, Somalia or Kosovo). A POW Coordinator has been designated at each VA regional office. The POW Coordinator can furnish former POWs with information about the benefits and services available to them.

Healthcare
Former POWs are recognized as a special category of veterans and will be placed on one of the top three Priority Groups established for VA healthcare by Congress. Those who have a service-connected disability are eligible for VA health care. This includes hospital, nursing home, and outpatient treatment. Former POWs who do not have a service- connected disability are eligible for VA hospital and nursing home care, regardless of the veterans' ability to provide payment. They are also eligible for outpatient care on a priority basis. Public Law 108-170, which became law December 6, 2003, eliminated prescription drug co-payments for former POWs.

Veterans who were POWs may receive complete dental care.

Former POWs are eligible for any prosthetic item, including eyeglasses and hearing aids. A VA physician must order the prosthetic items, when medically indicated, for eligible veterans.

Veterans may enroll in VA healthcare by completing VA Form 10-10EZR, "Application for Medical Benefits."

Disability Compensation
Veterans are encouraged to apply for VA Disability Compensation for any disabilities related to service, whether they were a POW or not. Some disabilities are presumptive, which means that if a veteran is diagnosed with certain conditions, and the veteran was a POW for at least 30 days (certain conditions do not include the 30-day requirement), the VA presumes the captivity caused the disability.

Presumptions of Service Connection Relating to Certain Diseases and Disabilities for Former Prisoners of War

Public Law 108-183, The Veterans Benefits Act of 2003, eliminated the requirement that a veteran was a POW for at least 30 days for certain conditions. Therefore, there are now two categories of presumptive conditions - those that include the 30-day requirement, and those that do not.

In the case of any veteran who is a former prisoner of war, and who was detained or interned for thirty days or more, any of the following which became manifest to a degree of 10% or more after active military, naval or air service, shall be considered to have been incurred in or aggravated by such service, notwithstanding that there is no record of such disease during the period of service:

- Avitaminosis
- Beriberi (Including beriberi heart disease, which includes ischemic heart disease-coronary artery disease-for former POWs who suffered during captivity from edema- swelling of the legs or feet- also known as "wet" beriberi)
- Chronic dysentery Helminthiasis
- Malnutrition (including optic atrophy associated with malnutrition)
- Pellagra
- Any other nutritional deficiency
- Peripheral neuropathy, except where directly related to infectious causes irritable bowel syndrome
- Peptic ulcer disease Cirrhosis of the liver
- Osteoporosis, on or after September 28, 2009

In the case of any veteran who is a former prisoner of war, and who was detained or interned for any period of time, any of the following which became manifest to a degree of 10% or more after active military, naval or air service, shall be considered to have been incurred in or aggravated by such service, notwithstanding that there is no record of such disease during the period of service:

- Psychosis
- Dysthymic disorder, or depressive neurosis
- Post-traumatic osteoarthritis
- Any of the Anxiety States Cold Injury
- Stroke and complications
- Heart Disease and complications
- Osteoporosis, on or after October 10, 2008, when Post Traumatic Stress Disorder is diagnosed

Former POWs should file a claim by completing "VA Form 21-526, Application for Compensation or Pension." The VA also offers a POW protocol exam. This is a one-time exam available to all former POWs and is conducted at a VA medical facility to help determine if any presumptive disabilities exist.

Survivors of POWS

There are benefits for survivors of POWs. The major benefit is Dependency and Indemnity Compensation (DIC) which is a monthly benefit payable to the surviving spouse (and the former POW's children and parents in some cases) when the former POW:

- Was a service member who died on active duty; or
- Died from service-related disabilities; or
- Died on or before September 30, 1999, and was continuously rated totally disabled for a service-connected condition (including individual unemployability) for at least 10 years immediately preceding death; or
- Died after September 30, 1999 and was continuously rated totally disabled for a

service-connected condition (including individual unemployability) for at least 1 year immediately preceding death.

DIC is terminated for a surviving spouse who remarries but can be resumed if the remarriage ends in death, divorce, or annulment. However, a surviving spouse who remarries on or after attaining age 57, and on or after December 16, 2003, can continue to receive DIC.

Prisoner of War Medal

A Prisoner of War Medal is available to any member of the U.S. Armed Forces taken prisoner during any armed conflict dating from World War I.

What to Remember
✓ Former POWs are veterans who were forcibly detained or interned in the line of duty during active military service by an enemy government or its agents, or a hostile force.
✓ Former POWs are eligible for a wide variety of benefits.
✓ Each VA Regional Office has a POW Veterans Outreach Coordinator.
✓ Former POWs are eligible for VA health care and disability compensation.
✓ Dependents and survivors may be eligible for compensation, pension, health care, education and home loan benefits.

CHAPTER 29

CAREGIVERS

The VA recognizes the role of family caregivers in supporting the health and wellness of veterans. As of October 1, 2022, the program is now open to family caregivers of all eligible veterans of all eras. This includes eligible veterans who served after May 7, 1975, and before September 11, 2001. Family caregivers of eligible veterans of all eras can now apply.

Eligibility

You may be eligible if both you and the veteran you're caring for meet all of the following requirements.

You must be at least 18 years old to be a caregiver, and at least one of the following must be true:

- You're a spouse, son, daughter, parent, stepfamily member or extended family member of the veteran, or
- You live full time with the veteran or you're willing to live full time with the veteran if the VA designates you as a family caregiver.

All of the following must be true for the veteran you're caring for:

- The veteran has a VA disability rating—individual or combined—of 70% or higher, and
- The veteran was discharged from the U.S. military or has a date of medical discharge, and
- The veteran needs at least six months of continuous in-person personal care services

Personal care services can include care or assistance to support health and well-being, everyday personal needs like feeding, bathing and dressing, and safety, protection or instruction in their living environment.

The veteran can appoint:

- 1 Primary Family Caregiver—this is the main caregiver
- Up to 2 Secondary Family Caregivers—these are people who serve as backup support to the primary caregiver when needed.

Eligible Primary and Secondary Family Caregivers can receive:

- Caregiver education and training
- Mental health counseling
- Travel, lodging, and financial assistance when traveling with the veteran to receive care

Eligible Primary Family Caregivers may also receive:

- A monthly stipend payment
- Access to health care benefits through the Civilian and Medical Program of the Department of Veterans Affairs (CHAMPVA)—if they don't already qualify for care or services under another health care plan

- At least 30 days of respite care per for the veteran.

Applying

A caregiver and veteran have to apply together for this program. Both have to be part of the application process to determine if they're eligible for the Program of Comprehensive Assistance for Family Caregivers. Both have to sign and date the application, and completely answer all questions for their role. You can do this process online.

Every time a veteran wants to add a new family caregiver, the veteran and new caregiver have to submit a new application.

Along with applying online, you can also apply by mail. You fill out a joint Application for the Program of Comprehensive Assistance for Family Caregivers (VA Form 10-10CG).

You mail the form and supporting documents to:

Program of Comprehensive Assistance for Family Caregivers
Health Eligibility Center
2957 Clairmont Road NE, Suite 200
Atlanta, GA 30329-1647

You can also bring your completed VA Form 10-10CG to your local VA medical center's Caregiver Support Coordinator. Contact the Caregiver Support Line at 855-260-3274 for more information.

You shouldn't send any medical records with the application. The VA will follow up after receiving the application.

If you're caring for a veteran who isn't eligible for the program, you may still be able to access support and resources. The VA Caregiver Support Program actually includes two programs. One is the Program of Comprehensive Assistance for Family Caregivers, and the other is called the Program of General Caregiver Support Services.

Program of General Caregiver Support Services

The Program of General Caregiver Support Services is a program providing peer support mentoring, skills training, online programs, coaching, telephone services and referrals to available resources for caregivers of veterans. The veteran has to be enrolled in VA health care and be receiving care from a caregiver. Each VA facility is staffed with a CSP Team/Caregiver Support Coordinator who can help you figure out what services and resources are available to you. There's no formal application to enroll in the program.

As a caregiver, the best thing you can do for people who depend on you is taking care of yourself. There are a variety of resources from the VA that provide general support services to family caregivers. These include:

Caregiver Support Line

The VA Caregiver Support Line is available at 1-855-260-3274. There are licensed professionals who staff the support line, and they can connect you with VA services, a Caregiver Support Coordinator at your nearest VA medical center or just listen. Caregivers can participate in monthly telephone education groups where they can discuss self-care tips and ask questions on different topics.

Peer Support Mentoring

Caregivers of veterans are eligible to participate in the VA Caregiver Peer Support Mentoring Program, both as mentors and mentees. Mentors and mentees communicate using email, telephone, and letter writing. Mentors receive training before being paired with another caregiver and are volunteers with their local VA medical center Voluntary Services department. Caregivers participating in the Caregiver Peer Support Mentoring Program agree to participate for six months, but many participate for much longer.

REACH VA
Mentoring in the areas of caregiving, stress management, mood management, and problem-solving is available through REACH VA. The program is available for caregivers of veterans diagnosed with ALS, dementia, MS, PTSD or spinal cord injury/disorder.

The REACH VA program is an opportunity for caregivers of veterans to take better care of themselves and their loved ones by providing them with important information in the areas of caregiving and building their skills in stress management and problem-solving.

The trained and certified REACH VA program coach usually provides four individual sessions with the caregiver over a period of two to three months, extending the session if both the coach and caregiver feel they need to do more work.

To be eligible for the REACH VA program, the caregiver must be caring for a veteran or a veteran caring for a loved one where the veteran is receiving services at the VA. Caregivers receive a Caregiver Notebook, which is the first resource for caregiver issues and challenges.

Approval and Designation of Primary and Secondary Family Caregivers
A veteran's eligibility for the Program of Comprehensive Assistance for Family Caregivers may be determined by the VA Caregiver Support Coordinator, who receives the application or the Caregiver Support Coordinator who will refer the application to the veteran's VA primary care team to complete eligibility determinations.

Alternatively, a VA Caregiver Support Program Multidisciplinary Clinical Eligibility Team or an individual VA provider may be designated to complete eligibility determinations with input from the veteran's primary care team.

Application Requirements
The veteran, and individuals who wish to be considered for designation by VA as primary or secondary family caregivers must complete, sign, and submit VA Form 10-10CG, or successor form, to VA. Individuals interested in serving as family caregivers must be identified as such on the joint application, and no more than 3 individuals may serve as family caregivers at one time for a veteran, with no more than one serving as the primary family caregiver. NOTE: Caregiver benefits, including the stipend provided to primary family caregivers, are effective as of the date the application was received by VA or when the veteran begins receiving care at home, whichever occurs later

Upon receiving the application, a VA Caregiver Support Coordinator will evaluate eligibility by identifying a potentially qualifying injury that was incurred or aggravated in the line of duty in the active military, naval, or air service on or after September 11, 2001, and assessing whether the potentially qualifying injury may render the Veteran in need of personal care services from a caregiver.

In performing this initial eligibility evaluation, the Caregiver Support Coordinator will consider input from the caregiver and the Veteran or Veterans' representative, as applicable. If the Caregiver Support Coordinator determines that the Veteran does not have a potentially qualifying injury that was incurred or aggravated in the line of duty in the active military, naval, or air service on or after September 11, 2001, or the potentially qualifying injury does not render the Veteran in need of personal care services from a caregiver, the Caregiver Support Coordinator will issue the applicant a determination that the applicant is not eligible for the Program of Comprehensive Assistance for Family Caregivers.

If the Coordinator does not issue such a determination, the Caregiver Support Coordinator will refer the application to the Veteran's VA primary care team, a VA Caregiver Support Program Multidisciplinary Clinical Eligibility Team or an individual VA provider to perform the required clinical evaluations, including evaluating the Veteran's level of dependency for purposes of determining the applicable stipend.

Caregiver Eligibility

In order to serve as a primary or secondary family caregiver and for VA to approve the application, the applicant must meet all of the following requirements:

- Be at least 18 years of age.
- Be either:
  o The veteran's spouse, son, daughter, parent, stepfamily member or extended family member; or
  o Someone who lives with the veteran full-time or will do so if designated as a family caregiver.
- There must be no determination by VA of abuse or neglect of the veteran by the family caregiver applicant.

Before VA approves an application for an applicant to serve as a primary or secondary family caregiver, all of the following requirements must be met:

- The applicant must be initially assessed by a VA primary care team as being able to complete caregiver education and training. NOTE: The Veterans Transportation Service may be available for purposes of this assessment in accordance with 38 CFR Part 70 Subpart B.
- The assessment must consider relevant information specific to the veteran as well. For example, whether the applicant can communicate and understand details of the treatment plan and specific instructions related to the care of the veteran and whether the applicant will be able to follow, without supervision, a treatment plan listing the specific care needs of the veteran.
- For purpose of the VA directive, caregiver training is part of a program of education and training designed and approved by VA that consists of issues generally applicable to family caregivers, as well as issues specific to the needs of the veteran.
- No later than ten business days after VA certifies completion of caregiver education and training, or should a veteran be hospitalized during this process, no later than ten days from the date the veteran returns home, a VA clinician or a clinical team will visit the veteran's home to assess the caregiver's completion of training and competence to provide personal care services and to measure the veteran's well-being.
- If the VA determines the veteran and at least one family caregiver applicant meets the applicable eligibility requirements, VA will approve the application and designate primary and/or secondary family caregivers. The approval and designation are a clinical decision.

Requirements of Primary Family Caregivers

Primary caregivers are expected to:

- Work closely with the veteran/servicemember's team to support, promote, and encourage the veteran/servicemember in attaining the highest level of independence possible.
- Promptly inform the veteran/servicemember's primary care team of any changes in their physical or mental health condition.
- Provide a written statement to the Caregiver Support Coordinator when the

veteran/servicemember's address changes, to avoid disruption of stipend payment. This notification should be made anytime the veteran/servicemember moves.

- Demonstrate flexibility in scheduling home visits --- be physically present and participate during home visits and monitoring required by the Program of Comprehensive Assistance for Family Caregivers.

- Promptly inform the Caregiver Support Coordinator if you're no longer able to serve as the Primary Family Caregiver, or if the veteran/servicemember is admitted to the hospital or a long-term treatment facility.

- The stipend is not an entitle—instead, it recognizes the care and support provided by the caregiver to the veteran/servicemember.

Caregiver Access to Military Commissaries, Exchanges and Recreation Facilities

According to the Department of Defense, starting January 1, 2020, all service-connected veterans Purple Heart recipients, former prisoners of war and individuals approved and designated as the primary family caregivers of eligible veterans under the Department of Veterans Affairs Program of Comprehensive Assistance for Family Caregivers can use commissaries, exchanges, and morale, welfare and recreational (MWR) retail facilities in-person and online.

Eligible caregivers will receive an eligibility letter from the VA Office of Community Care. Primary family caregivers who lose their eligibility letter can call 1-877-733-7927 to request a replacement.

Stipend Benefit for the Primary Family Caregiver

A Primary Family Caregiver stipend is monetary compensation paid for providing personal care services to an eligible veteran. The stipend benefit is not meant to replace career earnings, and the stipend payment doesn't create an employment relationship between the VA and the Primary Family Caregiver. Only the designated Primary Family Caregiver of an eligible veteran is entitled to receive a stipend. There can be only one Primary Family Caregiver designated at a time. The stipend is a VA enhanced service and is not considered taxable income.

The veteran's Patient Aligned Care Team at your assigned VA Medical Center makes a determination as far as the number of hours of personal care services a veteran requires. The PACT provides a clinical evaluation of the veteran's level of dependency, based on the degree to which the veteran is in need of supervision or protection based on symptoms or residuals of neurological or other impairment or injury.

As a result of the clinical evaluation and score, eligible veterans are rated as follows:
- High Tier: Equates to a maximum of 40 hours of care per week
- Medium Tier: Equates to a maximum of 25 hours of care per week
- Low Tier: Equates to a maximum of 10 hours of care per week

The stipend amount is based on the weekly number of hours of personal care services that an eligible veteran requires. It is calculated by multiplying the Bureau of Labor Statistics hourly wage for home health aides, Skill code 311011, for the geographic region in which the eligible veteran resides by the Consumer Price Index Cost of Living Adjustment (COLA), and then multiplying that total by the number of weekly hours of caregiver assistance required. This is then multiplied by the average number of weeks in each month.

An Example: If an eligible veteran requires 10 hours of personal care services weekly (Tier 1) and the caregiver's hourly wage (including COLA) is $10 per hour, then the monthly stipend would be: (10 hours x $10) X 4.35=435.

The stipend is paid monthly for personal care services that a Primary Family Caregiver provided in the prior month. Once an application is approved, the stipend will be retroactive to the date the application was received at the VAMC.

The Caregiver Support Line can provide more information at 1-855-260-3274 Monday through Friday, 8 a.m. to 8 p.m. Eastern Standard Time. You can learn more by mail by contacting:

VHA Office of Community Care
Caregiver Support Program
PO Box 460637
Denver CO 80246-0637

CHAMPVA Benefits for Primary Family Caregivers

CHAMPVA for the Primary Family Caregiver is a health care benefits program in which the Department of Veterans Affairs shares the cost of certain health care services and supplies with the Primary Family Caregiver, who is not entitled to care or services under a health-plan contract, including a health insurance plan, TRICARE, Medicare or Medicaid. CHAMPVA is managed by the VHA Office of Community Care.

Not all caregivers are eligible for CHAMPVA medical benefits. Only the designated Primary Family Caregiver who is without health insurance coverage is eligible for CHAMPVA benefits.

The Primary Family Caregiver is also eligible to receive health care at a VA facility through the CHAMPVA In-House Treatment Initiative Program if the VA facility has the excess capacity to provide care. Medical care and supplies received through a participating VA facility are not subject to cost shares or deductibles.

What to Remember
✓ As of October 1, 2022, the Program of Comprehensive Assistance for Family Caregivers is open to family caregivers of eligible veterans of all eras, including veterans who served after May 7, 1975, and before September 11, 2001.
✓ The family caregiver must be 18 years old, and be a spouse, son, daughter, parent, stepfamily member or extended family member of the veteran, or live with the veteran full-time or be willing to.
✓ To be eligible, the veteran receiving care must have a VA disability rating individual or combined of 70% or higher. The veteran needs at least six months of continuous, in-person, personal care services.
✓ A veteran can appoint one primary family caregiver and up to two secondary family caregivers.
✓ Eligible primary caregivers may receive a monthly stipend and access to health care benefits through CHAMPVA, as well as at least 30 days of respite care for the veteran.

CHAPTER 30

VETERAN FEDERAL HIRING

Veterans' Preference gives eligible veterans preference in appointment over many other applicants. Veterans' preference applies to almost all new appointments in both the competitive and excepted service. Veterans' preference does, not guarantee veterans a job and it does not apply to internal agency actions like promotions, transfers, reinstatements, and reassignments.

In accordance with Title 5, U.S. Code, Section 2108 (5 USC 2108), veterans' preference eligibility is based on dates of active-duty service, receipt of a campaign badge, Purple Heart, or a service-connected disability. Not all active-duty service may qualify for veterans' preference.

Only veterans discharged or released from active duty in the armed forces under honorable conditions are eligible for veterans' preference. That means you must have been discharged under honorable or general discharge. If you are a "retired member of the armed forces" you are not included in the definition of preference eligible unless you are a disabled veteran OR you retire below the rank of major or its equivalent.

There are three types of preference eligible, disabled (10-point preference eligible), non-disabled (5-point preference eligible), and sole survivorship preference (0-point preference eligible).

You are 0-point Preference eligible—no points are added to the passing score or rating of a veteran who is the only surviving child in a family in which the father or mother or one more sibling:

- Served in the armed forces; and
- Was killed, died as a result of wounds, accident or disease, is in captured or missing in action status, was permanently 100 percent disabled or hospitalized on a continuing basis and is not gainfully employed because of the disability or hospitalization where the death, status or disability did not result from the intentional misconduct or willful neglect of the parent or sibling and was not incurred during a period of unauthorized absence.

You are a 5-point preference eligible if your active-duty service meets any of the following:

- For more than 180 consecutive days, other than for training, any part of which occurred during the period beginning September 11, 2001, and ending on August 31, 2010, the last day of Operation Iraqi Freedom, OR
- Between August 2, 1990, and January 2, 1992, OR
- For more than 180 consecutive days, other than for training, any part of which occurred after January 31, 1955, and before October 15, 1976, OR
- In a war, campaign or expedition for which a campaign badge has been authorized or between April 28, 1952, and July 1, 1955.

You are 10-point preference eligible if you served at any time and you:
- Have a service-connected disability, OR
- Received a Purple Heart.

Preference eligibles are divided into five basic groups, as follows:
CPS - Disability rating of 30% or more (10 points)
CP - Disability rating of at least 10% but less than 30% (10 points)
XP - Disability rating less than 10% (10 points)
TP - Preference eligibles with no disability rating (5 points)
SSP – Sole Survivorship Preference (0 points)

Disabled veterans receive 10 points, regardless of their disability rating.

When agencies use a numerical rating and ranking system to determine the best-qualified applicants for a position, an additional 5 or 10 points are added to the numerical score of qualified preference eligible veterans.

When an agency does not use a numerical rating system, preference eligibles who have a compensable service-connected disability of 10 percent or more (CPS, CP) are placed at the top of the highest category on the referral list (except for scientific or professional positions at the GS-9 level or higher). XP and TP preference eligibles are placed above non-preference eligibles within their assigned category.

You must provide acceptable documentation of your preference or appointment eligibility. Acceptable documentation may be:
- A copy of your DD-214, "Certificate of Release or Discharge from Active Duty," which shows dates of service and discharge under honorable conditions.
- A "certification" that is a written document from the armed forces that certifies the service member is expected to be discharged or released from active-duty service in the armed forces under honorable conditions not later than 120 days after the date the certification is signed.

You may obtain a letter from the Department of Veterans Affairs reflecting your level of disability for preference eligibility by visiting a VA Regional Office, contacting a VA call center, or online.

NOTE: Prior to appointment, an agency will require the service member to provide a copy of the DD-214.

If claiming 10-point preference, you will need to submit a Standard Form (SF-15Adobe Acrobat Version [152 KB]) "Application for 10-point Veterans' Preference."

VOW (Veterans Opportunity to Work Act)
On November 21, 2011, the President signed the VOW (Veterans Opportunity to Work) Act. (Public Law 112-56). The VOW Act amends chapter 21 of title 5, United States Code (U.S.C.) by adding section 2108a, "Treatment of certain individuals as veterans, disabled veterans, and preference eligibles." This new section requires Federal agencies to treat certain active-duty service members as preference eligibles for purposes of an appointment in the competitive or excepted service, even though the service members have not been discharged or released from active duty.

Because many service members begin their civilian job search prior to being discharged or released from active-duty service, they may not have a DD Form 214, Certificate of Release or Discharge from Active Duty, when applying for Federal jobs. The VOW Act was enacted to ensure these individuals do not lose the opportunity to be considered for Federal service (and awarded their veterans' preference entitlements if applicable) despite not having a DD form 214 to submit along with their résumés.

Agencies are required to accept, process, and grant tentative veterans' preference to those active-duty service members who submit a certification (in lieu of a DD-form 214) along with their job application materials. Agencies must verify the individual meets the definition of 'preference eligible' under 5 U.S.C. 2108 prior to appointment.

A "certification" is any written document from the armed forces that certifies the service member is expected to be discharged or released from active-duty service in the armed forces under honorable conditions within 120 days after the certification is submitted by the applicant. The certification letter should be on letterhead of the appropriate military branch of the service and contain (1) the military service dates including the expected discharge or release date, and (2) the character of service.

If the certification has expired, an agency must request other documentation (e.g., a copy of the DD form 214) that demonstrates the service member is a preference-eligible per 5 U.S.C. 2108, before veterans' preference can be awarded.

Filling a Position Through the Competitive Examining Process

Announcing the Vacancy

To fill a vacancy by selection through the competitive examining process, the selecting official requests a list of eligibles from the examining office. The examining office must announce the competitive examining process through USAJOBS. OPM will notify the State employment service where the job is being filled. Subsequently, the examining office determines which applicants are qualified, rates and ranks them based on their qualifications, and issues a certificate of eligibles, which is a list of eligibles with the highest scores from the top of the appropriate register. A certificate of eligibles may be used for permanent, term, or temporary appointment.

Category Rating

Category rating is an alternative ranking and selection procedure authorized under the Chief Human Capital Officers Act of 2002 (Title XIII of the Homeland Security Act of 2002) and codified at 5 U.S.C. § 3319. Category rating is part of the competitive examining process. Under category rating, applicants who meet basic minimum qualification requirements established for the position and whose job-related competencies or knowledge, skills and abilities (KSAs) have been assessed are ranked by being placed in one of two or more predefined quality categories instead of being ranked in numeric score order. Preference eligibles are listed ahead of non-preference eligibles within each quality category. Veterans' preference is absolute within each quality category. For more detailed information on Category Rating please visit Chapter 5 of the Delegated Examining Operations Handbook.

The "Rule of Three" and Veteran pass overs

Selection must be made from the highest three eligibles on the certificate who are available for the job--the "rule of three." However, an agency may not pass over a preference eligible to select a lower-ranking nonpreference eligible or nonpreference eligible with the same or lower score.

Example: If the top person on a certificate is a 10-point disabled veteran (CP or CPS) and the second and third persons are 5-point preference eligibles, the appointing authority may choose any of the three.

Example: If the top person on a certificate is a 10-point disabled veteran (CP or CPS), the second person is not a preference eligible, and the third person is a 5-point preference eligible, the appointing authority may choose either of the preference eligibles. The appointing authority may not pass over the 10-point disabled veteran to select the nonpreference eligible unless an objection has been sustained.

Disqualification of Preference Eligibles

A preference eligible can be eliminated from consideration only if the examining office sustains the agency's objection to the preference eligible for adequate reason. These reasons, which must be recorded, include medical disqualification under 5 CFR Part 339, suitability disqualification under 5 CFR Part 731, or other reasons considered by the Office of Personnel Management (OPM) or an agency under delegated examining authority to be disqualifying.

OPM must approve the sufficiency of an agency reason to medically disqualify or pass over a preference eligible on a certificate based on medical reasons to select a nonpreference eligible. Special provisions apply to the proposed disqualification or pass over for any reason of a preference eligible with a 30 percent or more compensable disability. See Disqualification of 30 Percent or more Disabled Veterans below.

Agencies have delegated authority for determining suitability in accordance with 5 CFR Part 731.

The preference eligible (or his or her representative) is entitled on request to a copy of the agency's reasons for the proposed pass over and the examining office's response.

An appointing official is not required to consider a person who has three times been passed over with appropriate approval or who has already been considered for three separate appointments from the same or different certificates for the same position. But in each of these considerations, the person must have been within reach under the rule of three and a selection must have been made from that group of three. Further, the preference eligible is entitled to advance notice of discontinuance of certification.

U.S.C. 3317, 3318 and 5 CFR 332.402, 332.404, 332.405, 332.406, and Parts 339 and 731

Disqualification of a 30 Percent or More Disabled Veteran

- If an agency proposes to pass over a disabled veteran on a certificate to select a person who is not a preference eligible, or to disqualify a disabled veteran based on the physical requirements of the position, it must at the same time notify both the Office of Personnel Management (OPM) and the disabled veteran of the reasons for the determination and of the veteran's right to respond to OPM within 15 days of the date of the notification.
- The agency must provide evidence to OPM that the notice was timely sent to the disabled veteran's last known address.
- OPM must make a determination on the disabled veteran's physical ability to perform the duties of the position, taking into account any additional information provided by the veteran.
- OPM will notify the agency and the disabled veteran of its decision, with which the agency must comply. If OPM agrees that the veteran cannot fulfill the physical requirements of the position, the agency may select another person from the certificate of eligibles. If OPM finds the veteran able to perform the job, the agency may not pass over the veteran.
- OPM is prohibited by law from delegating this function to any agency. U.S.C. 3312, 3318.

Filing Late Applications

A veteran may file a late application under the following circumstances by contacting the employing agency. Agencies are responsible for accepting, retaining, and considering their applications as required by law and regulation regardless of whether the agency uses case examining or maintains a continuing register of eligibles.

- Applications from 10-point preference eligibles must be accepted, as described below, for future vacancies that may arise after a case examining register or continuing register is closed. Agencies must accept applications from other individuals who are eligible to file on a delayed basis only as long as a case examining register exists.

- A 10-point preference eligible may file a job application with an agency at any time. If the applicant is qualified for positions filled from a register, the agency must add the candidate to the register, even if the register is closed to other applicants. If the applicant is qualified for positions filled through case examining, the agency will ensure that the applicant is referred on a certificate as soon as possible. If there is no immediate opening, the agency must retain the application in a special file for referral on certificates for future vacancies for up to three years. The Office of Personnel Management's Delegated Examining Operations Handbook provides detailed instructions.

- A preference eligible is entitled to be reentered on each register (or its successor) where previously listed if he or she applies within 90 days after resignation without delinquency or misconduct from a career or career-conditional appointment.

- A preference eligible is entitled to be entered on an appropriate existing register if he or she applies within 90 days after furlough or separation without delinquency or misconduct from a career or career-conditional appointment or if found eligible to apply after successfully appealing a furlough or discharge from career or career-conditional appointment.

- A person who lost eligibility for appointment from a register because of active duty in the Armed Forces is entitled to be restored to the register (or its successor) and receive priority consideration when certain conditions are met. See 5 CFR 332.322 for more details.

- A person who was unable to file for an open competitive examination or appear for a test because of service in the Armed Forces or hospitalization continuing for up to 1 year following discharge may file after the closing date if the register of eligibles still exists.

- A Federal employee who was unable to file for an open competitive examination or appear for a test because of active Reserve duty continuing beyond 15 days may file after the closing date of an existing register. 5 U.S.C. 3305, 3314, 3315, and 5 CFR 332.311, 332.312, 332.321, 332.322.

Preference in Federal Employment

Veterans' preference is a tool to assist in the placement of veterans in federal government positions, providing a "first consideration." Veterans' preference was established by the Veterans' Preference Act of 1944, as amended and is found in certain provisions of 5 U.S.C. 2108.

Veterans' preference applies to permanent and temporary positions in both the competitive and excepted services, which are two class of jobs in the federal government. For the competitive service, applicants must compete with other individuals for positions that are posted on the USAJOBS website through a structured process. In the excepted service, applicants such as veterans with disabilities may be noncompetitively considered and hired through special appointing authorities that agencies may utilize to fill jobs. The excepted service contains certain agencies and entitles groups of individuals and positions that are outside the competitive service.

The methods used for the competitive and excepted services differ.

For competitive service, veterans' preference gives eligible veterans additional points toward passing examination score or rating. Eligible veterans are also placed at the top of hiring certificates for positions, except for professional and scientific positions at grate GS-09 and above. Eligible veterans who apply for professional or scientific positions still receive points and are listed ahead of other applicants having the same rating.

Regarding Excepted Service, veterans' preference allows eligible veterans to apply noncompetitively under special appointing authorities.

Veterans must be discharged from active duty under an honorable or general discharge to be eligible for veteran's preference. Veterans preference does not guarantee a veteran will be selected for employment, apply to internal agency agencies, or apply to Senior Executive Service, positions in the legislative and judicial branches of the federal government, or positions in certain exempted agencies like the Central Intelligence Agency.

VA partners with the Vocational Rehabilitation and Employment Office and its Regional Veteran Employment Coordinators (RVECs) to support the Department's National Veterans Employment Program. VESO was designed to be a strategic management program that oversees VA Veteran employment initiatives and manages VA for Vets, its flagship initiative. VA for Vets is a comprehensive, career support and management program for Veterans, National Guard, Reserve members and VA employees. Services include employment counseling, assistance in identifying transferable military skills (skills matching), qualifications and career assessment, assistance in drafting competitive resumes, instruction in developing comprehensive job search strategies, and direct job placement assistance. RVECs also advocate on behalf of Veterans, promoting the values, work ethic, leadership, dedication, skills, and qualifications Veterans possess, all of which make them ideal candidates to fill any position in the federal sector.

VR&E staff members must be thoroughly familiar and current with Veterans' preference regulations and documentation requirements.

Vocational Rehabilitation Counselors (VRCs) and Employment Coordinators (ECs) act as resources to perform the following tasks:
*   Provide guidance on Veterans' preference and assist job ready Veterans in completing federal vacancy applications.
*   Review applications and ensure Veterans have complete packages including appropriate Veterans' preference and required documentation.
*   Educate Human Resources (HR) personnel and managers on the importance of hiring Veterans with disabilities and Veterans' preference regulations and rules. Hiring managers have at their discretion the ability to select applicants from various lists, some of which may not contain any Veteran applicants.
*   Develop a basic understanding of the Office of Personnel Management's (OPM) Delegated Examining Operating Handbook (DEOH) and its regularly occurring memoranda, which provide updated information on Veterans' preference. The DEOH provides operational procedures for agencies to use in the staffing and placement process for competitive examining of positions.

OPM Role
The Office of Personnel Management is responsible for prescribing and enforcing regulations in the administration of veterans' preference in the competitive and excepted services. OPM is the deciding agency in requests for selecting non-veterans over veterans in the job selection process.

Required Documentation
Veterans who claim veterans' preference must submit the following documentation with their federal job applications. If a veteran doesn't submit the required documentation, the veterans' preference doesn't apply.

For Claiming 5-Points Preference
DD214, Certificate of Release or Discharge from Active Duty (must show Veteran's character of service upon discharge).

For Claiming 10-Points Preference
- DD214, Certificate of Release or Discharge from Active Duty (must show Veteran's character of service upon discharge).
- Purple Heart Recipients, which is listed on the DD214 or other official documentation
- SF-15, Application for 10-Point Veteran' Preference form
- Letter from the VA Regional Office or on eBenefits stating the veteran's percentage of disability

For Claiming To Be a Spouse or Child of a Qualifying Veteran (10-Points Preference)
- DD214, Certificate of Release or Discharge from Active Duty (must
- show the Veteran's character of service upon discharge)
- SF-15, Application for 10-Point Veterans' Preference form
- Letter from the VA Regional Office or on ebenefits showing that the veteran is unemployable or 100-percent service-connected
- If veteran is deceased, a copy of the death certificate

Veterans must contact their local VA Regional Office to obtain a veterans' preference letter at 900-827-1000, or the letter can be obtained on eBenefits.

Special Appointing Authorities

Special appointing authorities are non-competitive and excepted service appointing authorities, which federal agencies can use entirely at their discretion. These authorities provide flexibility in staffing hard-to-fill positions and overcoming underrepresentation and allow for quick and easy hiring. The following is a chart showing the overview of authorities.

VRA	VRA gives agencies the discretion to appoint eligible Veterans to positions in the federal government without competition. Veterans may be appointed to any grade level in the General Schedule through GS-11 or equivalent. This authority also allows applicants with disabilities rated at 30 percent or more or rated at 10 or 20 percent and determined to have a "serious employment handicap" to be employed by VA as Veterans benefits counselors, Veterans claims examiners, Veterans representatives at educational institutions and counselors at readjustment centers. Applicants must meet the basic qualifications for the position to be filled.
30 Percent or More Disabled Veterans' Authority	Veterans may be initially appointed noncompetitively to a temporary or term position. Then, as early as day 61 of employment under this authority, hiring managers may convert veterans to a career or career-conditional appointment. There is no grade-level limitation for this authority. Applicants must meet all qualification requirements for the position to be filled.
Disabled Veterans Enrolled in a VA Training Authority	This authority is the equivalent to the Non-Paid Work Experience (NPWE) program. Veterans eligible for training through VR&E may enroll in training or work experience under an agreement between any government agency (local, state or federal) and VA. Veterans are not considered government employees for most purposes. Training is tailored to the individual's needs and

	goals. Certificates of Training are provided at the end of the training/work experience, which allows agencies to appoint Veterans noncompetitively under status quo appointments. Those appointments may be converted to career or career conditional at any time
VEOA	This authority, unique to the competitive service, allows Veterans to apply to positions under merit promotion procedures (inside the federal government) when the agency is recruiting outside of its own workforce. Veterans' preference is not a consideration when selections are made for these appointments.
Schedule A for Persons with Disabilities	This excepted service authority is an alternative to authorities specifically designed for Veterans. Schedule A provides a way to hire individuals with physical, psychiatric or cognitive impairments without competition. Schedule A employees can be converted to permanent positions in the competitive service after completing two years on the job demonstrating satisfactory performance, with or without reasonable accommodation
Employment of Veterans with Disabilities Who Have Completed a Training Course Under Chapter 31	This authority, unique to the competitive service, allows any agency to appoint a Veteran with a disability noncompetitively to positions or class of positions for which he/she is trained. Veterans with disabilities must satisfactorily complete an approved course of training prescribed by VR&E.

Noncompetitive Hiring

Eligible veterans may be appointed to federal positions without competing with the general public and federal agencies can hire veterans without posting a vacancy announcement. Veterans who are eligible for Special Hiring Authorities may be noncompetitively hired if they meet eligibility and qualification requirements for the position. The following highlights the differences between competitive and noncompetitive hiring.

Competitive Hiring	Noncompetitive Hiring
Positions must be announced to a pool of job seekers in USAJOBS	Positions don't require public announcement
The applicant doesn't have to meet the same eligibility requirements as a noncompetitive job posting	The applicant must meet the requirements for noncompetitive status and be able to perform the essential duties of the job with or without reasonable accommodation
The applicant is rated based on qualifications	The applicant will not be subject to the usual requirement to determine the most qualified candidate and rating of qualification
All veterans have the option to apply	Veterans must provide proof of eligibility for Veterans' Preference or special appointments in order to be considered for noncompetitive placement. Documents may include SF-15, Application for 10-Point

	Veterans Preference, DD214, Certification of Job Readiness, and other medical documentation as requested

Veterans Recruitment Appointment (VRA) Authority

The VRA is a special authority by which agencies can, if they wish, appoint eligible veterans without competition to positions at any grade level through General Schedule (GS) 11 or equivalent. (The promotion potential of the position is not a factor.) VRA appointees are hired under excepted appointments to positions that are otherwise in the competitive service. There is no limitation to the number of VRA appointments an individual may receive, provided the individual is otherwise eligible.

If the agency has more than one VRA candidate for the same job and one (or more) is a preference eligible, the agency must apply the Veterans' preference procedures prescribed in 5 CFR Part 302 in making VRA appointments. A veteran who is eligible for a VRA appointment is not automatically eligible for Veterans' preference.

After two years of satisfactory service, the agency must convert the veteran to career or career-conditional appointment, as appropriate.

Eligibility Criteria

The Jobs for Veterans Act, Public Law 107-288, amended title 38 U.S.C. 4214 by making a major change in the eligibility criteria for obtaining a Veterans Recruitment Appointment (VRA). Those who are eligible:

- Disabled veterans; or
- Veterans who served on active duty in the Armed Forces during a war, or in a campaign or expedition for which a campaign badge has been authorized; or
- Veterans who, while serving on active duty in the Armed Forces, participated in a United States military operation for which an Armed Forces Service Medal was awarded; or
- Recently separated veterans.
- Veterans claiming eligibility on the basis of service in a campaign or expedition for which a medal was awarded must be in receipt of the campaign badge or medal.

In addition to meeting the criteria above, eligible veterans must have been separated under honorable conditions (i.e., the individual must have received either an honorable or general discharge).

Note: Under the eligibility criteria, not all 5-point preference eligible veterans may be eligible for a VRA appointment. For example, a veteran who served during the Vietnam era (i.e., for more than 180 consecutive days, after January 31, 1955, and before October 15, 1976) but did not receive a service-connected disability or an Armed Forces Service medal or campaign or expeditionary medal would be entitled to 5 pt. veterans' preference. This veteran, however, would not be eligible for a VRA appointment under the above criteria.

As another example, a veteran who served during the Gulf War from August 2, 1990, through January 2, 1992, would be eligible for veterans' preference solely on the basis of that service. However, service during that time period, in and of itself, does not confer VRA eligibility on the veteran unless one of the above VRA eligibility criteria is met.

Lastly, if an agency has 2 or more VRA candidates and 1 or more is a preference eligible, the agency must apply Veterans' preference. For example, one applicant is VRA eligible on the basis of receiving an Armed Forces Service Medal (this medal does not confer veterans' preference eligibility). The second applicant is VRA eligible on the basis of being a disabled veteran (which does confer veterans' preference eligibility). In this example, both individuals are VRA eligible, but only one of them is eligible for Veterans' preference. As a result, agencies must apply the procedures of 5 CFR 302 when considering VRA candidates for appointment.

Making Appointments

Ordinarily, an agency may simply appoint any VRA eligible who meets the basic qualifications requirements for the position to be filled without having to announce the job or rate and rank applicants. However, as noted, Veterans' preference applies in making appointments under the VRA authority. This means that if an agency has 2 or more VRA candidates and 1 or more is a preference eligible, the agency must apply Veterans' preference. Furthermore, an agency must consider all VRA candidates on file who are qualified for the position and could reasonably expect to be considered for the opportunity; it cannot place VRA candidates in separate groups or consider them as separate sources in order to avoid applying preference or to reach a favored candidate.

Terms and Conditions of Employment

A VRA appointee may be promoted, demoted, reassigned, or transferred in the same way as a career employee. As with other competitive service employees, the time in grade requirement applies to the promotion of VRAs. If a VRA-eligible employee is qualified for a higher grade, an agency may, at its discretion, give the employee a new VRA appointment at a higher grade up through GS-11 (or equivalent) without regard to time-in-grade.

Agencies must establish a training or education program for any VRA appointee who has less than 15 years of education. This program should meet the needs of both the agency and the employee.
Appeal Rights

During their first year of employment, VRA appointees have the same limited appeal rights as competitive service probationers, but otherwise, they have the appeal rights of excepted service employees. This means that VRA employees who are preference eligibles have adverse action protections after one year (see Chapter 7). VRA's who are not preference eligibles do not get this protection until they have completed 2 years of current continuous employment in the same or similar position.

Nonpermanent Appointment Based on VRA Eligibility

Agencies may make a noncompetitive temporary or term appointment based on an individual's eligibility for VRA appointment. The temporary or term appointment must be at the grades authorized for VRA appointment but is not a VRA appointment itself and does not lead to conversion to career conditional. Refer To: 38 U.S.C. 4214; Pub. L. 107-288; 5 CFR Part 307; 5 CFR 752.401 (c)(3)

Protected Veteran Status

Protected veteran status was created through the Vietnam Era Veterans' Readjustment Act in 1974. It protects veterans from discrimination based on their military service. The act specifically protects workers working for federal contractors and other companies that do business with the U.S. government. As a protected veteran, workers are able to request reasonable accommodation to perform their job duties and other things. Protected veterans are provided with Affirmative Action-type requirements for certain employers to hire vets who fall under protected status. A protected veteran is considered to be:

- Disabled veterans
- Veterans who served on active duty during a war, campaign, or expedition for which a campaign badge has been authorized
- Veterans with an Armed Forces Service Medial pursuant to Executive Order 12985

- Recently separated veterans

All of the above must have a military discharge not characterized as dishonorable.

Under VEVRAA, employers doing business with the federal government must recruit, hire and provide upward mobility for those with protected status. This applies to both contractors and subcontractor s doing business with the government. An employer can't, under VEVRAA, refuse protected veterans any reasonable accommodations in the workplace for disabilities or related issues.

Those who feel they have been discriminated against as protected veterans should file a complaint with the Office of Federal Contract Compliance Programs (OFCCP). You cannot be legally retaliated against for filing a complaint including harassment, threats, coercion, etc. Companies that violate protected veteran laws risk being prohibited from receiving future contracts with the federal government, and that includes violation of the anti-retaliation laws designed to protect veterans who file complaints.

What to Remember
✓ Veterans Preference gives eligible veterans preference in federal hiring over many other applicants.
✓ Veterans' preference applies to all new appointments in the competitive service and many in the excepted service.
✓ Preference doesn't guarantee a job, and it doesn't apply to internal actions like promotions and reassignments.
✓ Preference eligibility can be based on a variety of factors include active-duty service, receipt of a Purple Heart or a service-connected disability.
✓ Not all active-duty service may qualify for veterans' preference.

CHAPTER 31

HOMELESS VETERANS

Approximately one-third of the adult homeless population living on the streets or in shelters has served their country in the armed services. Many other veterans are considered at risk because of poverty, lack of support from family and friends and precarious living conditions in overcrowded or substandard housing.

The VA describes its mission to end veteran homelessness as having three core goals and approaches. First, the VA conducts outreach to seek out veterans proactively if they might need assistance. The VA also works to connect homeless and at-risk veterans with housing, health care, community employment services and other support services. The VA also collaborates with federal, state and local agencies, employers, housing providers, faith-based and community non-profits with the goal of expanding employment and affordable housing options for veterans who are exiting homelessness.

In 2012 as part of Public Law 112-154, there were a number of changes in homeless and housing programs for veterans. These include:

*   Authorization of grant funds for the construction of transitional housing for homeless veterans. It also ensures the matching of funds from private and public sources for transitional housing.
*   Extension of funding for programs designed to assist homeless veterans, including homeless veterans' reintegration programs and support services for low-income veteran families.
*   Public Law 112-154 increases case management and coordination of care for homeless veterans through state and local agencies.

Homeless Providers Grant and Per Diem Program

VA's Homeless Providers Grant and Per Diem Program are offered annually (as funding permits) by the Department of Veterans Affairs Health Care for Homeless Veterans (HCHV) Programs to fund community agencies providing services to homeless veterans. The purpose is to promote the development and provision of supportive housing and/or supportive services with the goal of helping homeless veterans achieve residential stability, increase their skill levels and/or income, and obtain greater self-determination.

Hotline for Homeless Veterans

In March 2010, VA announced a new hotline for homeless veterans: (877) 4AID-VET.

Well-trained expert responders will staff the hotline 24 hours a day, seven days a week. Responders can provide emergency support and resources to homeless veterans, as well as family members, community agencies and non-VA providers.

Loan Guarantee Program for Homeless Veterans Multifamily Housing

This initiative authorizes VA to guarantee no more than 15 loans with an aggregate value of

$100 million within 5 years for construction, renovation of an existing property and refinancing of existing loans, facility furnishing or working capital. No more than 5 loans could be guaranteed under this program prior to November 11, 2001. The amount financed is a maximum of 90% of project costs.

Legislation allows the Secretary to issue a loan guarantee for large-scale self-sustaining multifamily loans. Eligible transitional project are those that:
1) Provide supportive services including job counseling;
2) Require veteran to seek and maintain employment;
3) Require veteran to pay reasonable rent;
4) Require sobriety as a condition of occupancy; and,
5) Serves other veterans in need of housing on a space-available basis.

Stand Downs

Stand Downs are one part of the Department of Veterans Affairs' efforts to provide services to homeless veterans. Stand Downs are typically one-to-three-day events providing services to homeless veterans such as food, shelter, clothing, health screenings, VA and Social Security benefits counseling, and referrals to a variety of other necessary services, such as housing, employment and substance abuse treatment. Stand Downs are collaborative events, coordinated between local VAs, other government agencies, and community agencies who serve the homeless.

For additional information on Stand Down dates and locations, please contact the Homeless Veterans Programs Office at (202) 273-5764.

Compensated Work Therapy/Transitional Residence (CWT/TR) Program

In VA's Compensated Work Therapy/Transitional Residence (CWT/TR) Program, disadvantaged, at-risk, and homeless veterans live in CWT/TR community-based supervised group homes while working for pay in VA's Compensated Work Therapy Program (also known as Veterans Industries). Veterans in the CWT/TR program work about 33 hours per week, with approximate earnings of $732 per month, and pay an average of $186 per month toward maintenance and up-keep of the residence. The average length of stay is about 174 days. VA contracts with private industry and the public sector for work done by these veterans, who learn new job skills, relearn successful work habits, and regain a sense of self- esteem and self-worth.

CHALENG

Community Homelessness Assessment, Local Education, and Networking Groups (CHALENG) for veterans is a nationwide initiative in which VA medical center and regional office directors work with other federal, state, and local agencies and nonprofit organizations to assess the needs of homeless veterans, develop action plans to meet identified needs, and develop directories that contain local community resources to be used by homeless veterans.

More than 10,000 representatives from non-VA organizations have participated in Project CHALENG initiatives, which include holding conferences at VA medical centers to raise awareness of the needs of homeless veterans, creating new partnerships in the fight against homelessness, and developing new strategies for future action.

For more information, please contact CHALENG at (404) 327-4033.

DCHV

The Domiciliary Care for Homeless Veterans (DCHV) Program provides biopsychosocial treatment and rehabilitation to homeless veterans. The program provides residential treatment to approximately 5,000 homeless veterans with health problems each year and the average length of stay in the program is 4 months. The domiciliaries conduct outreach and referral; vocational counseling and rehabilitation; and post-discharge community support.

Drop-In Centers

These programs provide a daytime sanctuary where homeless veterans can clean up, wash their clothes, and participate in a variety of therapeutic and rehabilitative activities. Linkages with longer-term assistance are also available.

Comprehensive Homeless Centers

VA's Comprehensive Homeless Centers (CHCs) place the full range of VA homeless efforts in a single medical center's catchment area and coordinate administration within a centralized framework. With extensive collaboration among non-VA service providers, VA's CHCs in Anchorage, AK; Brooklyn, NY; Cleveland, OH; Dallas, TX; Little Rock, AR; Pittsburgh, PA; San Francisco, CA; and West Los Angeles, CA, provide a comprehensive continuum of care that reaches out to homeless veterans and helps them escape homelessness.

VBA-VHA Special Outreach and Benefits Assistance

VHA has provided specialized funding to support twelve Veterans Benefits Counselors as members of HCMI and Homeless Domiciliary Programs as authorized by Public Law 102-590. These specially funded staff provides dedicated outreach, benefits counseling, referral, and additional assistance to eligible veterans applying for VA benefits. This specially funded initiative complements VBA's ongoing efforts to target homeless veterans for special attention. To reach more homeless veterans, designated homeless veterans' coordinators at VBA's 58 regional offices annually make over 4,700 visits to homeless facilities and over 9,000 contacts with non-VA agencies working with the homeless and provide over 24,000 homeless veterans with benefits counseling and referrals to other VA programs. These special outreach efforts are assumed as part of ongoing duties and responsibilities. VBA has also instituted new procedures to reduce the processing times for homeless veterans' benefits claims.

VBA'S Acquired Property Sales for Homeless Veterans

This program makes all the properties VA obtains through foreclosures on VA-insured mortgages available for sale to homeless provider organizations at a discount of 20 to 50 percent, depending on time of the market.

VA Excess Property for Homeless Veterans Initiative

This initiative provides for the distribution of federal excess personal property, such as hats, parkas, footwear, socks, sleeping bags, and other items to homeless veterans and homeless veteran programs. A Compensated Work Therapy Program employing formerly homeless veterans has been established at the Medical Center in Lyons, NJ to receive, warehouse, and ship these goods to VA homeless programs across the country.

Supportive Services for Veteran Families (SSVF)

In July 2011, the VA announced the award of $60 million in grants to serve homeless and at-risk veterans and their families as part of the Supportive Services for Veteran Families (SSVF) program. The SSVF program awards grants to private non-profits and consumer cooperatives that provide a range of services to eligible low-income veteran families.

VA has been authorized to offer community-based grants through the Supportive Services for Veteran Families (SSVF) Program, which will provide supportive services to very low-income Veteran families in or transitioning to permanent housing. Funds are granted to private non- profit organizations and consumer cooperatives who will assist very low-income veteran families by providing a range of supportive services designed to promote housing stability.

Through the SSVF Program, VA aims to improve very low-income Veteran families' housing stability. Grantees (private non-profit organizations and consumer cooperatives) will provide eligible Veteran families with outreach, case management, and assistance in obtaining VA and other benefits, which may include:

- Health care services
- Daily living services
- Personal financial planning services
- Transportation services
- Fiduciary and payee services
- Legal services
- Childcare services
- Housing counseling services

In addition, grantees may also provide time-limited payments to third parties (e.g., landlords, utility companies, moving companies, and licensed childcare providers) if these payments help Veterans' families stay in or acquire permanent housing on a sustainable basis.

Healthcare for Homeless Veterans Program (HCHV)

HCHV programs now serve as the hub for a myriad of housing and other services which provide VA a way to outreach and assist homeless Veterans by offering them entry to VA care. Outreach is the core of the HCHV program. The central goal is to reduce homelessness among Veterans by conducting outreach to those who are the most vulnerable and are not currently receiving services and reengaging them in treatment and rehabilitative programs.

Contract Residential Treatment Program ensures that Veterans with serious mental health diagnoses can be placed in community-based programs that provide quality housing and services.

Program Monitoring and Evaluation

VA has built program monitoring and evaluation into all of its homeless veterans' treatment initiatives and it serves as an integral component of each program. Designed, implemented, and maintained by the Northeast Program Evaluation Center (NEPEC) at VAMC West Haven, CT, these evaluation efforts provide important information about the veterans served and the therapeutic value and cost-effectiveness of the specialized programs. Information from these evaluations also helps program managers determine new directions to pursue in order to expand and improve services to homeless veterans.

What to Remember
✓ Veterans who are homeless or at imminent risk of homelessness are encouraged to contact the National Call Center for Homeless Veterans at (877) 4AID-VET.
✓ A collaborative program between HUD and VA combines HUD housing vouchers with VA supportive services to help veterans who are homeless, and their families find and sustain permanent housing.
✓ State, local and tribal governments and nonprofits receive capital grants and per diem payments to develop and operate transitional housing and service centers for homeless veterans.
✓ For low-income veterans, Supportive Services for Veteran Families (SSVF) provides case management and supportive services.
✓ VA can connect veterans with health care and community employment services.

CHAPTER 32

BENEFITS FOR FEMALE VETERANS

Female veterans are entitled to the same benefits that are available to male veterans (see below list), and the VA has also designed a number of programs aimed specifically at female veterans, including healthcare benefits and programs.

Female veterans are entitled to all of the benefits available to male veterans, including:
- Disability Compensation For Service-Related Disabilities.
- Disability Pension For Non-Service-Related Disabilities.
- Education Assistance Programs.
- Work-Study Allowance.
- Vocational Rehabilitation And Counseling.
- Insurance.
- Home Loan Benefits.
- Medical Inpatient and Outpatient Care.
- Substance Abuse Treatment and Counseling.
- Sexual Trauma and Assault Counseling.
- Nursing Home Care.

Burial Benefits:
- Burial in A VA National Cemetery.
- Employment Assistance.
- Survivors' Benefit Programs.

Healthcare for Female Veterans

In the spring of 2012, the VA announced a new partnership with the American Heart Association to combat heart disease in female veterans. The program, entitled "Go Red for Women," is working to actively address the issues associated with heart disease in women, including smoking, poor diet and obesity, all of which are risk factors. Heart disease has been identified as the leading cause of death among women, and a third of female veterans suffer from high cholesterol or high blood pressure.

Women Veterans Health Care works to make certain that all eligible women Veterans requesting VA care are assured of:
- Comprehensive primary care by a proficient and interested primary care provider
- Privacy, safety, dignity, and sensitivity to gender-specific needs
- The right care in the right place and time
- State-of-the-art health care equipment and technology
- High-quality preventive and clinical care, equal to that provided to male veterans

Services Provided to Women Veterans Include:

Primary Care
General care includes health evaluation and counseling, disease prevention, nutrition counseling, weight control, smoking cessation, and substance abuse counseling and treatment as well as gender-specific primary care, e.g., cervical cancer screens (Pap smears), breast cancer screens (mammograms), birth control, preconception counseling, Human Papillomavirus (HPV) vaccine, menopausal support (hormone replacement therapy).

Mental Health Care
Mental health includes evaluation and assistance for issues such as depression, mood, and anxiety disorders; intimate partner and domestic violence; sexual trauma; elder abuse or neglect; parenting and anger management; marital, caregiver, or family-related stress; and post-deployment adjustment or post-traumatic stress disorder (PTSD).

Sexual Trauma
Military Sexual Trauma (MST). Women—and men as well—may experience repeated sexual harassment or sexual assault during their military service. Special services are available to women who have experienced MST. VA provides free, confidential counseling and treatment for mental and physical health conditions related to MST.

Specialty Care
Management and screening of chronic conditions include heart disease, diabetes, cancer, glandular disorders, osteoporosis, and fibromyalgia as well as sexually transmitted diseases such as HIV/AIDS and hepatitis.

Reproductive health care includes maternity care, infertility evaluation, and limited treatment, sexual problems, tubal ligation, urinary incontinence, and others. VA is prohibited by legislative authority from providing either in-vitro fertilization or abortion services.

Rehabilitation, homebound, and long-term care. VA referrals are given to those in need of rehabilitation therapies such as physical therapy, occupational therapy, speech-language therapy, exercise therapy, recreational therapy, and vocational therapy. Homebound and long-term care services are available as well, limited to those meeting specific requirements

The VA has placed a strategic priority on six pillars, which are designed to give women veterans the best possible health care services. These include:
Comprehensive Primary Care: Women Veterans Health Care works closely with Primary Care Services to redesign the way primary care is delivered, which tailors it to the specific needs of women.

Women's Health Education: Women Veterans Health Care has partnered with the VA Employee Education Services to create mini residencies in women's health, which will allow providers to be more educated in the advanced topics regarding women's health.

Reproductive Health: Women Veterans Health Care is working to implement safe prescribing measures for female veterans of childbearing age. Other efforts include improving follow-up of abnormal mammograms, tracking the timeliness of breast cancer treatment, and ensuring the care women receive from non-VA, maternity and emergency department care.

Communication and Partnerships: Women Veterans Health Care is working to develop a VA-wide communication plan to enhance the language, practice, and culture of the VA to be more inclusive of female veterans.

Women's Health Research: Women Veterans Health Care has partnered with the Women's Health Evaluation Initiative, based in Palo Alto, CA to develop a series of Sourcebooks with key descriptive information about women Veterans including demographics, population growth over time, diagnoses, utilization and cost of care.

Special Women Veteran Populations: The Women's Veteran Health Strategic Healthcare Group (WVHSHG) is working to ensure that the needs of all women Veterans are addressed, including those populations that require special attention.

Rural and homebound veterans can benefit from emerging technology that will deliver care remotely through "e-clinics," mobile clinics, and home-based care services.

Women veterans with mental illnesses can benefit through integration of mental health services within primary care so that necessary treatment is provided in a comprehensive and coordinated way. Women Veterans Health Care is also working to enhance the availability of woman-safe inpatient psychiatric acuteunits.

Aging women veterans can benefit from the latest advances in medical science and technology to identify and address cardiovascular disease as well as advances in treatments for diabetes, osteoporosis, and menopause.

Referrals are made for services that VA is unable to provide. Women Veterans Program Managers are available in a private setting at all VA facilities to assist women veterans seeking treatment and benefits.

VA has also established a division within the National Center for Post-Traumatic Stress Disorder, the Women's Health Science Division. The center is based at the Boston VA Medical Center and conducts clinical research addressing trauma-related problems of female veterans. Veterans may contact the Center at:

VA Medical Center Women's Health Sciences Division
150 South Huntington Avenue
Boston, MA 02130
(617) 232-9500

Women Veterans Comprehensive Health Centers

Eight Women Veterans Comprehensive Health Centers have been established to develop and enhance programs focusing on the gender-specific healthcare needs of female veterans. The locations are as follows:

Boston VA Medical Center Boston, Massachusetts (617) 232-9500, extension 4276
Chicago Area Network (Hines, Lakeside, North Chicago, and West Side VA Medical Centers) (312) 569-6168
Durham VA Medical Center Durham, NC (919) 286-0411
Minneapolis VA Medical Center Minneapolis, Minnesota (612) 725-2030
Southeast Pennsylvania Network (Coatesville, Lebanon, Philadelphia and Wilmington VA Medical Centers) (215) 823-44496
San Francisco VA Medical Center San Francisco, California (415) 221-4810, extension 2174
Sepulveda / West Los Angeles VA Medical Centers Sepulveda, California and Los Angeles, California (415) 221-4810
Tampa VA Medical Center Tampa, Florida (813) 972-2000, extension 3678

State Women Veterans Coordinators

In addition to the Department of Veterans Affairs' Women Veterans Coordinators that are located at local, regional offices and medical centers, there may be an Office of State Veterans Affairs or a State Commission for Veterans Coordinator available within the State government veterans' program. These State offices are part of the veteran-advocate community, and their staff may assist female veterans in accessing State and Federal entitlements. For specific information on availability in your state, contact your local VA office.

What to Remember
✓ Women served in the U.S. Military as early as the Revolutionary War
✓ Women veterans may be eligible for a wide variety of benefits available to all veterans such as disability compensation, pension, education and training
✓ VA Center for Women Veterans monitors and coordinates the administration of benefit services and programs for women veterans.
✓ There are Women Veteran Coordinators in every regional office who are the primary point of contact for female veterans.
✓ At each VA medical center nationwide, there is a Women Veterans Program Manager who advises and advocates for women veterans

CHAPTER 33

★ ★ ★ ★ ★ ★ ★ ★

BENEFITS FOR GULF WAR VETERANS

The first Gulf War, an offensive led by U.S. and coalition troops in January 1991, followed the August 1990 Iraqi invasion of Kuwait. The war was over on February 28, 1991, and an official cease-fire was signed in April 1991. The last U.S. troops who participated in the ground war returned home on June 13, 1991. In all, about 697,000 U.S. troops had been deployed to the Gulf region during the conflict.

Operation Iraqi Freedom, a military campaign that began on March 20, 2003, and Operation New Dawn, named for the reduced role of U.S. troops in Iraq, has created a new group of Gulf War veterans.

With variation in exposures and veterans' concerns ranging from oil well fire smoke to possible contamination from Iraqi chemical/biological agents, VA has initiated wide-ranging research projects evaluating illnesses as well as environmental risk factors.

The Department of Veterans Affairs (VA) offers Gulf War veterans physical examinations and special eligibility for follow-up care, and it operates a toll-free hotline at (800) 749-8387 to inform these veterans of the program and their benefits. Operators are trained to help veterans with general questions about medical care and other benefits. It also provides recorded messages that enable callers to obtain information 24 hours a day. Beginning in 2012, it was announced Veterans of the Persian Gulf War, with an undiagnosed illness, will have an additional five years to qualify for benefits from the VA. These changes also affect veterans of the Southwest Asia conflict. The change will apply to those veterans who may be eligible to claim VA disability compensation, as well as the ability of survivors to qualify for Dependency and Indemnity Compensation. Under previous VA guidelines, any undiagnosed illness used to establish eligibility for VA benefits had to be apparent by December 31, 2011. The change in rules pushed this date to December 31,2016.

Gulf War Service
Gulf War service is active duty in any of the following areas in Southwest Asia during the Gulf War, which began in 1990 and continues to the present:
Iraq
Kuwait
Saudi Arabia
The neutral zone (between Iraq and Saudi Arabia)
Bahrain
Qatar
The United Arab Emirates Oman
Gulf of Aden Gulf of Oman
Waters of the Persian Gulf, the Arabian Sea, and the Red Sea
The airspace above these locations

The PACT Act (2022 Update)

The PACT Act is a new law that expands VA healthcare and benefits for veterans exposed to burn pits and other toxic substances. The PACT Act expands and extends eligibility for VA health care for veterans with toxic exposures and veterans of the Vietnam, Gulf War and post-911 eras. It adds more than 20 presumptive conditions for burn pits and other toxic exposures. It also requires VA to provide a toxic exposure screening to ever veteran enrolled in VA health care.

The VA is encouraging veterans to apply no matter their separation date.

Please see Chapter 26: Toxic Exposures for More Information

Persian Gulf Registry Program

A free, complete physical examination with basic lab studies is offered to every Gulf War veteran, whether or not the veteran is ill. Veterans do not have to be enrolled in VA healthcare to participate in registry examinations. Results of the examinations, which include review of the veteran's military service and exposure history, are entered into special, computerized registries. The registries enable VA to update veterans on research findings or new compensation policies through periodic newsletters. The registries could also suggest areas to be explored in future scientific research. Registry participants are advised of the results of their examinations in personal consultations and by letters.

Special Access to Follow-Up Care

VA has designated a physician at every VA medical center to coordinate the special registry examination program and to receive updated educational materials and information as experience is gained nationally. Where an illness possibly related to military service in the Southwest Asia Theater of operations during the Gulf War is detected during the examination, follow-up care is provided on a higher-eligibility basis than most non-service-connected care.

Exam Protocols

VA has expanded its special registry examination protocol as more experience has been gained with the health of Gulf veterans. The protocol elicits information about symptoms and exposures, calls the clinician's attention to diseases common to the Gulf region, and directs baseline laboratory studies including chest X-ray (if one has not been done recently), blood count, urinalysis, and a set of blood chemistry and enzyme analyses that detect the "biochemical fingerprints" of certain diseases. In addition to this core laboratory work for every veteran undergoing the Gulf War program exam, physicians order additional tests and specialty consults as they would normally in following a diagnostic trail -- as symptoms dictate. If a diagnosis is not apparent, facilities follow the "comprehensive clinical evaluation protocol." The protocol suggests 22 additional baseline tests and additional specialty consultations, outlining dozens of further diagnostic procedures to be considered, depending on symptoms.

Gulf War Illness

Gulf War veterans have reported a variety of medically unexplained symptoms, such as fatigue, headache, joint pains, sleep disturbances and memory problems since serving in the Gulf. VA has recognized certain health problems as associated with Gulf War service or military service.

VA has recognized that certain health problems for Gulf War Veterans are associated with Gulf War service or military service. These Veterans may be eligible for disability compensation and health care for these illnesses. Surviving spouses, children and dependent parents of Gulf War Veterans who died as the result of illnesses associated with Gulf War service may be eligible for survivors' benefits.

Please see Chapter 26: Toxic Exposures for More Information

Obtaining Copies of Hospital Records

A program is in place to help Gulf War veterans obtain copies of their in-patient hospital records from hospitals established during the Persian Gulf War.

Although these records were always located in the National Personnel Records Center in St. Louis, MO, they were stored only by the name of the hospital and the date of treatment.

An electronic database has been created to cross-reference patient names and social security numbers with their theater hospitals and admission dates.

Veterans may call (800) 497-6261 to find out if their inpatient record has been added to the database and to obtain the paperwork necessary to request a copy.

What to Remember
✓ Anyone who served on active duty from August 2, 1990, to present is considered a Gulf War veteran
✓ Any veteran who served on active-duty military service for any period from August 2, 1990, to the present meets the wartime service requirement
✓ Certain illnesses are presumed by VA to be related to military service in designated areas of Southwest Asia and may entitle a veteran to disability compensation
✓ Presumptive illnesses for Gulf War veterans include medically unexplained illnesses, often called Gulf War Syndrome, certain infectious diseases and ALS diagnosed in all veterans who had 90 days or more of continuous active-duty military service
✓ Gulf War veterans may still be able to establish service connection individually for other non-presumptive diseases and illnesses related to Gulf War Service

CHAPTER 34

★ ★ ★ ★ ★ ★ ★ ★

SERVICEMEMBERS' CIVIL RELIEF ACT

Congress and state legislatures have long recognized that military service can often place an economic and legal burden on servicemembers. The Soldiers' and Sailors' Civil Relief Act of 1918 was passed in order to protect the rights of service members while serving on active duty.

Servicemembers were protected from such things as repossession of property, bankruptcy, foreclosure or other such actions while serving in the military. This Act remained in effect until shortly after World War I when it expired. The Soldiers' and Sailors' Civil Relief Act of 1940 (SSCRA) was passed in order to protect the rights of the millions of service members activated for World War II. The SSCRA has remained in effect until the present day and has been amended many times since 1940 to keep pace with the changing military.

In December 2003, Congress passed legislation renaming SSCRA as the Servicemembers' Civil Relief Act (SCRA). The SCRA updates and strengthens the civil protections enacted during World War II.

The SCRA is designed to protect active-duty military members, reservists who are in active federal service, and National Guardsmen who are in active federal service. Some of the benefits under the SCRA extend to dependents of active duty military members as well.

Public Law 107-330 extended protections under the SCRA to members of the National Guard who are called to active service authorized by the President or the Secretary of Defense, or a state governor for a period of more than 30 consecutive days, for purposes of responding to a national emergency declared by the President and supported by Federal funds.

The SCRA can provide many forms of relief to military members. Below are some of the most common forms of relief.

Protection from Eviction

If a military member is leasing a house or apartment and his or her rent is below a certain amount, the SCRA can protect the individual from being evicted for a period of time, usually three months. The dwelling place must be occupied by either the active-duty member or his or her dependents and the rent on the premises cannot exceed $4,089.62 month (rate as of 2021). This rent ceiling will be adjusted annually for consumer price index (CPI) changes. Additionally, the military member must show that military service materially affects his or her ability to pay rent. If a landlord continues to try to evict the military member or does actually evict the member, he or she is subject to criminal sanctions such as fines or even imprisonment.

331

Termination of Pre-Service Residential Leases

The SCRA also allows military members who are just entering active-duty service to lawfully terminate a lease without repercussions. To do this, the service member needs to show that the lease was entered into prior to the commencement of active-duty service, that the lease was signed by or on behalf of the service member, and that the service member is currently in military service or was called to active-duty service for a period of 180 days or more. Proper written notice with a copy of orders must be provided to the landlord.

Termination of Residential Leases During Military Service

The SCRA allows military members who receive permanent change of station (PCS) orders or are deployed for a period of 90 days or more to terminate a lease by providing written notice to the landlord along with a copy of the military orders. The termination of a lease that provides for monthly payment of rent will occur 30 days after the first date on which the next rental payment is due and payable after the landlord receives proper written notice.

Mortgages

The SCRA can also provide military members temporary relief from paying their mortgage. To obtain relief, a military member must show that their mortgage was entered into prior to beginning active duty, that the property was owned prior to entry into military service, that the property is still owned by the military member, and that military service materially affects the member's ability to pay the mortgage.

Mortgage foreclosures or lien actions initiated during your military service or within nine months after the end of your military service must be stayed upon your request. This applies only to obligations that you have undertaken before entering active duty. A court may instead "adjust the obligations in a way that preserves the interests of all parties." Note that the nine-month period reverts to 90 days as of January 1, 2013, unless further extended.

No sale, foreclosure, or seizure of property for a breach of a pre-service mortgage-type obligation is valid if made during or within 9 months after the period of active duty, unless pursuant to a valid court order. This provides the service member tremendous protections from foreclosure in the many states that permit foreclosures to proceed without involvement of the courts. Service members who miss any mortgage payments should immediately see a legal assistance attorney.

Maximum Rate of Interest

Under the SCRA, a military member can cap the interest rate at 6% for all obligations entered into before beginning active duty if the military service materially affects his or her ability to meet the obligations. This can include interest rates on credit cards, mortgages, and even some student loans (except for Federally guaranteed student loans), to name a few. To qualify for the interest rate cap the military member has to show that he or she is now on active duty, that the obligation or debt was incurred prior to entry on active duty, and that military service materially affects the members' ability to pay. To begin the process, the military member needs to send a letter along with a copy of current military orders to the lender requesting relief under the SCRA. The interest rate cap lasts for the duration of active duty service. The interest rate cap will apply from the first date of active-duty service. The military member must provide written notice to the creditor and a copy of military orders not later than 180 days after the servicemember's termination or release from military service.

Termination of Automobile Leases During Military Service

The SCRA allows military members to terminate pre-service automobile leases if they are called up for military service of 180 days or longer. Members who sign automobile leases while on active duty may be able to terminate an automobile lease if they are given orders for a permanent change of station outside the continental United States or to deploy with a military unit for a period of 180 days or longer.

Stay of Proceedings

If a military member is served with a complaint indicating that they are being sued for some reason, they can obtain a "stay" or postponement of those proceedings if the military service materially affects their ability to proceed in the case. A stay can be used to stop the action altogether or to hold up some phase of it. According to the SCRA, military members can request a "stay" during any stage of the proceedings. However, the burden is on the military member to show that their military service has materially affected their ability to appear in court. In general, individuals can request a stay of the proceedings for a reasonable period of time (30-60 days). For example, if they are being sued for divorce, they can put off the hearing for some period of time, but it is unlikely that a court will allow the proceedings to be put off indefinitely. The stay can be granted in administrative proceedings.

Default Judgments

A default judgment is entered against a party who has failed to defend against a claim that has been brought by another party. To obtain a default judgment, a plaintiff must file an affidavit (written declaration of fact) stating that the defendant is not in the military service and has not requested a stay. If someone is sued while on active duty and fails to respond, and as a result a default judgment is obtained against them, they can reopen the default judgment by taking several steps. First, they must show that the judgment was entered during their military service or within 30 days after they've left the service. Second, they must write to the court requesting that the default judgment be reopened while they are still on active duty or within 90 days of leaving the service. Third, they must not have made any kind of appearance in court, through filing an answer or otherwise, prior to the default judgment being entered. Finally, they must indicate that their military service prejudiced their ability to defend their case and show that they had a valid defense to the action against them.

Insurance

Under SCRA, the U.S. Department of Veterans Affairs (VA) will protect, from default for nonpayment of premiums, up to $250,000 of life insurance for servicemembers called to active duty. (This amount was previously $10,000.) The protection provided by this legislation applies during the insured's period of military service and for a period of two years thereafter. The following are conditions for eligibility for protection:

- The policy must be whole life, endowment, universal life or term insurance.
- The policy must have been in force on a premium-paying basis for at least six months at the time the servicemember applies for benefits.
- Benefits from the policy cannot be limited, reduced or excluded because of military service.
- Policies for which an additional amount of premium is charged due to military service are not eligible for protection under SCRA.

The servicemember must apply for protection of their life insurance by filing "VA Form 29-380 "Application For Protection Of Commercial Life Insurance Policy" with his/her insurance company and forwarding a copy of the application to VA.

Benefits of SCRA Life Insurance Protection

Once the servicemember has applied for protection of their life insurance policy and VA determines that the policy is eligible for protection under SCRA:

*   The servicemember is still responsible for making premium payments. However, the policy will not lapse, terminate, or be forfeited because of the servicemember's failure to make premium payments or to pay any indebtedness or interest due during their period of military service or for a period of two years thereafter.
*   The rights of the servicemember to change their beneficiary designation or select an optional settlement for a beneficiary are not affected by the provisions of this Act.

Limitations of SCRA Life Insurance Protection

Once the servicemember has applied for protection of their life insurance policy and VA determines that the policy is eligible for protection under SCRA:

*   Premium payments are deferred only, not waived. During this period, the government does not pay the premiums on the policy but simply guarantees that the premiums will be paid at the end of the servicemember's period of active duty.
*   A servicemember cannot receive dividends, take out a loan, or surrender the policy for cash without the approval of VA. (Dividends or other monetary benefits shall be added to the value of the policy and will be used as a credit when final settlement is made with the insurer.)
*   If the policy matures as a result of the insured's death, or any other means, during the protected period, the insurance company will deduct any unpaid premiums and interest due from the settlement amount.

Termination of Period Under SCRA

The servicemember has up to two years after their military service terminates to repay the unpaid premiums and interest to the insurer. If the amount owed is not paid before the end of the two years, then:

*   The insurer treats the unpaid premiums as a loan against the policy.
*   The government will pay the insurer the difference between the amount due and the cash surrender value (if the cash surrender value of the policy is less than the amount owed.)
*   The amount the United States government pays to the insurance company under the SCRA Act becomes a debt due the government by the insured.
*   If the policy matures as a result of the insured's death, or any other means, during the protected period, the insurance company will deduct any unpaid premiums and interest due from the settlement amount.

Taxation

A service member's state of legal residence may tax military pay and personal property. A member does not lose residence solely because of a transfer to another state pursuant to military orders.

For example, if an Illinois resident who is a member of Illinois Army National Guard is activated to federal military service and sent to California for duty, that person remains an Illinois resident while in California. The service member is not subject to California's authority to tax his/her military income. However, if the service member has a part-time civilian job in California, California will tax his/her non-military income earned in the state.

The Servicemembers Civil Relief Act also contains a provision preventing servicemembers from a form of double taxation that can occur when they have a spouse who works and is taxed in a state other than the state in which they maintain their permanent legal residence. The law prevents states from using the income earned by a servicemember in determining the spouse's tax rate when they do not maintain their permanent legal residence in that state. Public Law 111-98, which became law on November 11, 2009, provides that when a service member leaves his or her home state in accord with military or naval orders, the service member's spouse may retain residency in his or her home state for voting and tax purposes, after relocating from that state to accompany the service member.

Right to Vote

In addition to the protections involving debt payments and civil litigation, the act guarantees service members the right to vote in the state of their home of record and protects them from paying taxes in two different states.

Caution

The SSCRA does not wipe out any of an individual's obligations. Rather, it temporarily suspends the right of creditors to use a court to compel an individual to pay, only if the court finds that the inability to pay is due to military service. The obligation to honor existing debts remains, and someday the individual must "pay up."

It is important to remember that the SSCRA affords no relief to persons in the Service against the collection of debts or other obligations contracted or assumed by them after entering such Service.

The Servicemembers' Civil Relief Act is highly technical. The above summary is intended only to give a general overview of the protection available. The specific nature of all the relief provided under the law is a matter about which an individual may need to contact an attorney. The Act is designed to deal fairly with military personnel and their creditors. While relief is very often available, individuals are expected and required to show good faith in repayment of all debts.

Veterans Benefits Act of 2010

Amends the Servicemembers Civil Relief Act (SCRA) to prohibit lessors from charging early termination fees with respect to residential, business, agricultural, or motor vehicle leases entered into by servicemembers who subsequently enter military service or receive orders for a permanent change of station or deployment in support of a military operation.

Allows a servicemember to terminate a contract for cellular or home telephone service at any time after receiving military orders to deploy for at least 90 days to a location that does not support the contract. Requires the return of any advance payments made by a deploying servicemember under such a contract.

Allows the Attorney General to bring a civil suit against any violator of the SCRA. Also gives servicemembers a "private right of action" to file their own lawsuits against those who violate their legal rights.

What to Remember
✓ Military members who are called up to active duty may be able to request mortgage relief.
✓ The Servicemembers Civil Relief Act is meant to ease economic and legal burdens on military personnel during active service.
✓ The Act may provide a lower interest rate or prevent eviction or foreclosure for up to nine months following the period of military service.
✓ To qualify, the borrower must request protection under the Act.
✓ The loan must have originated before the current period of active military service.

CHAPTER 35

★ ★ ★ ★ ★ ★ ★ ★

DECISION REVIEWS AND APPEALS

Veterans and other claimants for VA benefits have the right to appeal decisions made by a VA regional office or medical center. The legacy VA appeals process changed and if you disagree with a VA decision dated on or after February 2019, you can choose from 3 decision review options to continue your case. If you disagree with the results of the initial option you choose, you can try another eligible option.

Appeals Modernization

On August 23, 2017, the President signed the Veterans Appeals Improvement and Modernization Act of 2017 (Appeals Modernization Act), to create a decision review process aimed at allowing VA to improve the delivery of benefits and services.

The process allows veterans to seek a faster resolution when they disagree with a VA decision.

If you received an initial claim after February 2019 and you disagree, you can choose one of three lanes to have your disagreement reviewed.

Supplemental Claim

You can file a Supplemental Claim to add new evidence relevant to your case or identity new evidence you want the VA to gather for you. A reviewer determines if the new evidence will change the decision.

Filing a Supplemental Claim for Pact Act-Related Conditions

If VA denied your claim in the past but now consider your condition presumptive, it encourages you to file a Supplemental Claim. The VA will review the claim again.

To file a Supplemental Claim, you have to add evidence that's new—not provided to the VA previously, and relevant to your case. You can file a Supplemental Claim anytime, but the VA recommends filing within a year from the date on your decision letter.

You can't file a Supplemental Claim if you have a contested claim.

New evidence means the VA didn't have it before the last decision. Relevant means it can prove or disprove something in your claim. You can submit this evidence yourself or ask the VA to get evidence for you. Evidence can include medical records from a VA medical center, from another federal health facility, or your private health care provider. VA won't accept a Supplemental Claim without new and relevant evidence.

To file a Supplemental Claim, fill out Decision Review Request: Supplemental Claim (VA Form 20-0995).

You can submit your application by mail or in person. If you're sending it by mail, send it to the VA regional office that matches the benefit type you selected on the form. For compensation, mail it to:

Department of Veterans Affairs
Claims Intake Center
PO Box 4444
Janesville, WI 53547-4444

If you're submitting in person, you bring the completed form and supporting documents to a VA regional office.

If you get a Supplemental Claim decision you don't agree with, you can then request a Higher-Level review of the Supplemental Claim decision, or file for a Board Appeal. If you file for a Board Appeal, your case will be reviewed by a Veterans Law Judge. You also have the option to file another Supplemental Claim if you have additional evidence you'd like to submit for review.

Higher-Level Reviews

If you don't agree with a VA decision, you or your representative can request a new review to be done by a senior reviewer. The senior reviewer determines if there is an error or difference of opinion that would change the decision. You can't submit new evidence with a Higher-Level review.

You can request a Higher-Level Review of an initial claim or Supplement Claim decision. You have a year from the date on your decision letter to request a Higher-Level Review. You can't request a Higher-Level review after a previous Higher-Level Review or Board Appeal on the same claim. You can't request a Higher-Level review if you have a contested claim.

Informal Conferences

An informal conference is a call with the senior reviewer working on your case. If you choose to have an informal conference, the senior reviewer contacts you by phone to schedule a time to talk about your case. You or your representative during this time can discuss why you think the decision should change and identify evidence. If the senior reviewer doesn't get an answer when they call, they'll leave a voicemail. If they can't make contact after two attempts, they review and decide your case without talking to you.

If you request a Higher-Level review online, you can use the online form in step 3 to ask for an informal conference.

If you request a Higher-Level review by mail or in person, you can ask for an informal conference by marking the circle in item 16A on VA Form 20-0996.

You can only have one informal conference for every Higher-Level review.

If the reviewer finds an error, that might change the original decision. Then you'll get a new decision sent to you. If the reviewer finds the VA didn't help you get all the evidence needed for your claim, this is called a duty-to-assist error. To fix this error, the VA closes your Higher-Level review and opens a new Supplemental Claim. The VA will send a letter outlining the steps they're taking to fix the error. Then the VA will help you get missing evidence, and decide your case based on the new evidence.

If you get a Higher-Level Review decision you don't agree with, you can then file for a Board Appeal. When you file for a Board Appeal, your case will get reviewed by a Veterans Law Judge. You'll also have the option to file a Supplemental Claim if you have more evidence you want to submit for review.

Board Appeals

When you choose the Board Appeals option, you're appealing to a Veterans Law Judge at the Board of Veterans' Appeals in Washington D.C. A judge who's an expert in veterans law will review the case.

You can request a Board Appeal after an initial claim, Supplemental Claim or higher-Level Review decision. You can't request two Board Appeals in a row for the same claim. You have a year from the date on your decision letter to request a Board Appeal, unless you have a contest claim.

If you select a Board Appeal, you have three options.

Option 1: Request a Direct Review
When you choose the direct review option, a judge reviews your appeal based on evidence you already submitted. You can't submit new evidence, and you can't have a hearing. The direct review option takes an average of a year for the Board to complete.

Option 2: Submit New Evidence
If you choose this option, you submit new evidence for a Veterans Law Judge to review. You have to submit the evidence within 90 days of the date the VA receives your request for a Board Appeal. This option takes an average of 1 ½ years to complete.

Option 3: Request a Hearing
If you request a hearing with a Veterans Law Judge, you can add new and relevant evidence and submit it at the hearing or within 90 days after the hearing. Adding evidence is optional. You can speak to the Veterans Law Judge by a virtual hearing at home, a videoconference hearing at a VA location near your home, or an in-person hearing at the Board in D.C. where you have to pay your travel costs. It takes an average of two years for the Board to complete the hearing option.

You can request a Board appeal online, or by mail. To do it by mail, you fill out a Decision Review Request: Board of Appeal (Notice of Disagreement) (VA Form 10182). Send the completed form to:

Board of Veterans' Appeals
PO Box 27063
Washington, D.C. 20038

You can also bring your completed form to a VA regional office.

If you don't agree with the Board's decision and you have new and relevant information supporting your case, you can file a Supplemental Claim. You also have the option to appeal to the U.S. Court of Appeals for Veterans Claims.

After You Request a Review

VA will mail you a decision packet when your review is complete, which will have details about the decision for your case.

While you wait, you don't need to do anything unless the VA asks for more information. Don't request another review if you haven't heard back. Instead contact VA.

You may be able to choose a different review option after submitting a form.

To switch after you submitted a Higher-Level Review or Supplemental Claim, both must be true:
- The VA hasn't decided your Supplemental Claim or Higher-Level review, and
- You're still within one year from the date on your original VA decision letter

To switch to a different review option, submit a new request for the option you want. Include a signed letter that indicates you want to withdraw your original decision review request.

To switch after you've already submitted a Board Appeal, all of the following must be true:
- You haven't submitted new evidence or had a hearing, and
- The Board hasn't decided your case, and
- You're still within one year from the date on your original VA decision letter

To switch to a Supplemental Claim or Higher-Level Review, submit a new request for the review option you want, and include a signed letter that says you want to withdraw your original Board Appeal request.

To switch to a different type of Board Appeal, all of the following must be true:
- You haven't submitted new evidence or had a hearing, and
- The Board hasn't decided your case, and
- You're still within one year from the date on your original VA decision letter or 60 days from your original Board Appeal request, whichever is later

To switch to a different type of Board Appeal, submit a new Board Appeal request and choose the type you want. You won't have to withdraw your original request.

What to Remember
✓ The legacy VA appeals process changed to the decision review process.
✓ If you disagree with a VA decision dated on or after February 19, 2019, you can choose from three decision review options.
✓ The three decision review options include Supplemental Claim, Higher-Level Review or Board Appeal.
✓ If you aren't happy with the results of the first option you choose, you can try another eligible option.
✓ With Supplemental Claim reviews, a reviewer decides if new and relevant evidence changes the prior decision. With a Higher-Level Review, a senior claims adjudicator reviews the decision using the same evidence from the prior decision. With the Board Appeal review, a Veterans Law Judge will review the decision.

CHAPTER 36

VA REGIONAL OFFICES & BENEFITS OFFICES

Department of Veterans Affairs
Headquarters
810 Vermont Avenue NW
Washington, DC 20420
(202) 273-5400

Alabama (SDN 5) Montgomery RO
345 Perry Hill Road
Montgomery, AL 36109

Alaska (SDN 8) Anchorage RO
1201 North Muldoon Road
Anchorage, AK 99504

Arizona (SDN 9) Phoenix RO
3333 North Central Avenue
Phoenix, AZ 85012

Arkansas (SDN 7)
North Little Rock RO
2200 Fort Roots Drive Building 65
North Little Rock, AR 72114

California (SDN 9)
Los Angeles RO Federal Building
11000 Wilshire Boulevard
Los Angeles, CA 90024

Oakland RO
1301 Clay Street
Room 1400 North
Oakland, CA 94612

San Diego RO
8810 Rio San Diego Drive
San Diego, CA 92108

Colorado (SDN 8) Denver RO
155 Van Gordon Street
Lakewood, CO 80228

Connecticut (SDN 1) Hartford RO
555 Willard Avenue
Newington, CT 06111
Mailing Address:
P.O. Box 310909
Newington, CT 06131

Delaware (SDN 3) Wilmington RO
1601 Kirkwood Highway
Wilmington, DE 19805

Florida (SDN 5) St. Petersburg RO
9500 Bay Pines Boulevard Bay Pines, FL
33708 Mailing Address:
P.O. Box 1437
St. Petersburg, FL 33731

Georgia (SDN 4) Atlanta RO
1700 Clairmont Road
Decatur, GA 30033
Mailing Address:
P.O. BOX 100026
Decatur, GA 30031-7026

Hawaii (SDN 9) Honolulu RO
459 Patterson Road E-Wing
Honolulu, HI 96819-1522

Idaho (SDN 8) Boise RO
444 W. Fort Street
Boise, ID 83702-4531

Illinois (SDN 6) Chicago RO
2122 West Taylor Street
Chicago, IL 60612

Indiana (SDN 2) Indianapolis RO
575 North Pennsylvania Street
Indianapolis, IN 46204

Iowa (SDN 6) Des Moines RO
210 Walnut Street
Des Moines, IA 50309

Kansas (SDN 6) Wichita RO
5500 East Kellogg
Wichita, KS 67211

Kentucky (SDN 3) Louisville RO
321 West Main Street, Suite 390
Louisville, KY 40202

Louisiana (SDN 7) New Orleans RO
1250 Poydras Street
New Orleans, LA 70113

Maine (SDN 1) Togus VA Med/RO
1 VA Center
Augusta, ME 04330

Maryland (SDN 3) Baltimore RO
31 Hopkins Plaza Federal Bldg
Baltimore, MD 21201

Massachusetts (SDN 1) Boston RO
JFK Federal Building
15 New Sudbury St. Government Center
Boston, MA 02114

Michigan (SDN 2) Detroit RO
Federal Building
477 Michigan Avenue
Detroit, MI 48226

Minnesota (SDN 6) St. Paul RO
1 Federal Drive
Fort Snelling
St. Paul, MN 55111-4050

Mississippi (SDN 5) Jackson RO
1600 East Woodrow Wilson Ave.
Jackson, MS 39216

Missouri (SDN 6) St. Louis RO
Federal Building
400 South, 18th Street
St. Louis, MO 63103

Montana (SDN 8) Fort Harrison RO
3633 Veterans Drive
Fort Harrison, MT 59636-0188

Nebraska (SDN 6) Lincoln RO
3800 Village Drive
Lincoln, NE 68516
Mailing Address: PO Box 85816
Lincoln, NE 68501-5816

New Hampshire (SDN 1) Manchester RO
Norris Cotton Federal Bldg
275 Chestnut Street
Manchester, NH 03101

New Jersey (SDN 2) Newark RO
20 Washington Place
Newark, NJ 07102

New Mexico (SDN 8) Albuquerque RO
Dennis Chavez Federal Building
500 Gold Avenue,
S.W. Albuquerque, NM 87102

Nevada (SDN 9) Reno RO
5460 Reno Corporate Dr.
Reno, NV 89511-2250

New York (SDN1) Buffalo RO
130 South Elmwood Avenue
Buffalo, NY 14202

New York RO
245 West Houston Street
New York, NY 10014

North Carolina (SDN 4) Winston-Salem RO
Federal Building
251 North Main Street
Winston-Salem, NC 27155

North Dakota (SDN 6) Fargo RO
2101 Elm Street
Fargo, ND 58102

Ohio (SDN 2) Cleveland RO
A.J. Celebrezze Federal Building
1240 East 9th Street
Cleveland, OH 44199

Oklahoma (SDN 7) Muskogee RO
125 South Main Street
Muskogee, OK 74401

Oregon (SDN 8) Portland RO
100 SW Main Street, Floor 2
Portland, OR 97204

Mailing Address:
100 SW Main St FL 2
Portland, OR 97204

Pennsylvania (SDN 2) Philadelphia RO
5000 Wissahickon Avenue
Philadelphia, PA 19101

Pittsburgh RO
1000 Liberty Avenue
Pittsburgh, PA 15222

Rhode Island (SDN 1) Providence RO
380 Westminster St.
Providence, RI 02903

South Carolina (SDN 4) Columbia RO
6437 Garners Ferry Road
Columbia, SC 29209

South Dakota (SDN 6) Sioux Falls RO
2501 West 22nd Street
Sioux Falls, SD 57117

Tennessee (SDN 4) Nashville RO
110 9th Avenue South
Nashville, TN 37203

Southern Area Office
3322 West End, Suite 408
Nashville, TN 37203

Texas (SDN 7) Houston RO
6900 Almeda Road
Houston, TX 77030

Waco RO
1 Veterans Plaza
701 Clay Avenue
Waco, TX 76799

Utah (SDN 8) Salt Lake City RO
550 Foothill Drive
Salt Lake City, UT 84158

Vermont (SDN 1) White River Junction
RO
215 N. Main St.
White River Junction, VT 05009

Virginia (SDN 3) Roanoke RO
116 N. Jefferson St.
Roanoke, VA 24016

Washington (SDN 8) Seattle RO Federal
Building
915 2nd Avenue
Seattle, WA 98174

West Virginia (SDN 3) Huntington RO
640 Fourth Avenue
Huntington, WV 25701

Wisconsin (SDN 6) Milwaukee RO
5400 West National Avenue
Milwaukee, WI 53214

Wyoming (SDN 8) Cheyenne RO
2360 East Pershing Blvd.
Cheyenne, WY 82001

District of Columbia (SDN 3)
Washington DC RO
1722 I Street N.W.
Washington, DC 20421

Puerto Rico (SDN 5) San Juan RO
50 Carr 165
Guaynabo, PR 00968-8024

Philippines (SDN 9) Manila RO
1501 Roxas Boulevard
Pasay City, PI 1302
Note: SDN = Service Delivery Network

*In order to reach any of the above
offices by phone, call (800) 827-1000
and your call will be directed
appropriately.

CHAPTER 37

STATE BENEFITS

Many states offer services and benefits to veterans in addition to those offered by the Department of Veterans' Affairs. To find out more about a particular state's programs, individuals should contact the following

Alabama VA
RZA Plaza
770 Washington Ave,
Suite 530
Montgomery, AL 36102-1509
(334) 242-5077

Alaska VA
P.O. Box 5800
Fort Richardson, AK 99505- 5800
(907) 428-6031

Arizona VA Services
3839 North Third
Street, Suite 209
Phoenix, AZ 85012
(602) 248-1554

Arkansas VA
2200 Fort Roots Drive
Room 119 – Bldg 65
North Little Rock, AR 72114
(501) 370-3820

California VA
1227 O Street, Suite 300
Sacramento, CA 95814
(800) 952-5626

Colorado VA
6848 South Revere Parkway
Centennial, CO 80112
(720) 250-1500

Connecticut VA
287 West Street
Rocky Hill, CT 06067
(860) 616-3603

Delaware VA
Robbins Building
802 Silverlake Blvd, Suite 100
Dover, DE 19904
(302) 739-2792
(800) 344-9900 (in-state only)

Florida VA
Mary Grizzle Building, Room 311-K
11351 Ulmerton Road
Largo, FL 33778
(727) 518-3202

Georgia VA
Floyd Veterans Memorial Building Suite
E-970
Atlanta, GA 30334
(404) 656-2300

Hawaii VA
 Mailing Address:
459 Patterson Road
E-Wing, Room 1-A103
Honolulu, HI 96819-1522
Location:
Tripler Army Med Center (Ward Road)
VAMROC, E-Wing, Room 1-A103
Honolulu, HI 96819
(808) 433-0420

Idaho VA
351 Collins Road
Boise, ID 83702
(208) 577-2310

Illinois VA
362 North Linwood Rd.
Galesburg, IL 61401
(309) 343-2510

Indiana VA
302 West Washington, Room E120
Indianapolis, IN 46204-2738
(317) 232-3910

Iowa Commission of Veterans' Affairs
7105 N.W. 70th Avenue Camp
Dodge - Building 3663
Johnston, IA 50131-1824
(515) 242-5331

Kansas Commission on Veterans' Affairs
Jayhawk Towers Suite 1004
700 S.W. Jackson Street
Topeka, KS 66603-3714
(785) 296-3976

Kentucky VA
1111 Louisville Road NGAKY Building
Frankfort, KY 40601
(502) 564-9203

Louisiana VA
P.O. Box 94095 Capitol Station
1885 Wooddale Blvd. 10th Floor,
Room 1013
Baton Rouge, LA 70806
(225) 219-5000

Maine Bureau of Veterans' Services
117 State House Station
Camp Keyes Building 7, Room 115
Augusta, ME 04333
(207) 430-6035

Maryland VA
16 Francis St. 4th Floor
Annapolis, MD 21401
(410) 260-3838

Massachusetts Department of Veterans'
Services
600 Washington Street,
Suite 1100
Boston, MA 02111
(617) 210-5480

Michigan VA
7109 W. Saginaw
Lansing, MI 48913
(517) 335-6523

Minnesota VA
State Veterans Service Building
20 West 12th Street, 2nd Floor Rm. 206
St. Paul, MN 55155-2079
(651) 296-2562

Mississippi State Veterans' Affairs
Board
3466 Highway 80 East
P.O. Box 5947
Pearl, MS 39288-5947
(601) 576-4850

Missouri Veterans' Commission Mailing
Address:
205 Jefferson Street
12th Floor Jefferson Building
P.O. Drawer 147
Jefferson City, MO 65102
(573) 751-3779

Montana Veterans' Affairs Division
P.O. Box 5715
1900 Williams Street
Helena, MT 59604
(406) 324-3740

Nebraska VA
State Office Building
301 Centennial Mall South
P.O. Box 95083
Lincoln, NE 68509-5083
(402) 471-2458

Nevada Commission for Veterans'
Affairs
5460 Reno Corporate Dr.
Suite 131
Reno, NV 89511
(775) 688-1656

New Hampshire Office of Veterans'
Service
275 Chestnut Street Room 517
Manchester, NH 03101-2411
(603) 624-9230
1-800-622-9230 (in-state only)

New Jersey VA
Eggert Crossing Road PO Box 340
Trenton, NJ 08625-0508
(800) 624-0508

New Mexico Veterans' Service
Commission
Bataan Memorial Building
407 Galisteo St. Rm. 142
Santa Fe, NM 87504
(866) 433-8387

New York Division of Veterans' Affairs
2 Empire State Plaza 17th Floor
Albany, NY 12223-1551
(518) 474-6114

North Carolina Division of Veterans'
Affairs
325 N. Salisbury Street
Raleigh, NC 27601
(919) 807-4250

North Dakota VA
4201 38th St S # 104
Fargo, ND 58104
(701) 239-7165

Ohio Governor's Office of Veterans'
Affairs
77 South High Street
Columbus, OH 43215
(614) 644-0898

Oklahoma VA
2311 N. Central Avenue
Oklahoma City, OK 73105
(405) 521-3684

Oregon VA
100 South Main St.
Portland, OR 97204
(503) 412-4777

Pennsylvania VA
Fort Indiantown Gap Building S-O-47
Annville, PA 17003-5002
(717) 861-2000

Rhode Island Division of Veterans'
Affairs
480 Metacom Avenue
Bristol, RI 02809
(401) 462-0324

South Carolina Office of Veterans'
Affairs
1205 Pendleton Street, Suite 463
Columbia, SC 29201
(803) 734-0200

South Dakota Division of Veterans'
Affairs
425 East Capitol Avenue
c/o 500 East Capitol Avenue
Pierre, SD 57501
(605) 773-3269

Tennessee VA
312 Rosa Parks Ave.
Nashville, TN 37243-1010
(615) 741-2931

Texas Veterans' Commission
1700 Congress Ave.
Austin, TX 78701
(512) 463-5538

Utah Division of Veterans' Affairs
550 Foothill Blvd, Room 202
Salt Lake City, UT 84113
(801) 326-2372

Vermont State Veterans' Affairs
118 State Street
Montpelier, VT 05602
(802) 828-3379

Virginia VA
900 East Main Street
Richmond, VA 23219
(804) 786-0286

Washington VA
P.O. Box 41150
1011 Plum Street
Olympia, WA 98504-1150
(800) 562-0132

West Virginia Division of Veterans'
Affairs
1514-B Kanawha Blvd.
Charleston, WV 25311
(304) 558-3661

Wisconsin VA
201 West Washington Avenue
Madison, WI 53703
(800) 947-8387

Wyoming Veterans' Affairs Commission
Wyoming ANG Armory
5905 CY Avenue - Room 101
Casper, WY 82604
(307) 265-7372

American Samoa Veterans' Affairs
P.O. Box 982942
Pago Pago, American Samoa 96799
(001) 684-633-4206

Guam Veterans' Affairs Office Mailing
Address:
770 East Sunset Blvd., Suite 165
Tamuning, GU 96913

(671) 475-4222

Puerto Rico Public Advocate for
Veterans' Affairs
Mailing Address:
Apartado 11737 Fernandez Juncos
Station
San Juan, PR 00910-1737
Location:
Mercantile Plaza Bldg, Fourth Floor,
Suite 4021
Hato Rey, PR 00918-1625
(787) 758-5760

Government of the Virgin Islands
Division of Veterans' Affairs
1013 Estate Richmond Christiansted
St. Croix VI 00820-4349
(340) 773-6663

CHAPTER 38

MILITARY PERSONNEL RECORDS

Note: The U.S. Department of Veterans Affairs does not maintain veterans' military service records.

The personnel records of individuals currently in the military service, in the reserve forces, and those completely separated from military service are located in different offices. A nominal fee is charged for certain types of service. In most instances service fees cannot be determined in advance. If your request involves a service fee you will be notified as soon as that determination is made. A veteran and spouse should be aware of the location of the veteran's discharge and separation papers. If a veteran cannot locate discharge and separation papers, duplicate copies may be obtained (further information regarding who to contact is included later in this chapter).

Use "Standard Form 180, Request Pertaining to Military Records," which is available from VA offices and veterans' organizations. Specify that a duplicate separation document or discharge is needed. The veteran's full name should be printed or typed so that it can be read clearly, but the request must also contain the signature of the veteran or the signature of the next of kin if the veteran is deceased. Include branch of service, service number or Social Security number and exact or approximate date and years of service.

It is not necessary to request a duplicate copy of a veteran's discharge or separation papers solely for the purpose of filing a claim for VA benefits. If complete information about the veteran's service is furnished on the application, VA will obtain verification of service from the National Personnel Records Center or the service department concerned.

Who to Contact
The various categories of military personnel records are described in the tables below. Please read the following notes carefully, to make sure an inquiry is sent to the right address. Please note especially that the record is not sent to the National Personnel Records Center as long as the person retains any sort of reserve obligation, whether drilling or non-drilling.

Locations of Medical Records—National Personnel Records Center
Veterans who plan to file a claim for medical benefits with the **Department of Veterans Affairs (VA)** do not need to request a copy of their military health record from the NPRC. After a claim is filed, the VA will obtain the original health record from the NPRC. In addition, many health records were lent to the Department of Veterans Affairs prior to the **1973 Fire**. Veterans who filed a medical claim should contact the Department of Veterans Affairs (VA) in order to determine if their record is already on file. The VA Toll Free number is: 1-800-827-1000 - it will connect the caller to the nearest VA office.

In the 1990s, the military services discontinued the practice of filing health records with the personnel record portion at the NPRC. In 1992, the Army began retiring most of its former members' health records to the Department of Veterans Affairs (VA). Over the next six years, the other services followed suit:

Branch	Status	Health Record to VA
Army	Discharged, retired or separated from any component	October 16, 1992
Navy	Discharged, retired or separated from any component	January 31, 1994
Air Force	Discharged, retired or separated from Active Duty/Discharged or retired from Reserves or National Guard	May 1, 1994 June 1, 1994
Marine Corps	Discharged, retired or separated from any component	May 1, 1994
Coast Guard	Discharged, retired or separated from Active Duty/Reservists with 90 days active duty for training	April 1, 1998

After the dates listed above, the Department of Veterans Affairs (VA), Records Management Center, in St. Louis, MO, became responsible for maintaining active duty health records and managing their whereabouts when on loan within the VA. Call the VA toll free number at 1-800-827-1000 to identify the current location of specific health records and to find out how to obtain releasable documents or information.

Location of Army Personnel Records (Includes Army Air Corps and Army Air Forces)

Dates of Service	Rank	Personnel Record Location	Health Record Location
1789 to November 1, 1912	Enlisted	NARA, Washington, DC	N/A
1789 to July 1, 1917	Officer	NARA, Washington, DC	N/A
November 1, 1912, to October 15, 1992 *Note: Many records were destroyed by the 1973 fire	Enlisted	National Personnel Records Center *Note: Personnel records are Archival 62 years after the service member's separation	
July 1, 1917, to October 15,1992	Officer	National Personnel Records Center *Note: personnel records are Archival 62 years after the service member's separation	
October 16, 992 to September 30,	All Personnel	National Personnel Records Center	Department of Veterans Affairs

2002			
Discharged, deceased or retired on or after October 1, 2002, to December 31, 2013	All Personnel	U.S. Army Human Resources Command *Note: all records are stored electronically at AHRC, but requests are serviced by National Personnel Records Center	Department of Veterans Affairs
Discharged, deceased or retired on or after January 1, 2014	All Personnel	U.S. Army Human Resources Command *Note: records are stored electronically at AHRC, but requests are served by National Personnel Records Center	AMEDD Record Processing Center
All active duty, including active Army Reserve	All Personnel	U.S. Army Human Resources Command	
All active and non-active-duty National Guard	All Personnel	The Adjutant General (of the appropriate state, DC or Puerto Rico)	

Location of Air Force Personnel Records

Dates of Service	Rank	Personnel Record Location	Health Record Location
September 24, 1947, to May 1, 1994 *Note: Many records were destroyed by the 1973 Fire	All Personnel	National Personnel Records Center *Note: Personnel records are Archival 62 years after the service member's separation	
May 1, 1994, to September 30, 2004	All Personnel	National Personnel Records Center	Department of Veterans Affairs
Discharged, deceased or retired from active duty on or after October 1, 2004, to December 31, 2013	All Personnel	Air Force Personnel Center HQ AFPC/DPSSRP	Department of Veterans Affairs
Discharged, deceased or retired from active duty on or after January 1, 2014	All Personnel	Air Force Personnel Center HQ AFPC/DPSSRP	AF STR Processing Center
Active (including National Guard on	All Personnel	Air Force Personnel Center HQ	

active duty in the Air Force), TDRL, or general officers retired with pay		AFPC/DPSSRP	
Reserve, retired reserve in non-pay status, current National Guard officers not on active duty in the Air Force or National Guard released from active duty in the Air Force	Various Personnel	Air Reserve Personnel Center HQ ARPC/DPTOCW (Contact Center)	
Current National Guard enlisted not on active duty in the Air Force	All Personnel	The Adjutant General (of the appropriate state, DC, or Puerto Rico)	

Location of Navy Personnel Records

Dates of Service	Rank	Personnel Record Location	Health Record Location
1798 to 1885	Enlisted	NARA, Washington DC	N/A
1798 to 1902	Officer	NARA, Washington DC	N/A
1885 to January 30, 1994	Enlisted	National Personnel Records Center *Note: Personnel records are Archival 62 years after the servicemember's separation	
1902 to January 30, 1994	Officer	National Personnel Records Center *Note: Personnel records are Archival 62 years after the service member's separation	
January 31, 1994, to December 31, 1994	All Personnel	National Personnel Records Center	Department of Veterans Affairs
Discharged, deceased or retired on or after January 1, 1995, to December 31, 2013	All Personnel	Navy Personnel Command	Department of Veterans Affairs
Discharged, deceased or retired on or after January 1, 2014	All Personnel	National Personnel Records Center	BUMED Navy Medicine Records Activity
Active, reserve or TDRL	All Personnel	Navy Personnel Command	

Location of Marine Corps Personnel Records

Dates of Service	Rank	Personnel Record Location	Health Record Location
1798 to 1904	All Personnel	NARA, Washington DC	N/A
1905 to April 30, 1994	All Personnel	National Personnel Records Center *Note: Personnel records are Archival 62 years after the servicemember's separation	
May 1, 1994, to December 31, 1998	All Personnel	National Personnel Record Center	Department of Veterans Affairs
Discharged, deceased or retired on or after January 1, 1999, to December 31, 2013	All Personnel	HQ U.S. Marine Corps	Department of Veterans Affairs

Discharged, deceased or retired on or after January 1, 2014	All Personnel	HQ U.S. Marine Corps	BUMED Navy Medicine Records Activity
Individual Ready Reserve	All Personnel	Marine Forces Reserve	
Active, Selected Marine Corps Reserve TDRL	All Personnel	HQ U.S. Marine Corps	

Location of Coast Guard Personnel Records—Including Revenue Cutter Service, Life-Saving Service and Lighthouse Service

Dates of Service	Rank	Personnel Record Location	Health Record Location
Discharged, deceased or retired prior to December 31, 1897	All Personnel	NARA, Washington DC	N/A
January 1, 1898, to March 31, 1998	All Personnel	National Personnel Records Center *Note: Personnel records are Archival 62 years after the service member's separation	
Discharged, deceased or retired from active duty from April 1, 998 to September 30, 2014	All Personnel	National Personnel Records Center	Department of Veterans Affairs
Discharged deceased or retired from active duty on or after October 1, 2014	All Personnel	National Personnel Records Center	
Active, Reserve or TDRL	All personnel	USCG Personnel Command	

Address List
Army

U.S. Army Human Resources Command
Attn: AHRC-PDR-VIB
1600 Spearhead Division Avenue Dept 420
Fort Knox, KY 40122-5402

Telephone: 1-888-276-9472

AMEDD Record Processing Center
3370 Nacogdoches Road, Suite 116
San Antonio, TX 78217

Air Force

Air Force Personnel Center
HQ AFPC/DPSSRP
550 C Street West, Suite 19
Randolph AFB, TX 78150-4721

Telephone: 1-800-525-0102

Air Reserve Personnel Center
HQ ARPC/DPTOCW (Contact Center)
18420 E Silver Creek Ave Bldg 390 MS 68
Buckley AFB, CO 80011

Telephone: 1-800-525-0102

AF STR Processing Center
ATTN: Release of Information
3370 Nacogdoches Road, Suite 116
San Antonio, TX 78217

Navy

Navy Personnel Command
(PERS-312E)
5720 Integrity Drive
Millington, TN 38055-3120

Telephone: 901-874-4885

Navy Medicine Records Activity (NMRA)
BUMED Detachment St. Louis
4300 Goodfellow Blvd., Building 103
St. Louis, MO 63120

Marine Corps

Headquarters U.S. Marine Corps
Personnel Management Support Branch
(MMSB-10)
2008 Elliot Road
Quantico, VA 22134-5030
Telephone: 1-800-268-3710

Marine Forces Reserve
4400 Dauphine St.
New Orleans, LA 70146-5400

Telephone: 1-800-255-5082

Coast Guard

Commander CG PSC-bobs-mr
USCG Personnel Command
4200 Wilson Blvd., Suite 1100

Arlington, VA 22203-1804

Telephone: 1-703-872-6392

Other Agencies

The Adjutant General
(of the appropriate state, DC, or Puerto Rico)

Department of Veterans Affairs
Records Management Center (VARMC)
P.O. Box 5020
St. Louis, MO 63115-5020
Telephone: 1-888-533-4558

Division of Commissioned Personnel and Readiness
ATTN: Records Officer
1101 Wootton Parkway, Plaza Level, Suite 100
Rockville, MD 20852
Telephone: 240-453-6041

National Archives & Records Administration
Old Military and Civil Records (NWCTB-Military)
Textual Services Division
700 Pennsylvania Ave., N.W.
Washington, DC 20408-0001
Telephone: 202-357-5000
Customer Service Center Telephone: 1-866-272-6272

National Personnel Records Center, NARA
1 Archives Drive
St. Louis, MO 63138
Request records online, or
Telephone: 314-801-0800

Selective Service System
Data Management Center
P.O Box 94638
Palatine, IL 60094-4638
Telephone: 847-688-6888 or toll-free: 1-888-655-1825

Facts About the 1973 St. Louis Fire and Lost Records

A fire at the NPRC in St. Louis on July 12, 1973, destroyed about 80 percent of the records for Army personnel discharged between November 1, 1912, and January 1, 1960. About 75 percent of the records for Air Force personnel with surnames from "Hubbard" through "Z" discharged between September 25, 1947, and January 1, 1964, were also destroyed

What Was Lost
It is hard to determine exactly what was lost in the fire, because:
- There were no indices to the blocks of records involved. The records were merely filed in alphabetical order for the following groups:
  - World War I: Army November 1, 1912 - September 7, 1939
  - World War II: Army September 8, 1939 - December 31, 1946
  - Post-World War II: Army January 1, 1947 - December 31, 1959
  - Air Force September 25, 1947 - December 31, 1963
- Millions of records, especially medical records, had been withdrawn from all three

groups and loaned to the Department of Veterans Affairs (VA) before the fire. The fact that one's records are not in NPRC files at a particular time does not mean the records were destroyed in the fire.

Reconstruction of Lost Records

If veterans learn that their records may have been lost in the fire, they may send photocopies of any documents they possess -- especially separation documents -- to the NPRC at:

National Personnel Records Center
Military Personnel Records
9700 Page Blvd.
St. Louis, MO 63132-5100

Alternate Sources

When veterans don't have copies of their military records and their NPRC files may have been lost in the St. Louis fire, essential information about their military service may be available from a number of other sources including:

The Department of Veterans Affairs (VA) maintains records on veterans whose military records were affected by the fire if the veteran or a beneficiary filed a claim before July 1973.

Service information may also be found in various kinds of "organizational" records such as unit morning reports, payrolls and military orders on file at the NPRC or other National Archives and Records Administration facilities.
There also is a great deal of information available in records of the State Adjutants General, and other state "veterans services" offices.

By using alternate sources, NPRC often can reconstruct a veteran's beginning and ending dates of active service, the character of service, rank, time lost on active duty, and periods of hospitalization. NPRC can issue NA Form 13038, "Certification of Military Service," considered the equivalent of a Form DD-214, "Report of Separation from Active Duty," to use in establishing eligibility for veterans' benefits.

Necessary Information for File Reconstruction

The key to reconstructing military data is to give the NPRC enough specific information so the staff can properly search the various sources. The following information is normally required:

- Full name used during military service.
- Place of entry into service.
- Branch of service.
- Last unit of assignment.
- Approximate dates of service.
- Place of discharge.
- Service number or Social Security number

What to Remember
✓ The easiest way to request your military service records is through the milConnect website (https://milconnect.dmdc.osd.mil/milconnect/)
✓ You can use milConnect to get your DD214 and request other records
✓ You can also request your military records by mailing or faxing a Request Pertaining to Military Records (Standard Form SF 180) to the National Personnel Records Center
✓ If you're a family member planning a burial for a veteran in a VA national cemetery, you can call 800-535-1117, which is the National Cemetery Scheduling Office.
✓ If you're the next of kin of a veteran who has passed away, you can mail or fax Standard Form SF 180 to the National Personnel Records Center.

CHAPTER 39

CORRECTION OF RECORDS BY CORRECTION BOARDS

Retirees may feel that their records need correcting or amending for any number of reasons. Correction boards consider formal applications for corrections of military records. Each service department has a permanent Board for Correction of Military (Naval) records, composed of civilians, to act on applications for correction of records.

In order to justify the correction of a military record, the applicant must prove to a Corrections Board that the alleged entry or omission in the record was in error or unjust. This board considers all applications and makes recommendations to the appropriate branch Secretary.

An application for correction of record must be filed within three years after discovering the error or injustice. If filed after the three-year deadline, the applicant must include in the application reasons the board should find it in the interest of justice to accept the late application. Evidence may include affidavits or signed testimony executed under oath, and a brief of arguments supporting the application. All evidence not already included in one's record must be submitted. The responsibility for securing new evidence rests with the applicant.

To justify any correction, it is necessary to show to the satisfaction of the board that the alleged entry or omission in the records was in error or unjust. Applications should include all available evidence, such as signed statements of witnesses or a brief of arguments supporting the requested correction. Application is made with DD Form 149, available at VA offices, from veterans' organizations or from the internet. Each of the military services maintains a discharge review board with authority to change, correct or modify discharges or dismissals that are not issued by a sentence of a general courts-martial. The board has no authority to address medical discharges.

The veteran or, if the veteran is deceased or incompetent, the surviving spouse, next of kin or legal representative may apply for a review of discharge by writing to the military department concerned, using DoD Form 293. This form may be obtained at a VA regional office, from veterans' organizations or from the Internet. However, if the discharge was more than 15 years ago, a veteran must petition the appropriate service Board for Correction of Military Records using DoD Form 149, which is discussed in the "Correction of Military Records" section of this booklet. A discharge review is conducted by a review of an applicant's record and, if requested, by a hearing before the board.

Discharges awarded as a result of a continuous period of unauthorized absence in excess of 180 days make persons ineligible for VA benefits regardless of action taken by discharge review boards unless VA determines there were compelling circumstances for the absence.

Boards for the correction of military records also may consider such cases.

Veterans with disabilities incurred or aggravated during active military service may qualify for medical or related benefits regardless of separation and characterization of service. Veterans separated administratively under other than honorable conditions may request that their discharge be reviewed for possible recharacterization, provided they file their appeal within 15 years of the date of separation. Questions regarding the review of a discharge should be addressed to the appropriate discharge review board at the address listed on DoD Form 293.

Jurisdiction

Correction boards are empowered to deal with all matters relating to error or injustice in official records. The boards cannot act until all other administrative avenues of relief have been exhausted. Discharges by sentence of Special Court-Martial and administrative discharges cannot be considered by correction boards unless:

- Application to the appropriate Discharge Review Board has been denied and rehearing is barred; or
- Application cannot be made to the Discharge Review Board because the time limit has expired.

Application

DD Form 149, Application for Correction of Military or Naval record, must be used to apply for the correction of military records. It should be submitted, along with supporting evidence, to one of the review boards listed below:

Army
Army Review Boards Agency (ARBA)
ATTN: Client Information and Quality Assurance
251 18th Street South, Suite 385
Arlington, VA 22202

Navy & Marine Corps
Board for Correction of Naval Records
701 S. Courthouse Road
Bldg 12, Suite 1001
Arlington, VA 22204-2490
(703) 604-6884

Coast Guard
DHS Office of the General Counsel
Board for Correction of Military Records
245 Murray Lane, Stop 0485
Washington, DC 20528
(202) 447-4099

Air Force
Board for Correction of Air Force Records
SAF/MRBR
550-C Street West, Suite 40
Randolph AFB, TX 78150-4742

Decisions

In the absence of new and material evidence the decision of a correction board, as approved or modified by the Secretary of the Service Department, is final. Adverse decisions are subject to judicial review in a U.S. District Court. Decisions of the Boards for Correction of Military or Naval Records must be made available for public inspection. Copies of the decisional documents will be provided on request.

What to Remember
✓ The National Archives and Records Administration can't make changes or corrections to records or discharge status other than minor administration corrections.
✓ You need to apply to the review board for your service branch for corrections or changes.
✓ The secretary of a military department through a board for correction of military records can change records to remove an injustice or correct an error.
✓ Generally, the veteran, survivor or legal representative must file a request for correction within three years after discovery of the error or injustice.
✓ If the discharge was more than 15 years ago, a veteran has to petition using DoD Form 149

CHAPTER 40

DISCHARGE REVIEW

Discharge Review Boards
Each branch of service has discharge review boards to review the discharge or dismissal of former service members. (The Navy Board considers Marine Corps cases.)

Online Tool for Discharge Upgrade Process
The Department of Defense (DoD), through a joint program with the VA, has launched a new web-based tool that provides customized guidance to veterans who want to upgrade or change the conditions of their military discharge.

By answering questions, veterans can receive information on the specific armed services board to contact, the forms to fill out, special guidance applicable to their case, where to send their application, and helpful tips for appealing their discharge.

The military has estimated that tens of thousands of veterans with less than honorable discharges are especially likely to have unjust discharges deserving of upgrades. These are veterans who were discharged due to incidents relating to post-traumatic stress disorder, traumatic brain injury or sexual orientation. Fragmented and confusing information has historically deterred veterans from obtaining crucial information and — in many cases — necessary benefits.

The discharge upgrade tool is available at https://www.vets.gov/discharge-upgrade-instructions.

Consideration for Veterans' Discharge Upgrade Requests
The Defense Department released guidance in 2017 to clarify the liberal consideration given to veterans who request upgrades of their discharge saying they had mental health conditions or were victims of sexual assault or sexual harassment.

The new guidance clarifies that the liberal consideration policy includes conditions resulting from post-traumatic stress disorder, traumatic brain injury, sexual assault or sexual harassment. The policy is meant to ease the burden on veterans and give them a reasonable opportunity to establish the extenuating circumstances of their discharge. Under new guidance, the following are some of the key things involved in discharge relief:
- Evidence may come from sources other than a veteran's service record and may include records from the DoD Sexual Assault Prevention and Response Program (DD Form 2910, Victim Reporting Preference Statement), and/or DD Form 2911 (DoD Sexual Assault Forensic SAFE Report), law enforcement authorities, rape crisis centers, mental health counseling centers, hospitals, physicians, pregnancy tests, tests for sexually transmitted diseases, and statements from family members, friends, roommates, co-workers, fellow servicemembers, or clergy.
- Evidence may also include changes in behavior, requests for transfer to another military duty assignment, deterioration in performance, inability of the individual to conform their behavior to the expectations of a military

environment, substance abuse, periods of depression, panic attacks or anxiety without an identifiable cause, unexplained economic or social behavior changes, relationship issues or sexual dysfunction

- Evidence of misconduct, including any misconduct underlying a veteran's discharge may be evidence of a mental health condition, including PTSD, TBI or of behavior consistent with experiencing sexual assault or sexual harassment
- The veteran's testimony alone, written or oral, may establish the existence of a condition or experience, that the condition or experience existing during or was aggravated by military service, and that the condition or experience causes or mitigates the discharge
- Absent clear evidence to the contrary, a diagnosis from a licensed psychiatrist or psychologist is evidence the veteran had a condition that could excuse or mitigate the discharge
- Evidence that may reasonably support more than one diagnosis should be liberally considered as supporting a diagnosis, where applicable, that could excuse or mitigate the discharge
- A veteran asserting a mental health condition without a corresponding diagnosis of such condition from a licensed psychiatrist or psychologist will receive liberal consideration of evidence that may support the evidence of such a condition
- Review Boards are not required to find that a crime of sexual assault or an incident of sexual harassment occurred in order to grant liberal consideration to a veteran that the experience happened during military service, was aggravated by military service, or that it excuses or mitigates the discharge

Authority

Discharge review boards can be based on the official records and such other evidence as may be presented, upgrade a discharge or change the reason and authority for discharge. Discharge review boards cannot grant disability retirement, revoke a discharge, reinstate any person in the service, recall any person to active duty, act on requests for re-enlistment code changes or review a discharge issued by sentence of a general court-martial. Discharge review boards have no authority to address medical discharges.

Discharges awarded as a result of a continuous period of unauthorized absence in excess of 180 days make persons ineligible for VA benefits regardless of action taken by discharge review boards, unless VA determines there were compelling circumstances for the absence. Boards for the correction of military records also may consider such cases.

Application

DD Form 293, "Application for Review of Discharge or Dismissal from the Armed Forces of the United States," is used to apply for review of discharge. (If more than 15 years have passed since discharge, DD Form 149 should be used.) The individual or, if legal proof of death is provided, the surviving spouse, next-of-kin, or legal representative can apply. If the individual is mentally incompetent, the spouse, next-of-kin, or legal representative can sign the application, but must provide legal proof of incompetence. The instruction for completing DD Form 293 must be read and complied with.

Time Limitation

Initial application to a discharge review board must be made within 15 years after the date of discharge.

Personal Appearance

A personal appearance before the Discharge Review Board is a legal right. A minimum 30-day notice of the scheduled hearing date is given unless the applicant waives the advance notice in writing. Reasonable postponements can be arranged if circumstances preclude appearance on the scheduled date. All expenses of appearing before the board must be paid by the applicant. If no postponement of a scheduled hearing date is requested and the applicant does not appear on the date scheduled, the right to a personal hearing is forfeited and the case will be considered on the evidence of record.

Hearings

Discharge review boards conduct hearings at various locations in the U.S. Information concerning hearing locations and availability of counsel can be obtained by writing to the appropriate board at the address shown on DD Form 293. Those addresses are listed at the end of this chapter.

Published Uniform Standards for Discharge Review

A review of discharge is conducted to determine if an individual was properly and equitably discharged. Each case is considered on its own merits. A discharge is considered to have been proper unless the discharge review determines:

- That there is an error of fact, law, procedures, or discretion which prejudiced the rights of the individual, or
- That there has been a change of policy which requires a change of discharge

A discharge is considered to have been equitable unless the discharge review determines:

- That the policies and procedures under which the individual was discharged are materially different from current policies and procedures and that the individual probably would have received a better discharge if the current policies and procedures had been in effect at the time of discharge; or
- That the discharge was inconsistent with the standards of discipline; or
- That the overall evidence before the review board warrants a change of discharge. In arriving at this determination, the discharge review board will consider the quality and the length of the service performed, the individual's physical and mental capability to serve satisfactorily, abuses of authority which may have contributed to the character of the discharge issued and documented discriminatory acts against the individual.

An authenticated decisional document is prepared, and a copy provided to each applicant and council. A copy of each decisional document, with identifying details of the applicant and other persons deleted to protect personal privacy, must be made available for public inspection and copying. These are located in a reading room in the Pentagon, Washington, DC. To provide access to the documents by persons outside the Washington, D.C. area, the documents have been indexed. The index includes case number of each case; the date, authority and reason for, and character if the discharge, and the issues addressed in the statement of findings, conclusions and reasons.

Interested parties may contact the DVARO or the State veterans Agency for the location of an index. A copy of the index will be made available at the sites of traveling board hearings during the period the board is present. An individual can go through the index and identify cases in which the circumstances leading to discharge are similar to those in the individual's case. A copy of these case decisional documents can be requested by writing to:

DA Military Review Boards Agency
ATTN: SFBA (Reading Room) Room 1E520
The Pentagon Washington, D.C. 20310

Examination of decisional documents may help to identify the kind of evidence that was used in the case and may indicate why relief was granted or denied. Decisional documents do not set precedence - each case is considered on its own merits.

Reconsideration

An application that has been denied can be reopened if:

- The applicant submits newly discovered evidence that was not available at the time of the originalconsideration.

- The applicant did not request a personal hearing in the original application and now desires to appear before the board. If the applicant fails to appear at the hearings, the case will be closed with no further action.
- The applicant was not represented by counsel in the original consideration and now desires counsel and the application for reconsideration is submitted within 15 years following the date of discharge.
- Changes in policy, law or regulations have occurred or federal court orders have been issued which substantially enhances the rights of the applicant.

Service Department Discharge Review Board Addresses

Army Discharge Review Board
Attention: SFMR-RBB
251 18th St. South Suite 385
Arlington, VA 22202-4508

Navy & USMC
Secretary of the Navy Council of Review Boards
720 Kennon St. SE Suite 309
Washington, DC 20374

Air Force
Air Force Military Personnel Center Attention: DP-MDOA1
Randolph AFB, TX 78150-6001

Coast Guard
DHS Office of the General Counsel Board for Correction of Military Records
Mailstop # 485
245 Murray Lane
Washington, DC 20528

What to Remember
✓ Even with a less than honorable discharge you may be able to access some VA benefits.
✓ You may have access through the Character of Discharge review process.
✓ When you apply, VA will review your record to determine if your service was honorable for VA purposes.
✓ The process can take up to a year.
✓ Someone who wants this review should provide documents supporting their case, similar to what you'd send for a discharge upgrade.

CHAPTER 41

PENAL AND FORFEITURE PROVISIONS

The first section of this chapter outlines basic information for veterans with questions concerning the effect of incarceration on VA benefits. The later sections of this chapter provide detailed information regarding misappropriation by fiduciaries, fraudulent acceptance of payments, forfeiture for fraud, forfeiture for treason, and forfeiture for subversive activities.

Basic Information
VA benefits are restricted if a veteran, surviving spouse, child, or dependent parent is convicted of a felony and imprisoned for more than 60 days. VA may still pay certain benefits, however, the amount paid depends on the type of benefit and reason for imprisonment. Following is information about the benefits most commonly affected by imprisonment.

Please note that overpayments due to failure to notify VA of a veteran's incarceration results in the loss of all financial benefits until the overpayment is recovered.

VA Disability Compensation
VA disability compensation payments are reduced if a veteran is convicted of a felony and imprisoned for more than 60 days. Veterans rated 20 percent, or more are limited to the 10 percent disability rate. For a veteran whose disability rating is 10 percent, the payment is reduced by one-half. Once a veteran is released from prison, compensation payments may be reinstated based upon the severity of the service-connected disability(ies) at that time.

The disability compensation paid to a veteran incarcerated because of a felony is limited to the 10% disability rate, beginning with the 61st day of imprisonment. For a surviving spouse, child, dependent parent or veteran whose disability rating is 10%, the payment is at the 5% rate. This means that if a veteran was receiving $188 or more prior to incarceration, the new payment amount will be $98. If a veteran was receiving $98 before incarceration, the new payment amount will be $49.

If a veteran resides in a halfway house, participates in a work release program, or is on parole, compensation payments will not be reduced.

The amount of any increased compensation awarded to an incarcerated veteran that results from other than a statutory rate increase may be subject to reduction due to incarceration.

VA Disability Pension
VA will stop a veteran's pension payments beginning on the 61st day of imprisonment for conviction of either a felony or misdemeanor. Payments may be resumed upon release from prison if the veteran meets VA eligibility requirements.

VA Medical Care

While incarcerated veterans do not forfeit their eligibility for medical care, current regulations restrict VA from providing hospital and outpatient care to an incarcerated veteran who is an inmate in an institution of another government agency when that agency has a duty to give the care or services.

However, VA may provide care once the veteran has been unconditionally released from the penal institution. Veterans interested in applying for enrollment into the VA health care system should contact the nearest VA health care facility upon their release.

Educational Assistance / Subsistence Allowance

Beneficiaries incarcerated for other than a felony can receive full monthly benefits, if otherwise entitled. Convicted felons residing in halfway houses (also known as "residential re-entry centers") or participating in work-release programs also can receive full monthly benefits. Claimants incarcerated for a felony conviction can be paid only the costs of tuition, fees, and necessary books, equipment, and supplies.

VA cannot make payments for tuition, fees, books, equipment, or supplies if another Federal State, or local program pays these costs in full.

If another government program pays only a part of the cost of tuition, fees, books, equipment, or supplies, VA can authorize the incarcerated claimant payment for the remaining part of the costs.

Clothing Allowance

In the case of a veteran who is incarcerated in a Federal, State, or local penal institution for a period in excess of 60 days and who is furnished clothing without charge by the institution, the amount of any annual clothing allowance payable to the veteran shall be reduced by an amount equal to 1/365 of the amount of the allowance otherwise payable under that section for each day on which the veteran was so incarcerated during the 12-month period preceding the date on which payment of the allowance would be due.

Payments to Dependents

VA may be able to take part of the amount that the incarcerated veteran is not receiving and pay it to his or her dependents, if they can show need. Interested dependents should contact the nearest VA regional office for details on how to apply. They will be asked to provide income information as part of the application process.

VA will inform a veteran whose benefits are subject to reduction of the right of the veteran's dependents to an apportionment while the veteran is incarcerated, and the conditions under which payments to the veteran may be resumed upon release from incarceration.

VA will also notify the dependents of their right to an apportionment if the VA is aware of their existence and can obtain their addresses.

No apportionment may be made to or on behalf of any person who is incarcerated in a Federal, State, or local penal institution for conviction of a felony.

An apportionment of an incarcerated veteran's VA benefits is not granted automatically to the veteran's dependents. The dependent(s) must file a claim for an apportionment.

Restoration of Benefits

When a veteran is released from prison, his or her compensation or pension benefits may be restored. Depending on the type of disability, the VA may schedule a medical examination to see if the veteran's disability has improved or worsened.

Misappropriation by Fiduciaries

Whoever, being a guardian, curator, conservator, committee, or person legally vested with the responsibility or care of a claimant or a claimant's estate, or any other person having charge and custody in a fiduciary capacity of money heretofore or hereafter paid under any of the laws administered by the VA for the benefit of any minor, incompetent, or other beneficiary, shall lend, borrow, pledge, hypothecate, use, or exchange for other funds or property, except as authorized by law, or embezzle or in any manner misappropriate any such money or property derived wherefrom in whole or in part, and coming into such fiduciary's control in any matter whatever in the execution of such fiduciary's trust, or under color of such fiduciary's office or service as such fiduciary, shall be fined in accordance with Title 18, or imprisoned not more than 5 years, or both.

Any willful neglect or refusal to make and file proper accountings or reports concerning such money or property as required by law shall be taken to be sufficient evidence prima facie of such embezzlement or misappropriation.

Fraudulent Acceptance of Payments

Any person entitled to monetary benefits under any of the laws administered by the VA whose right to payment ceases upon the happening of any contingency, who thereafter fraudulently accepts any such payment, shall be fined in accordance with Title 18, or imprisoned not more than one year, or both.

Whoever obtains or receives any money or check under any of the laws administered by the VA without being entitled to it, and with intent to defraud the United States or any beneficiary of the United States, shall be fined in accordance with Title 18, or imprisoned not more than one year, or both.

Forfeiture for Fraud

Whoever knowingly makes or causes to be made or conspires, combines, aids, or assists in, agrees to, arranges for, or in any way procures the making or presentation of a false or fraudulent affidavit, declaration, certificate, statement, voucher, or paper, concerning any claim for benefits under any of the laws administered by the VA (except laws pertaining to insurance benefits) shall forfeit all rights, claims, and benefits under all laws administered by the VA (except laws pertaining to insurance benefits).

Whenever a veteran entitled to disability compensation has forfeited the right to such compensation under this chapter, the compensation payable but for the forfeiture shall thereafter be paid to the veteran's spouse, children, and parents. Payments made to a spouse, children, and parents under the preceding sentence shall not exceed the amounts payable to each if the veteran had died from service-connected disability. No spouse, child, or parent who participated in the fraud for which forfeiture was imposed shall receive any payment by reason of this subsection. Any apportionment award under this subsection may not be made in any case after September 1, 1959.

Forfeiture of benefits by a veteran shall not prohibit payment of the burial allowance, death compensation, dependency and indemnity compensation, or death pension in the event of the veteran's death.

After September 1, 1959, no forfeiture of benefits may be imposed under the rules outlined in this chapter upon any individual who was a resident of, or domiciled in, a State at the time the act or acts occurred on account of which benefits would, but not for this subsection, be forfeited unless such individual ceases to be a resident of, or domiciled in, a State before the expiration of the period during which criminal prosecution could be instituted. The paragraph shall not apply with respect to:

- Any forfeiture occurring before September 1, 1959; or
- An act or acts that occurred in the Philippine Islands before July 4, 1946.

The VA is authorized and directed to review all cases in which, because of a false or fraudulent affidavit, declaration, certificate, statement, voucher, or paper, a forfeiture of gratuitous benefits under laws administered by the VA was imposed, pursuant to this section or prior provisions of the law, on or before September 1, 1959. In any such case in which the VA determines that the forfeiture would not have been imposed under the provisions of this section in effect after September 1, 1959, the VA shall remit the forfeiture, effective June 30, 1972.

Benefits to which the individual concerned becomes eligible by virtue of any such remission may be awarded, upon application for, and the effective date of any award of compensation, dependency and indemnity compensation, or pension made in such a case shall be fixed in accordance with the facts found, but shall not be earlier than the effective date of the Act or administrative issue. In no event shall such award or increase be retroactive for more than one year from the date of application, or the date of administrative determination of entitlement, whichever is earlier.

Forfeiture for Treason

Any person shown by evidence satisfactory to the VA to be guilty of mutiny, treason, sabotage, or rendering assistance to an enemy of the United States or its allies shall forfeit all accrued or future gratuitous benefits under laws administered by the VA.

The VA, in its discretion, may apportion and pay any part of benefits forfeited under the preceding paragraph to the dependents of the person forfeiting such benefits. No dependent of any person shall receive benefits by reason of this subsection in excess of the amount to which the dependent would be entitled if such person were dead.

In the case of any forfeiture under this chapter, there shall be no authority after September 1, 1959, to:

- Make an apportionment award pursuant to the preceding paragraph; or
- Make an award to any person of gratuitous benefits based on any period of military, naval, or air service commencing before the date of commission of the offense.

Forfeiture for Subversive Activities

Any individual who is convicted after September 1, 1959, of any offense listed below shall, from and after the date of commission of such offense, have no right to gratuitous benefits (including the right to burial in a national cemetery) under laws administered by the VA based on periods of military, naval, or air service commencing before the date of commission of such offense, and no other person shall be entitled to such benefits on account of such individual. After receipt of notice of the return of an indictment for such an offense, the VA shall suspend payment of such gratuitous benefits pending disposition of the criminal proceedings. If any individual whose rights to benefits has been terminated pursuant to this section, is granted a pardon of the offense by the President of the United States, the right to such benefits shall be restored as of the date of such pardon.

The offenses referred to in the previous paragraph are:
Sections 894, 904 and 906 of Title 10 (articles 94, 104, and 106 of the Uniform Code of Military Justice).
Sections 792, 793, 794, 798, 2381, 2382, 2383, 2384, 2385, 2387, 2388, 2389, 2390, and chapter 105 of Title 18.

Sections 222, 223, 224, 225 and 226 of the Atomic Energy Act of 1954 (42 U.S.C. 2272, 2273, 2274, 2275, and 2276).
Section 4 of the Internal Security Act of 1950 (50 U.S.C. 783).

The Secretary of Defense, the Secretary of Transportation, or the Attorney General, as appropriate, shall notify the VA in each case in which an individual is convicted of an offense mentioned in this chapter.

What to Remember
✓ VA can pay certain benefits to veterans who are incarcerated in a federal, state or local penal institution; however, the amount depends on the type of benefit and the reason for incarceration.
✓ VA disability compensation payments are reduced if a veteran is convicted of a felony and imprisoned for more than 60 days.
✓ Veterans rated 20% or more are limited to the 10% disability rate.
✓ For a veteran whose disability rating is 10%, the payment is reduced by one-half.
✓ When a veteran is released from prison, compensation payments may be reinstated based upon the severity of the service-connected disability at that time.

CHAPTER 42

★ ★ ★ ★ ★ ★ ★ ★

PERIODS OF WAR

Military service is classified either as wartime or peacetime service. This is important because there are significant advantages specifically accruing only to veterans with wartime service. The following list sets out the periods of wartime designated by Congress for pension purposes. To be considered by the VA to have served during wartime, a veteran need not have served in a combat zone, but simply during one of these designated periods.

Indian Wars:
The period January 1, 1817, through December 31, 1898. Service must have been rendered with the United States military forces against Indian tribes or nations.

Spanish-American War:
The period April 21, 1898, through July 4, 1902. In the case of a veteran who served with the United States military forces engaged in hostilities in the Moro Province the ending date is July 15, 1903.

Mexican Border Period:
The period May 9, 1916, through April 5, 1917, in the case of a veteran who during such period served in Mexico, on the borders thereof, or in the waters adjacent thereto.

World War I:
The period April 6, 1917, through November 11, 1918. In the case of a veteran who served with the United States military forces in Russia, the ending date is April 1, 1920. Service after November 11, 1918, and before July 2, 1921 is considered World War I for compensation or pension purposes, if the veteran served in the active military, naval, or air service after April 5, 1917 and before November 12, 1918.

World War II:
The period December 7, 1941, through December 31, 1946. If the veteran was in service on December 31, 1946, continuous service before July 26, 1947, is considered World War II service.

Korean Conflict:
The period June 27, 1950, through January 31, 1955.

Vietnam Era:
The period August 5, 1964 (February 28, 1961, for Veterans who served "in country" before August 5, 1964) and ending May 7, 1975.

Persian Gulf War:
The period August 2, 1990, through a date to be set by law or Presidential Proclamation.

Future Dates:
The period beginning on the date of any future declaration of war by the Congress and ending on a date prescribed by Presidential Proclamation or concurrent resolution of the Congress. (Title U.S.C. 10)

CHAPTER 43

★★★★★★★★

COMMON FORMS

General Administration Forms	
Form #	Purpose
20-572	Request For Change of Address / Cancellation of Direct Deposit
Compensation and Pension Forms	
Form #	Purpose
21-22	Appointment Of Veterans Service Organization As Claimant's Representative
21-0304	Application For Spina Bifida Benefits
21P-534EZ	Application for DIC, Death Pension, and/or Accrued Benefits
21-526EZ	Application for Disability Compensation Benefits
21-526c	Pre-Discharge Compensation Claim
21P-527EZ	Application for Pension
21P-530	Application for Burial Benefits
21P-601	Application for Accrued Amounts Due a Deceased Beneficiary
21-686c	Declaration of Status of Dependents
Form 9	Appeal to Board of Veterans Appeals
21-0958	Notice of Disagreement
21-0966	Intent to File a VA Claim
21-2680	Request for Aid and Attendance/Housebound Status
21-4138	Statement in Support of Claim
21-4142	Authorization for Release of Information
21-8940	Application for Increased Compensation Due Unemployability
Vocational Rehabilitation and Employment Forms	
Form #	Purpose
22-1190	Application for VA Education Benefits
22-5490	Dependent's Application for VA Education Benefits
28-1900	Disabled Veterans Application for Vocational Rehabilitation
VA Insurance Forms	
Form #	Purpose

29-336	Designation Of Beneficiary – Government Life Insurance
29-1546	Application For Cash Surrender Value / Application for Policy Loan
29-4125	Claim For One Sum Payment
29-4364	Application For Service-Disabled Insurance

Finance and Budget Forms

Form #	Purpose
24-0296	Direct Deposit Enrollment

Miscellaneous Forms

Form #	Purpose
DD 149	Application For Correction of Military Record
DD 214	Report of Separation from Active Duty
DD 293	Application For the Review Of Discharge OR Dismissal From The Armed Forces Of The United States
SF 15	Application For 10-Point Veteran's Preference
SF 180	Request Pertaining to Military Records
SGLV8283	Claim For Death Benefits – Form Returned to Office Of Servicemembers' Group Life Insurance
SGLV8285	Request For Insurance (Servicemembers' Group Life Insurance)
SGLV8286	Servicemembers' Group Life Insurance Election and Certificate
SGLV8714	Application For Veterans' Group Life Insurance
VAF 8	Certification To Appeal
VAF 9	Appeal To Board of Veterans' Appeals
10-10ez	Instructions For Completing Application for Health Benefits
VAF 3288	Request For and Consent To Release Of Information From Claimant's Records
VAF 4107	Notice Of Procedural and Appellate Rights
VAF 4107b	Notice Of Procedural and Appellate Rights (Spanish Version)
40-1330M	Application For Standard Government Headstone or Marker For Installation In A Private Or State Veterans' Cemetery

Discharge Documents
DD214 and Other Department of Defense Documents

Form	Title
DD 214	Certificate of Release or Discharge from Active Duty
DD 13	Statement of Service
DA 1569	Transcript of Military Record
DD 2A	Armed Forces Identification Card (Active)
DD 2AF	Armed Forces Identification Card (Active)
DD 2CG	Armed Forces Identification Card (Active)
DD 2MC	Armed Forces Identification Card (Active)
DD 2N	Armed Forces Identification Card (Active)
DD 2NOAA	Armed Forces Identification Card (Active)
DD 2 (Retired)	US Uniformed Services Identification Card
DD 214	Certificate of Release or Discharge from Active Duty
DD 217	Discharge Certificate
DD 303	Certificate in Lieu of Lost or Destroyed Discharge
DD 303AF	Certificate in Lieu of Lost or Destroyed Discharge
DD 303CG	Certificate in Lieu of Lost or Destroyed Discharge
DD 303MC	Certificate in Lieu of Lost or Destroyed Discharge
DD 303N	Certificate in Lieu of Lost or Destroyed Discharge
DD 1300	Report of Casualty

CHAPTER 44

★★★★★★★

APPLICATION AND CHARACTER OF DISCHARGE

Applying for Benefits and Your Character of Discharge

Generally, in order to receive VA benefits and services, the Veteran's character of discharge or service must be under other than dishonorable conditions (e.g., honorable, under honorable conditions, general). However, individuals receiving undesirable, bad conduct, and other types of dishonorable discharges may qualify for VA benefits depending on a determination made by VA.

Basic eligibility for Department of Veterans Affairs (VA) benefits depends upon the type of military service performed, the duration of the service, and the character of discharge or separation. VA looks at the "character of discharge" to determine whether a person meets the basic eligibility requirements for receipt of VA benefits under title 38 of the United States Code. Any discharge under honorable conditions satisfies the character of discharge requirement for basic eligibility for VA benefits. Certain types of discharges, along with the circumstances surrounding those discharges, bar an individual from basic eligibility for VA benefits. Other types of discharges require VA to make a character of discharge determination in order to assess basic eligibility for VA benefits.

Under the law (38 U.S.C. § 5303), a release or discharge for any of the following reasons constitutes a statutory bar to benefits, unless it is determined that the Servicemember was insane at the time he/she committed the offense that resulted in the discharge:

- Sentence of a general court-martial
- Being a conscientious objector who refused to perform military duty, wear the uniform, or otherwise comply with lawful orders of competent military authority
- Desertion
- Absence without official leave (AWOL) for a continuous period of 180 days or more, without compelling circumstances to warrant such prolonged unauthorized absence (as determined by VA)
- Requesting release from service as an alien during a period of hostilities, or This means that if an individual is discharged for any of the above reasons, the law prohibits VA from providing any benefits
- Resignation by an officer for the good of the service

VA reviews military service records, including facts and circumstances surrounding the incident(s) leading to the discharge. VA also considers the following when making its determination:

- Any mitigating or extenuating circumstances presented by the claimant

- Any supporting evidence provided by third parties who were familiar with the circumstances surrounding the incident(s) in question
- Length of service
- Performance and accomplishments during service
- Nature of the infraction(s), and
- Character of service preceding the incident(s) resulting in the discharge.

VA considers whether an individual was insane when determining whether a statutory bar to benefits exists. When no statutory bar to benefits exists, the impact of disabilities may be considered during the analysis of any mitigating or extenuating circumstances that may have contributed to the discharge.

Specific Benefit Program Character of Discharge Requirements

Compensation Benefits
To receive VA compensation benefits and services, the Veteran's character of discharge or service must be under other tan dishonorable conditions (e.g., honorable, under honorable conditions, general).

Education Benefits
To receive VA education benefits and services through the Montgomery GI Bill program or Post-9/11 GI Bill program, the Veteran's character of discharge or service must be honorable.

To receive VA education benefits and services through any other VA educational benefits program, including the Survivors' and Dependents' Educational Assistance (DEA) program, the Veteran's character of discharge or service must be under other than dishonorable conditions (e.g., honorable, under honorable conditions, general).

Home Loan Benefits
To receive VA home loan benefits and services, the Veteran's character of discharge or service must be under other than dishonorable conditions (e.g., honorable, under honorable conditions, general).

Insurance Benefits
Generally, there is no character of discharge bar to benefits to Veterans' Group Life Insurance. However, for Service-Disabled Veterans Insurance and Veterans' Mortgage Life Insurance benefits, the Veteran's character of discharge must be other than dishonorable.

Pension Benefits
To receive VA pension benefits and services, the Veteran's character of discharge or service must be under other than dishonorable conditions (e.g., honorable, under honorable conditions, general).

Review of Discharge from Military Service
Each of the military services maintains a discharge review board with authority to change, correct or modify discharges or dismissals not issued by a sentence of a general court-martial. The board has no authority to address medical discharges.

The Veteran or, if the Veteran is deceased or incompetent, the surviving spouse, next of kin or legal representative, may apply for a review of discharge by writing to the military department concerned, using DD Form 293, "Application for the Review of Discharge from the Armed Forces of the United States." This form may be obtained at a VA regional office, or from Veterans organizations.

However, if the discharge was more than 15 years ago, a Veteran must petition the appropriate Service's Board for Correction of Military Records using DD Form 149, "Application for Correction of Military Records Under the Provisions of Title 10, U.S. Code, Section 1552." A discharge review is conducted by a review of an applicant's record and, if requested, by a hearing before the board.

Discharge Related to Mental Health Conditions, Sexual Assault, or Sexual Harassment

In December 2016, the Department announced a renewed effort to ensure that veterans are aware of the opportunity to have discharges and military records reviewed. Following a review, it was determined that clarifications were needed regarding mental health conditions, sexual assault and sexual harassment. New guidance was issued.

Clarifying Guidance to Military Discharge Review Boards and Boards for Correction Requests by Veterans for Modification of Their Discharge Due to Mental Health Conditions, Traumatic Brain Injury, Sexual Assault or Sexual Harassment

Requests for discharge relief usually involve four questions, which include:

- Did the veteran have a condition or experience that may excuse or mitigate the discharge?
- Did that condition exist/experience occur during military service?
- Does that condition or experience actually excuse or mitigate the discharge?
- Does that condition or experience outweigh the discharge?

Liberal consideration will be given to veterans for petitioning for discharge relief when the application for relief is based in whole or in part on matters relating to mental health conditions including PTSD, TBI, sexual assault or sexual harassment.

Evidence may come from sources other than a veteran's service record and may include records from the DoD Sexual Assault Prevention and Response Program (DD Form 2910, Victim Reporting Preference Statement) and/or DD Form 2911, DoD Sexual Assault Forensic Examination (SAFE) Report, law enforcement authorities, rape crisis centers, mental health and counseling centers, hospitals, physicians, pregnancy tests, tests for sexually transmitted diseases and statements from family members, friends, roommates, co-workers, and fellow servicemembers or clergy.

Evidence may also include changes in behavior, requests for transfer to another military assignment, deterioration in work performance, inability of the individual to conform their behavior to expectations of a military environment, substance abuse, episodes of depression or panic attacks, anxiety without an identifiable cause, unexplained economic or social behavior changes, relationship issues, or sexual dysfunction.

Evidence of misconduct, including any misconduct underlying a veteran's discharge may be evidence of a mental health condition, including PTSD, TBI or behavior consistent with experience sexual assault or sexual harassment.

The veteran's testimony alone, oral or written, may establish the existence of a condition or experience, that the condition or experience existed during or was aggravated by military service, and that the condition or experience excuses or mitigates the discharge.

The Department of Defense today announced a renewed effort to ensure veterans are aware of the opportunity to have their discharges and military records reviewed. Through enhanced public outreach, engagement with Veterans Service Organizations (VSOs), Military Service Organizations (MSOs), and other outside groups, as well as direct outreach to individual veterans, the department encourages all veterans who believe they have experienced an error or injustice to request relief from their service's Board for Correction of Military/Naval Records (BCM/NR) or Discharge Review Board (DRB).

Additionally, all veterans, VSOs, MSOs, and other interested organizations are invited to offer feedback on their experiences with the BCM/NR or DRB processes, including how the policies and processes can be improved.

In the past few years, the department has issued guidance for consideration of post-traumatic stress disorder (PTSD), as well as the repealed "Don't Ask, Don't Tell" and its predecessor policies. Additionally, supplemental guidance for separations involving victims of sexual assault is currently being considered.

The department is reviewing and consolidating all of the related policies to reinforce the department's commitment to ensuring fair and equitable review of separations for all veterans. Whether the discharge or other correction is the result of PTSD, sexual orientation, sexual assault, or some other consideration, the department is committed to rectifying errors or injustices and treating all veterans with dignity and respect.

To request an upgrade or correction:
Veterans who desire a correction to their service record or who believe their discharge was unjust, erroneous, or warrants an upgrade, are encouraged to apply for review.

For discharge upgrades, if the discharge was less than 15 years ago, the veteran should complete DD Form 293 and send it to their service's DRB (the address is on the form). For discharges over 15 years ago, the veteran should complete the DD Form 149 and send it to their service's BCM/NR (the address is on the form).

For corrections of records other than discharges, veterans should complete the DD Form 149 and submit their request to their service's BCM/NR (the address is on the form).

Key information to include in requests:
There are three keys to successful applications for upgrade or correction. First, it is very important to explain why the veteran's discharge or other record was unjust or erroneous—for example, how it is connected to, or resulted from unjust policies, a physical or mental health condition related to military service, or some other explainable or justifiable circumstance.

Second, it is important to provide support, where applicable, for key facts. If a veteran has a relevant medical diagnosis, for example, it would be very helpful to include medical records that reflect that diagnosis.
Third, it is helpful, but not always required, to submit copies of the veteran's applicable service records. The more information provided, the better the boards can understand the circumstances of the discharge.

BCM/NRs are also authorized to grant relief on the basis of clemency. Veterans who believe their post-service conduct and contributions to society support an upgrade or correction should describe their post-service activity and provide any appropriate letters or other documentation of support.

Personnel records for veterans who served after 1997 should be accessible online and are usually retrievable within hours of a request through the Defense Personnel Records Information Retrieval System (DPRIS). Those who served prior to 1997 or for whom electronic records are not available from DPRIS, can request their records from the National Personnel Records Center (NPRC).

Frequently Asked Questions
The following are answers to some of the common questions people have about discharge upgrades.

Can I Get Benefits Without a Discharge Upgrade?
Even with a less than honorable discharge, you might be able to access some VA benefits throughout the Character of Discharge review process. When you apply for VA benefits, the VA will review your record to determine if your service was honorable for VA purposes. The review can take up to a year. You provide the VA with documents supporting your case, similar to what you'd send with an application to upgrade your discharge as evidence.

VA recommends finding someone to advocate on your behalf, such as a Veterans Service Organization (VSO) or lawyer.

You can ask for a VA character of discharge review while at the same time applying for a discharge upgrade from the Department of Defense (DoD) or the Coast Guard.

If you need mental health services related to PTSD or other mental health problems linked to your service including conditions related to an experience of military sexual trauma, you might qualify for VA benefits immediately, even without a VA Character of Discharge review or a discharge upgrade.

What I Already Applied for An Upgrade or Correction and Was Denied?
If your previous upgrade application was denied, you can apply again but may have to follow a different process. You're most likely to be successful in applying again if your application is significantly different from when you last applied. For example, you may have evidence not available to you when you last applied, or the DoD might have issued new rules about discharges. DoD changed the rules for discharges related to PTSD, TBI and mental health in 2014, military sexual harassment and assault in 2017, and sexual orientation in 2011.

What If I Have Discharges for More Than One Period of Service?
If the Department of Defense or Coast Guard determined you served honorably in one period of service, you may use that honorable characterization to establish eligibility for VA benefits, even if later on you received a less than honorable discharge. You earned your benefits during the period when you served honorably, and make sure you specifically mention your period of honorable service when you apply for VA benefits. The only exception is for service-connected disability benefits. You're only eligible if you suffered disabilities during your period of honorable discharge. You can't use an honorable discharge from one period of service to establish eligibility for a service-connected disability from a different period of service.

What to Remember
✓ Generally, to receive VA benefits and services, the veteran's character of discharge or service must be under other than dishonorable conditions.
✓ In some cases, veterans receiving undesirable, bad conduct and other types of dishonorable discharges may qualify for benefits, as determined by VA.
✓ Each of the military services maintain a discharge review board with the authority to change, correct or modify discharges or dismissals not issued by a sentence of a general court-martial.
✓ For VA compensation benefits and services, discharge must be other than dishonorable, and the same is true for pension.
✓ To receive education benefits through GI Bill programs, the character of discharge or service must be honorable.

CHAPTER 45

ACCREDITED REPRESENTATIVES

The Department of Veterans Affairs accredits three types of representatives. These are Veterans Service Organization (VSO) representatives, attorneys and agents.

This is to ensure claimants have access to responsible, qualified representation on VA benefit claims. VA-accredited representatives must have good moral character and be capable of providing competent representation. The VA's Office of General Counsel is responsible for making those determinations through its accreditation process. The accreditation process differs depending on the type of accreditation being sought. An accredited VSO representative is someone who's been recommended by a VSO for accreditation, recognized by VA to assist on benefit claims.

The VSO has certified to VA that the representative possesses good character and is fit to represent Veterans and their families as an employee or member of their organization. An attorney is someone who is a member in good standing of at least one State bar. When an attorney applies for VA accreditation, VA typically presumes that the attorney possesses the good character and fitness necessary to represent Veterans and their family members based on the attorney's state license to practice law. An accredited claims agent is someone who is not an attorney but who has undergone a character review by OGC and has passed a written examination about VA law and procedures.

What An Accredited Representative Does

VA-accredited representative can help you understand and apply for any VA benefits you may be entitled to including compensation, education, Veteran readiness and employment, home loans, life insurance, pension, health care, and burial benefits. A VA-accredited representative may also help you request further review of, or appeal, an adverse VA decision regarding benefits. The VA's Office of General Counsel is responsible for making these determinations through its accreditation process. The accreditation process varies depending on which type of accreditation is being sought. An accredited VSO is someone who's been recommended for accreditation by a VSO recognized by VA to help assist with VA benefit claims.

The VSO has certified to VA that the representative possesses good character and is fit to represent veterans and their families as an employee of their organization or member. An attorney is someone in good standing of at least one State bar. When an attorney applies for VA accreditation, VA presumes the attorney has good character and the necessary fitness needed to represented veterans and their family members based on the attorney's state license to practice law. An accredited claims agent is someone who's not an attorney but went through a character review by OGC and has passed a written exam on VA law and procedures.

The VA Office of General Counsel keeps a list of VA-recognized organizations and VA-accredited organizations who have authorization to help in the preparation, presentation

378

and prosecution of VA benefit claims.

The Role of an Accredited Representative
Most of the representation provided to claimants on initial benefit claims is performed by VA-recognized VSOs and their accredited representatives. A VSO can help with gathering needed evidence and submitting a Fully Developed Claim. VSOs can correspond with VA on behalf of you about your claim. VSOs always provide representation free of charge for VA claims.

Many VSOs also sponsor programs like providing transportation to and from medical appointments at VA facilities.

Accredited Attorneys and Claims Agents
VA-accredited attorneys and claims agents will do most of their representation after VA issues an initial decision on a claimant's claim. This is when attorneys and claims agents can charge fees for representation. During this stage of the process of adjudication, an attorney or claims agent might help you further develop the evidence to support your claim. They can also help you create persuasive and legal arguments to submit to VA, and help you navigate the VA appeals process.

Fee for Service
VA-recognized VSOs and their representatives always provide their services on benefit claims for free. Unlike a VSO, a VA-accredit attorney or agent can charge a fee for their representation in appealing or requesting an additional review of an adverse VA decision. Only a VA-accredited attorney or claims agent can charge a fee for assisting in a claim for VA benefits, and only after VA issues an initial decision on the claim, and they have complied with the power-of-attorney and the fee agreement requirements.

How to Challenge a Fee
If you were charged a fee and you believe it was unreasonable or too high, you can file a motion to challenge it. The Office of General Counsel (OGC) of the Department of Veterans Affairs will review the fee agreement if you file a motion with the office.

There is no requirement to use any particular format or writing style when writing a motion. It can be as simply as a letter you write to OGC. However, OGC won't review the fee agreement unless your motion meets all of these requirements:
- Your motion must be in writing. A telephone call won't satisfy the requirement.
- Your motion has to include your full name and your VA file number.
- Your motion must state the reason or reasons why the fee called for in the agreement is unreasonable.
- You have to attach your motion to any evidence you want OGC to consider.

You need to serve a copy of your motion on the attorney or claims agent involved in the matter by mailing or delivering it to them.

To begin the OGC review of your fee agreement, you must mail a motion and proof of service to:

Department of Veterans Affairs
Office of General Counsel (022D)
810 Vermont Ave. NW
Washington, DC 20420

Proof of service is a statement by the person who sent or delivered the motion that includes the date and manner of service, the name of the person served, and the address of the place of delivery.

You have 120 days from the date of the final VA action, which in most instances is 120 days from the date of the fee eligibility decision to file a motion for review of a fee agreement. That means that a motion meeting all requirements including proof of service, must be filed at the above address, before the 120-day time limit expires.

After you file a motion, the involved attorney or claims agent may file a response to your motion with OGC within 30 days from the date on which you serve them with your motion. They are required to serve you with a copy of their response.

You then have 15 days from the date you're served with a response to file a reply with OGC. You also have to serve the attorney or claims agent with a copy of your reply.

Fifteen days after the date the attorney or claims agent response or 30 days after you serve the attorney or claims agent if they don't respond, OGC will close the record in the proceedings and no further evidence will be accepted.

The General Counsel will issue the final decision, which is appealable to the Board of Veterans Appeals.

How to Find a Representative

VA Office of General Counsel maintains a list of VA-recognized organizations and VA-accredited individuals that are authorized to assist in the preparation, presentation and prosecution of VA benefit claims. This is available online.

You can also find a VSO office in your reginal benefit office.

If you want to appoint a VA-recognized VSO to represent you or manage your current representative, you can do so online using eBenefits. You should speak to your VSO first before submitting an online request.

You can also appoint a VA-recognized VSO by completing VA Form 21-22, Appointment of Veterans Service Organization as Claimant's Representative.

You can appoint an attorney, claims agent or a specific, individual VSO representative to represent you by completing VA Form 21-22a, Appointment of Individual as Claimant's Representative.

You can then mail the completed form to:

Department of Veterans Affairs
Claims Intake Center
PO Box 444
Janesville, WI 53547-444

You can discharge your attorney, claims agent or VSO representative at any time and for any reason by informing VA of your request in writing. You can also replace your representative with a new representative by filing a new VA Form 21-22, Appointment of Veterans Service Organization as Claimant's Representative or VA Form 21-22a, Appointment of Individual As Claimant's Representative.

If you believe your VA-accredited representative acted unethically or violated the law, you can file a complaint with the VA's Office of General Counsel.

What to Remember
✓ VA accredits three types of representatives.
✓ Accredited representatives include VSO, attorneys and agents.
✓ A VSO has to certify to VA that the representative has good character and is fit to represent veterans and their families.
✓ Veterans and their families should be cautious of people who are unaccredited and who are preparing, presenting or prosecuting VA claims or holding themselves as being authorized to do so. This violates federal law.
✓ A VA-accredited representative can help you understand and apply for any VA benefits.

CHAPTER 46

ACCRUED BENEFITS

When you file a claim for Veterans Pension, Survivors Pension, VA DIC or accrued benefits, VA reviews all available evidence to determine eligibility. Accrued benefits are benefits that are due, but not paid prior to a beneficiary's death. Examples include:

- A claim or appeal for a recurring benefit that was pending at the time of death, but all evidence needed for a favorable decision was in VA's possession.
- A claim for a recurring benefit had been allowed, but the beneficiary died before award.

At the time of death, one or more benefit checks weren't deposited or negotiated.

Eligibility

VA pays accrued benefits based on the claimant's relationship to the deceased beneficiary. If there is no eligible living person, VA pays accrued benefits based on reimbursement.

Accrued benefits are paid to the first living person listed below:

Relationship to the Deceased Veteran	Accrued Benefit
Surviving spouse	Full amount to surviving spouse
Dependent children, including those between the ages of 18 and 23 who are attending school and those who are found helpless	Equal shares among children
Parents (both)	Equal shares if parents are dependent at the time of veteran's death
Sole surviving parent	Full amount to surviving parent, if dependent at the time of veteran's death

If the death is that of a surviving spouse, the accrued benefit is payable to the veteran's children. However, accrued Dependent's Educational Assistance (DEA) is payable only as reimbursement on the expenses of last sickness and burial.

If the death is of a child, the accrued benefit is payable to the surviving children of the veteran. They must be entitled to death, compensation, dependency and indemnity compensation or death pension, with two exceptions:

- If the deceased child was entitled to an apportioned share of the surviving spouse's award, the accrued benefit is payable only as a reimbursement. It can reimburse expenses of the deceased child's last sickness or burial.
- If the deceased child was in receipt of death pension, compensation or DIC, a remaining child who has elected DEA benefits is only entitled to the unpaid benefits due prior to the commencement of DEA benefits.

The line of succession for accrued benefits is set by law. If a preferred beneficiary doesn't file or prosecute a claim, payment isn't permitted to the person with equal or lower preference. This applies to a waiver of right to payment as well.

Reimbursement
If there isn't an entitled living person based on relationship, VA may reimburse the person who paid for or is responsible for the last illness and burial. If payments were made from the estate of the deceased beneficiary, the executor or administrator of the estate should file the claim. The amount payable as reimbursement is limited to the actual expenses paid and it's limited to the accrued benefits available.

Substitution
Substitution allows a person eligible for accrued benefits to substitute a deceased beneficiary on a pending claim or appeal. The substitute claimant may submit evidence in support of the pending claim, or appeal for potential accrued benefits.

Applying for Accrued Benefits
The forms that you may need to complete to file for accrued benefits include:
- VA Form 21P-601, Application for Accrued Amounts Due a Deceased Beneficiary
- VA Form 21P-534EZ, Application for Dependency and Indemnity Compensation, Death Pension and Accrued Benefits
- VA Form 21P-535, Application for Dependency and Indemnity Compensation by Parent(s) including Accrued Benefits and Death Compensation When Applicable
- VA Form 21P-0847, Request for Substitution of Claimant Upon Death Claimant

VA must receive an accrued benefits claim within one year of the beneficiary's death and/or the date of notification to the beneficiary.

VA has to receive a substitution claim within one year of the original claimant's death. If the substitute dies, the next substitute has one year from the original substitute's death to file a claim.

Evidence
The evidence must show that both of the following are true:
- The VA owed the deceased claimant payments based on existing ratings, decisions or evidence that the VA had when the claimant died, but the VA didn't make the payments before the death of the claimant, and
- You're the surviving spouse, child or dependent parent of the deceased veteran

You will need to submit the veteran's DD214 or other separation documents and a copy of the veteran's death certificate showing cause of death. You can also give the VA permission to gather both.

If a representative of the beneficiary's estate has been assigned, the VA needs a certified copy of the letters of administration or letters testamentary with the signature and seal of the appointing court.

If you're submitting a reimbursement claim for the veteran's last illness and burial expenses, VA needs a copy of all billing and account statements for services and supplies connected to the expenses. The billing or account statement should be submitted on the regular billhead of the creditor. The statement needs to show:
- The dates, nature and costs of services or supplies provided
- The name of the deceased veteran who paid for the services or supplies
- Proof the expenses have already been paid, and if so, who made those payments

What to Remember
✓ Accrued benefits are due but not paid before the death of a beneficiary.
✓ VA will pay accrued benefits based on the relationship of the claimant to the deceased beneficiary.
✓ Benefits are paid based on reimbursement if there is no eligible living person.
✓ The line of succession is set by law.
✓ VA must receive an accrued benefits claim within a year of the beneficiary's death and/or the date of notification to the beneficiary.

269
CHALENG, 321
CHAMPVA, 1, 11, 138, 152, 161, 162,
    165, 166, 167, 303, 308
Character of discharge, 59, 215, 373,
    374
Chronic disability, 27
Chronic Disease, 52
Clothing allowance, 68, 365
College Fund, 220, 235
Combined Ratings Table, 58
Community care, 137, 147, 148, 149
Community living centers, 21, 155
Community residential care, 21
Concurrent receipt, 37, 38
Contaminated Drinking Water, 281
Copayment Rates, 139
Correction boards, 357, 358
Correction of record, 357
Cost of living allowance, 42
Countable Income, 73, 76, 89, 157
CRSC program, 38

D

DD Form 214, 225, 268, 269
DD214, 3, 5, 7, 8, 9, 10, 11, 12, 36, 68,
    98, 136, 161, 166, 167, 287, 288, 296,
    314, 315, 317, 356, 372, 383
DD215, 8
Death Gratuity, 87, 251
Denial of a claim, 101
dental care, 144, 161, 300
Dependency and indemnity
    compensation, 366, 367
Dependency and Indemnity
    Compensation, 301, 328
Desertion, 373
DIC, 82, 83, 84, 87, 301, 302, 370
Dignified Burial of Unclaimed
    Veterans Act of 2012, 100
Direct Service Connection, 39
Disability compensation, 19, 22, 23,
    24, 25, 27, 29, 30, 37, 52, 54, 55, 56,
    58, 60, 157, 187, 277, 328, 329, 364,
    366
Disability Compensation, 22, 25, 56,
    64, 77, 87, 157, 300, 324, 370
Disabled Veterans Application for
    Vocational Rehabilitation, 229,
    247, 370

E

ebenefits, 9, 10, 22, 23, 29, 269, 315
Effective Dates, 38, 39, 227, 244
Eligibility determinations, 98, 247
Emergency Care, 145, 146
Emergency transportation, 148
Employmentservices, 320
enrollment certification, 225, 226,
    243, 244
Enrollment Certification, 217, 225,
    226, 240, 243, 244
Extended care, 20
Extended Care Services, 20, 156

F

Family coverage, 256
Female Vietnam veterans, 188
Fiduciary Program, 177
Fiduciary selection, 178
Flat markers, 103
flight training, 215, 217, 220, 226,
    235, 239, 241, 243
Foreclosure, 265, 273, 274, 275, 276,
    331, 332
Foreign Medical Program, 277, 278
Former prisoners of war, 300
Funeral expenses, 98, 99

G

Geriatric care, 21
Geriatrics, 20, 156
GI Bill, 188, 220, 235, 236, 238, 374
Government headstones, 114
Grave marker medallion, 103
Grave markers, 111
Gravesite Locator, 115
Gulf War, 26, 27, 104, 268, 317, 328,
    329, 330
Gulf War period, 26
Gulf War Syndrome, 27
Gulf War veterans, 27, 328, 329, 330
Gulf War Veterans, 26, 329

156, 278, 300
Nursing Home, 22, 155, 156, 324
Nursing home care, 156, 300

O

Office of Rural Health, 149
Other than dishonorable
conditions, 267, 268, 373, 374
Overpayment, 222, 227, 244, 245,
364
Overpayment, 61
Overseas graves, 116

P

PACT Act, 1, 16, 18, 26, 33, 34, 79, 281,
282, 283, 284, 285, 288, 294, 296,
329
Paralympics, 188
Parkinson's disease, 18, 19, 285, 286,
291
Pension, 20, 22, 23, 26, 37, 68, 77, 87,
100, 147, 301, 324, 370
Pension benefits, 23, 37
Permanently and totally disabled,
77, 145, 146, 155
Persian Gulf War, 18, 134, 267, 268,
284, 300, 328, 330, 369
Personnel records, 347
Philippine veteran, 109
Plot allowances, 99
Plot interment allowance, 101
Post 9/11 GI-Bill, 31
Post-9/11 GI Bill, 6, 31, 33, 186, 207,
208, 210, 211, 213, 214, 230, 231,
232, 233, 234, 235, 236, 237, 374
POW, 23, 300, 301
Pre-Discharge, 370
Prescription drug coverage, 142, 143
Presidential Memorial Certificates,
116, 117
Prestabilization Rating, 70
Presumption of Sound Condition,
52, 55
Presumptions Relating to Certain
Diseases, 54, 55
Presumptive Service Connection,
14, 39
Priority groups, 156

Priority Groups, 133
Prisoner of War, 302
Program of Comprehensive
Assistance for Family Caregivers,
17, 33, 63, 138, 304, 305, 306, 307,
308
Project SHAD, 294
PTSD, 54, 325, 361, 375, 376
Purchase Loans, 32, 264
Purple Heart, 63, 133, 134, 144, 230,
231, 236, 265, 307, 309, 310, 315,
319

Q

Quick Start, 30, 31

R

Radiation exposure, 24
Radiation Exposure, 281
Radiation-risk activities, 23, 24
REAP, 235
Reinstatement, 259
Release from liability, 274
REMOTE Act, 13
Request Pertaining To Military
Records, 347, 371
Reserve Component, 28, 106
Respiratory Cancers, 14
Restoration of entitlement, 274
Rural Initiative Plan, 106

S

Same-sex married couples, 57
SBA, 183, 184, 185, 266
SBP. *See* Survivor Benefit Plan
SCRA, 276, 331, 332, 333, 334, 335
S-DVI, 249, 250
Selected Reserve, 101, 102, 114, 163,
215, 217, 225, 238, 239, 241, 242,
244, 268, 269
September 11, 2001, 31, 37, 61, 99,
103, 309
Service dog, 147
Service-connected burial, 97, 101
Service-connected disabilities, 77,
152, 272

Service-connected disability, 22, 31,
58, 59, 60, 66, 68, 78, 84, 145, 147,
148, 152, 153, 155, 156, 159, 188,
190, 225, 228, 229, 231, 246, 249,
264, 267, 268, 272, 275, 309, 310,
317, 366
Service-Disabled Veterans
Insurance, 249
Servicemembers Civil Relief Act,
276, 334, 335
Servicemembers group life
insurance, 87
Servicemembers' Group Life
Insurance, 250, 253, 254, 255, 256,
258, 260, 371
Sexual assault, 325, 360, 361, 375, 376
SGLI, 250, 251, 252, 253, 254, 256, 257,
258, 259, 260
Small Business Administration, 266
Social Security, 23, 42, 79, 321, 347,
355
Solid Start, 26
Special appointing authorities, 315
Special Compensation, 38, 52, 59
Special Hiring Authorities, 316
Special Housing Adaptations
Grants, 65
Special Monthly Compensation, 43,
44, 45
Specially Adapted Housing Grant,
65
Specially Adapted Housing Grants,
65
Specialty care, 57
SRMGIB, 242
Stand Downs, 321
State Veterans Home, 21, 156
State Veterans Homes, 21, 155
Student Verification of Enrollment,
226, 243
Subsistence allowance, 78, 186, 187,
188
Substitution, 383
Supplemental Claim, 9, 19
Supplemental S-DVI, 249
Supportive Services for Veteran
Families, 322
Surviving spouse, 82, 83, 109, 301,
302, 357, 361, 364, 374
Surviving spouses, 32, 99, 116, 164,
272
Survivor Benefit Plan, 90, 91, 93, 95,

96, 163, 164

T

TBI, 255, 361, 375
Temporary Disability Rating, 69, 70
Temporary Residence Adaptation,
65, 66
The Veterans Benefits Act of 2003,
66, 301
Thrift Savings Plan, 28
Total Disability Income Provision,
250
Transgender Veterans, 57
Transition GPS, 29
Traumatic brain injury, 360
Traumatic Brain Injury, 35
Travel Expenses, 138, 170
TRICARE, 137, 138, 163, 164, 165, 167,
168, 169, 170, 171, 172, 173, 174, 175,
176, 308
TSGLI, 251, 254, 255

U

Unemployability, 70, 71, 370
Urgent care, 6
USGLI, 250

V

VA Dental Insurance Program, 161,
162
VA direct loan, 275
VA Form 21-526EZ, 22
VA Form 22-1990, 225, 242
VA guaranteed loans, 265
VA health care, 22, 29, 143, 145, 149,
152, 155, 300, 365
VA loan guaranty, 265, 266, 268, 269,
274
VA national cemetery, 98, 99, 107,
110
VA payment of emergency care, 148
VA reimbursement, 148
VADIP, 162, 163, 164
VALife, 16, 248, 249, 263
VEAP, 220, 235, 236
Vet Centers, 31, 190, 191

W

Y

Would You Like Additional Copies Of
<u>What Every Veteran Should Know</u>?

Simply tear out this form, and:

Phone 309-757-7760 or Fax 309-278-5304
Mail To: VETERANS INFORMATION SERVICE
P.O. Box 111
East Moline, IL 61244-0111

ORDER ONLINE AT: www.vetsinfoservice.com

☐ Yes! Send me _____ copies of *"What Every Veteran Should Know"*, at $30.00 each (shipping & handling included). I request the _____ (specify: <u>current</u> <u>already released 2023</u>, or <u>pre-reserve 2024</u>) edition. New annual books are published every March 1st.

☐ Yes! I would like to subscribe to *"What Every Veteran Should Know"* monthly supplement (an 8-page newsletter which keeps your book up-to-date), and receive _____ copies of all 12 monthly issues (1 year) for $38.00 per subscription (shipping & handling included).

☐ Yes! I want to save money, and receive both the book and the monthly supplement. I would like _____ sets, at $65 per set (shipping & handling included).

Name   Current Customer ID?

Address

City / State / Zip Code

Telephone Number   E-mail Address

Amount Enclosed   Daytime Phone # (including area code)

Method of Payment:

☐ Check   ☐ Visa   ☐ MasterCard   ☐ Money Order

Credit Card #   3-digit CVV   Expiration Date
(Month/Year)

Signature

Thank you for your order!
If you have any questions, feel free to contact us at
(309) 757-7760
www.vetsinfoservice.com - Email: help@vetsinfoservice.com

Would You Like Additional Copies Of
What Every Veteran Should Know?

Simply tear out this form, and:

Phone 309-757-7760 or Fax 309-278-5304

Mail To: VETERANS INFORMATION SERVICE
P.O. Box 111
East Moline, IL 61244-0111

ORDER ONLINE AT: www.vetsinfoservice.com

☐ Yes! Send me _____ copies of *"What Every Veteran Should Know"*, at $30.00 each (shipping & handling included). I request the _____ (specify: current already released 2023, or pre-reserve 2024) edition. New annual books are published every March 1st.

☐ Yes! I would like to subscribe to *"What Every Veteran Should Know"* monthly supplement (an 8-page newsletter which keeps your book up-to-date), and receive _____ copies of all 12 monthly issues (1 year) for $38.00 per subscription (shipping & handling included).

☐ Yes! I want to save money, and receive both the book and the monthly supplement. I would like _____ sets, at $65 per set (shipping & handling included).

Name   Current Customer ID?

Address

City / State / Zip Code

Telephone Number   E-mail Address

Amount Enclosed   Daytime Phone # (including area code)

Method of Payment:

☐ Check   ☐ Visa   ☐ MasterCard   ☐ Money Order

Credit Card #   3-digit CVV   Expiration Date
(Month/Year)

Signature

Thank you for your order!
If you have any questions, feel free to contact us at
(309) 757-7760
www.vetsinfoservice.com - Email: help@vetsinfoservice.com

Would You Like Additional Copies Of
<u>What Every Veteran Should Know</u>?

Simply tear out this form, and:

Phone 309-757-7760 or Fax 309-278-5304

Mail To: VETERANS INFORMATION SERVICE
P.O. Box 111
East Moline, IL 61244-0111

ORDER ONLINE AT: www.vetsinfoservice.com

☐ Yes! Send me _____ copies of *"What Every Veteran Should Know"*, at $30.00 each (shipping & handling included). I request the _____ (specify: <u>current</u> <u>already released 2023,</u> or <u>pre-reserve 2024</u>) edition. New annual books are published every March 1st.

☐ Yes! I would like to subscribe to *"What Every Veteran Should Know"* monthly supplement (an 8 page newsletter which keeps your book up-to-date), and receive _____ copies of all 12 monthly issues (1 year) for $38.00 per subscription (shipping & handling included).

☐ Yes! I want to save money, and receive both the book and the monthly supplement. I would like _____ sets, at $65 per set (shipping & handling included).

Name   Current Customer ID?

Address

City / State / Zip Code

Telephone Number   E-mail Address

Amount Enclosed   Daytime Phone # (including area code)

Method of Payment:

☐ Check   ☐ Visa   ☐ MasterCard   ☐ Money Order

Credit Card #   3-digit CVV   Expiration Date
(Month/Year)

Signature

Thank you for your order!
If you have any questions, feel free to contact us at
(309) 757-7760
www.vetsinfoservice.com - Email: help@vetsinfoservice.com

Would You Like Additional Copies Of
What Every Veteran Should Know?

Simply tear out this form, and:

Phone 309-757-7760 or Fax 309-278-5304

Mail To: VETERANS INFORMATION SERVICE
P.O. Box 111
East Moline, IL 61244-0111

ORDER ONLINE AT: www.vetsinfoservice.com

☐ Yes! Send me _____ copies of *"What Every Veteran Should Know"*, at $30.00 each (shipping & handling included). I request the _____ (specify: current already released 2023, or pre-reserve 2024) edition. New annual books are published every March 1st.

☐ Yes! I would like to subscribe to *"What Every Veteran Should Know"* monthly supplement (an 8-page newsletter which keeps your book up-to-date), and receive _____ copies of all 12 monthly issues (1 year) for $38.00 per subscription (shipping & handling included).

☐ Yes! I want to save money, and receive both the book and the monthly supplement. I would like _____ sets, at $65 per set (shipping & handling included).

Name   Current Customer ID?

Address

City / State / Zip Code

Telephone Number   E-mail Address

Amount Enclosed   Daytime Phone # (including area code)

Method of Payment:

☐ Check   ☐ Visa   ☐ MasterCard   ☐ Money Order

Credit Card #   3-digit CVV   Expiration Date
(Month/Year)

Signature

Thank you for your order!
If you have any questions, feel free to contact us at
(309) 757-7760
www.vetsinfoservice.com - Email: help@vetsinfoservice.com